ADVANCES IN MESHFREE TECHNIQUES

Computational Methods in Applied Sciences

Volume 5

Series Editor

E. Oñate
International Center for Numerical Methods in Engineering (CIMNE)
Technical University of Catalunya (UPC)
Edificio C-1, Campus Norte UPC
Gran Capitán, s/n
0834 Barcelona, Spain
onate@cimne.upc.edu
www.cimne.com

Advances in Meshfree Techniques

Edited by

V.M.A. Leitão

Instituto Superior Técnico,
Lisboa, Portugal

C.J.S. Alves

Instituto Superior Técnico,
Lisboa, Portugal

and

C. Armando Duarte

University of Illinois at Urbana-Champaign,
Urbana, IL, USA

Springer

A C.I.P. Catalogue record for this book is available from the Library of Congress.

ISBN 978-90-481-7533-8
ISBN 978-1-4020-6095-3 (e-book)

Published by Springer,
P.O. Box 17, 3300 AA Dordrecht, The Netherlands.

www.springer.com

Printed on acid-free paper

Table of Contents

Preface

Over the past decade, mesh reduction techniques (meshless or meshfree methods) have emerged as effective numerical techniques for solving science and engineering problems. Meshing difficulties of existing numerical techniques like the finite element and boundary element methods, were the initial thrust for the development of meshfree methods. Several techniques devoted to the simulation of crack propagation, moving material interfaces, large deformation and shear band localization problems, among others, were proposed during the last decade. Problems with highly oscillatory solutions such as Helmholtz equations and multiscale problems have also spurred great interest on this class of methods. The variety of problems that are now being addressed by these techniques continues to expand and the quality of the results obtained demonstrates the effectiveness of many of the methods currently available.

Various scientific meetings have been organized in recent years that were totally or at least partially devoted to this area of knowledge and a large number of papers have been published in prestigious journals. The number of methods that have recently been proposed is also an evidence of the growing interest of the engineering and mathematics community worldwide on these types of numerical techniques.

The objective of this book is to collect state-of-the-art research, methods and new ideas on the subject of mesh reduction techniques, and to contribute to the development of this area of knowledge. The book contains 15 invited contributions written by participants to one of the Thematic Conferences sponsored by the European Community on Computational Methods in Applied Sciences, the ECCOMAS Thematic Conference on Meshless Methods held in Lisbon, Portugal, from July 11th to 14th 2005. The vitality and multi-disciplinarity of the research community in this field is reflected on the broad range of methods discussed in this book. Each contribution in this book is an extended and revised version of the paper presented at the conference. They present a sample of the state of the art in the field with methods that have reached a certain level of maturity while also addressing many open issues. The list of contributors reveals a fortunate mix of highly distinguished authors as well as quite young but very active and promising researchers.

It is the editors' wish that this book may constitute a valuable reference for researchers interested in the field of mesh reduction techniques, one that may help promoting the use of these techniques and to extend its application to an even broader variety of engineering and science problems.

The editors would like to take this opportunity to thank all authors for submitting their contributions.

V.M.A. Leitão
C.J.S. Alves
C.A. Duarte

A Global-Local Approach for the Construction of Enrichment Functions for the Generalized FEM and Its Application to Three-Dimensional Cracks

C. Armando Duarte[1], Dae-Jin Kim[1] and Ivo Babuška[2]

[1]*Department of Civil and Environmental Engineering, University of Illinois at Urbana-Champaign, 205 North Mathews Avenue, Urbana, IL 61801, U.S.A.; E-mail: caduarte@uiuc.edu, dkim46@uiuc.edu*
[2]*Institute for Computational Engineering and Sciences, The University of Texas at Austin, 1 University Station, C0200, Austin, TX 78712, U.S.A.; E-mail: babuska@ices.utexas.edu*

Abstract. Existing generalized or extended finite element methods for modeling cracks in three-dimensions require the use of a sufficiently refined mesh around the crack front. This offsets some of the advantages of these methods specially in the case of propagating three-dimensional cracks. In this paper, a strategy to overcome this limitation is investigated. The approach involves the development of enrichment functions that are computed using a new global-local approach. This strategy allows the use of a fixed global mesh around the crack front and is specially appealing for non-linear or time dependent problems since it avoids mapping of solutions between meshes. The resulting technique enjoys the same flexibility of the so-called meshfree methods for this class of problem while being more computationally efficient.

The proposed generalized FEM with global-local functions, by numerically constructing the enrichment functions, brings the benefits of existing generalized FEM to a broader class of problems. The procedure is applied to the solution of three-dimensional linear elastic fracture mechanics problems. Numerical experiments demonstrating the computational efficiency and accuracy of the method are presented.

Key words: Generalized finite element method, extended finite element method, meshfree method, global-local method, fracture mechanics.

1 Introduction

The generalized finite element method (GFEM) [4, 15, 31, 38] and the eXtended finite element method (XFEM) [7, 26] are examples of the so-called partition of unity method which originated in the works of Babuška et al. [3] and Duarte and

V.M.A. Leitão et al. (eds), Advances in Meshfree Techniques, 1–26.

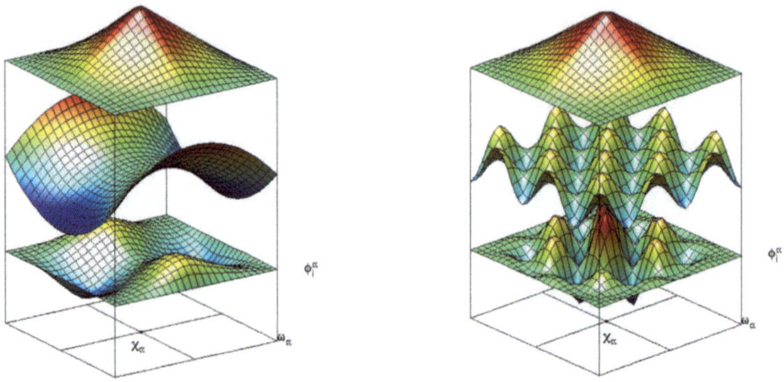

(a) (b)

Fig. 1. Construction of GFEM shape functions using (a) polynomial and (b) non-polynomial enrichment functions. Here, φ_α are the functions at the top, the enrichment functions, L, are the functions in the middle, and the generalized FE shape functions, ϕ_α, are the resulting bottom functions.

Oden [17, 19]. The shape functions in this class of methods are built from the product of a partition of unity, φ_α, and enrichment functions, L,

$$\phi_\alpha(\boldsymbol{x}) := \varphi_\alpha(\boldsymbol{x}) L(\boldsymbol{x}) \tag{1}$$

where φ_α, $\alpha = 1, \ldots, N$, N being the number of functions, constitute a partition of unity, i.e., a set of functions defined in a domain Ω with the property that $\sum_{\alpha=1}^{N} \varphi_\alpha(\boldsymbol{x}) = 1$ for all $\boldsymbol{x}$ in Ω. Figure 1 illustrates the construction of GFEM shape functions using a polynomial and a non-polynomial enrichment function.

The partition of unity property of functions φ_α, $\alpha = 1, \ldots, N$, implies that

$$\sum_{\alpha=1}^{N} \phi_\alpha(\boldsymbol{x}) = \sum_{\alpha=1}^{N} \varphi_\alpha(\boldsymbol{x}) L(\boldsymbol{x}) = L(\boldsymbol{x}) \tag{2}$$

Any enrichment function L can be represented *exactly* through linear combinations of generalized FE shape functions. Therefore, if the enrichment function L can approximate the solution $\boldsymbol{u}$ of a boundary value problem, the corresponding GFEM shape function also will. There is considerable freedom in the choice of the enrichment functions and the corresponding approximation spaces used in the GFEM. Figure 1b, for example, illustrates the construction of a GFEM shape function using a highly oscillatory non-polynomial enrichment function.

Customized enrichment functions can be used to model local features in a domain like cracks [11, 16, 26, 27, 30, 32, 42], edge singularities [15], boundary layers [13], inclusions [41], voids [37, 41], microstructures [25, 34], etc., instead of a strongly

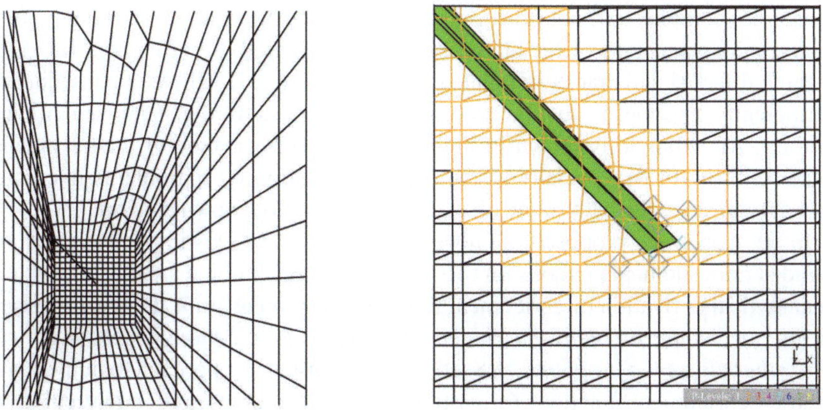

(a) (b)

Fig. 2. (a) Top view of a three-dimensional GFEM discretization for a cracked plate [16]. A finner mesh is required near the crack front to compensate the limitations of the enrichment functions. (b) Three-dimensional view of the crack front. Notice that the crack surface goes through the elements [16].

refined mesh, as required in the finite element method. This has led to a growing interest in this class of methods by the engineering community.

The GFEM and the XFEM have the ability to analyze crack surfaces arbitrarily located within the mesh (across finite elements). Figure 2 shows the representation of a three-dimensional crack surface using the generalized finite element method presented in [16].

The GFEM enjoys, for several classes of problems, the same level of flexibility and user friendliness as meshfree methods while being more computationally efficient. The GFEM and the XFEM are the most promising techniques to model, for example, propagating cracks in complex three-dimensional structures.

A crack can be represented by partition of unity methods using enrichment functions from the asymptotic expansion of the elasticity solution in the neighborhood of a crack. These expansions are well known for two-dimensional stress states, *but not for a general three-dimensional case*. Custom enrichment functions for two-dimensional linear problems are amenable to analytical derivation and *all* GFEM and XFEM references listed above rely on this approach. However, this is not the case for most non-linear and many three-dimensional linear problems. This has restricted the application of the GFEM to many important classes of problems. The linear elastic fracture mechanics theory in two- and three-dimensions is well developed. Asymptotic forms of the elasticity solution near the edges for three-dimensional problems are discussed in, for example, [21, 28]. There are terms in the three-dimensional expansions which are very similar to the two-dimensional case. Nevertheless, the theory is quite complex and not practical for engineering applications. Currently, two-dimensional expansions of the elasticity solution are used as enrichment functions for three-dimensional cracks in finite size domains [16, 27, 42]. As a consequence, a sufficiently fine mesh must be used around the crack front to achieve sufficient accuracy. An example is shown in Figure 2a. This offsets some of the advantages of the XFEM/GFEM, specially in the case of evolution type problems like crack propagation, since the path of the crack is, in general, not known a priori.

In this paper, we propose to remove the limitations of existing generalized and extended finite element methods for the solution of problems that are not amenable to the analytical derivation of enrichment functions. *The focus is on three-dimensional fracture mechanics problems but the methodology is broad enough to be used with many other classes of problems.* In Section 3, we present an approach to numerically construct enrichment functions for partition of unity methods and, in particular, for the generalized FEM.

The proposed enrichment functions are computed from the solution of local boundary valued problems defined in a neighborhood of a crack front in three-dimensions. The approach uses concepts from the classical global-local finite element method, where boundary conditions for local problems are extracted from the solution of a global problem [29]. However, unlike the classical global-local FEM, the proposed approach is able to account for local-global interactions and interactions among local features, like multiple cracks. The proposed approach allows the use of coarse global meshes around crack fronts while delivering accurate solutions and is specially appealing to evolution type problems like propagating cracks. Numerical experiments demonstrating the computational efficiency and accuracy of the method are presented in Section 4. The main conclusions of this work are presented in Section 5.

Strouboulis et al. [38–40] consider the problem of a scalar equation with micro-scale. A particular example is the Laplace equation on a domain with many holes. The main idea is to construct a space of functions which can approximate locally the exact solution. Because the exact solution is not smooth and any coarse solution can not approximate it well, it is not possible to use a crude solution as boundary condtions for the local problems. Hence a set of solutions with polynomial boundary conditions and a buffer zone is used. The solutions are then able to approximate well inside the local domains. The present paper does not address multiscale problems. It addresses instead three-dimensional fracture mechanics problems. In this case, the solutions are smooth away from crack fronts where it is possible to assume that the crude solution of a global problem can approximate the exact solution reasonably well. When the boundary of a local problem intersects the crack front, the crude global solution cannot approximate the exact solution well at this point. However, this point at the crack front will be *inside* of another local problem since the local domains overlap. Therefore, the solution of the other local problem will be able to approximate well the exact solution at the point even with non-exact boundary conditions because of the ellipticity of the problem under consideration. This allows us to utilize the crude solution of a global problem for the construction of local problems using the crude solution as boundary conditions.

2 The Generalized Finite Element Method

In this section, we briefly review the construction of generalized finite element shape functions. Further details can be found in, for example, [4, 14, 15, 18, 19, 24, 31, 36, 37]. The generalized FEM is one instance of the so-called partition of unity method. A partition of unity-based approximation of a scalar field $u(\boldsymbol{x})$ defined on a domain $\Omega \subset \mathbb{R}^n$, $n = 1, 2, 3$, can be written as

$$u_h(\boldsymbol{x}) = \sum_{\alpha=1}^{N} \varphi_\alpha(\boldsymbol{x}) u_{h\alpha}(\boldsymbol{x}) \tag{3}$$

where

(i) $\mathsf{PoU}_N = \{\varphi_\alpha\}_{\alpha=1}^{N}$ constitute a *partition of unity* (PoU) with N functions φ_α defined on Ω and with properties

$$\varphi_\alpha \in C_0^s(\omega_\alpha),\ s \geq 0, \qquad 1 \leq \alpha \leq N \tag{4}$$

$$\sum_{\alpha}^{N} \varphi_\alpha(\boldsymbol{x}) = 1 \qquad \forall \boldsymbol{x} \in \Omega \tag{5}$$

The support of φ_α, $\{\boldsymbol{x} : \varphi_\alpha(\boldsymbol{x}) \neq 0\}$, is denoted by ω_α (often called *cloud*) and $\boldsymbol{x}_\alpha$ denotes a node associated with function φ_α and its support. Examples of partition of unities are standard finite element shape functions, functions generated by moving least squares methods and Shepard functions [19, 23].

(ii) $u_{h\alpha}(\boldsymbol{x})$ denotes a local approximation of the field $u(\boldsymbol{x})$ defined on ω_α and belonging to the local space

$$\chi_\alpha(\omega_\alpha) = \text{span}\{L_{i\alpha}(\boldsymbol{x})\}_{i \in \mathcal{I}(\alpha)} \tag{6}$$

where the basis functions $L_{i\alpha}$, $i \in \mathcal{I}(\alpha)$, are also denoted by *enrichment functions* and $\mathcal{I}(\alpha)$ is an index set such that

$$L_{1\alpha} = 1 \tag{7}$$

Examples of linear and quadratic enrichment functions for a node $\boldsymbol{x}_\alpha = (x_\alpha, y_\alpha)$ in two-dimensions are

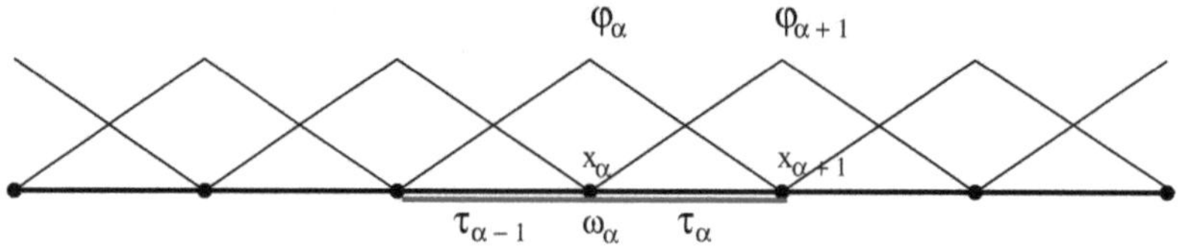

Fig. 3. One-dimensional finite element partition of unity.

$$\left\{1, \frac{(x-x_\alpha)}{h_\alpha}, \frac{(y-y_\alpha)}{h_\alpha}\right\} \quad \text{and}$$

$$\left\{1, \frac{(x-x_\alpha)}{h_\alpha}, \frac{(y-y_\alpha)}{h_\alpha}, \frac{(x-x_\alpha)^2}{h_\alpha^2}, \frac{(x-x_\alpha)}{h_\alpha}\frac{(y-y_\alpha)}{h_\alpha}, \frac{(y-y_\alpha)^2}{h_\alpha^2}\right\}, \quad (8)$$

respectively, where h_α is a scaling factor [15].

Using the definitions above, we can write $u_{h\alpha}(\boldsymbol{x})$ as

$$u_{h\alpha}(\boldsymbol{x}) = \sum_{i \in \mathcal{I}(\alpha)} a_{i\alpha} L_{i\alpha}(\boldsymbol{x}), \qquad a_{i\alpha} \in \boldsymbol{R} \quad (9)$$

The approximations used in *all* partition of unity methods like the *hp*-cloud method [18, 19], the generalized finite element method [4, 14, 15, 24, 31, 36, 37], the particle-partition of unity method [20], the extended finite element method [8, 26, 42], among others, are special cases of (3), the basic difference being the choice of the partition of unity functions, φ_α, $\alpha = 1, \ldots, N$, and/or the enrichment functions, $L_{i\alpha}$, $i \in \mathcal{I}(\alpha)$.

The partition of unity approximation $u_h(\boldsymbol{x})$ can be written as

$$u_h(\boldsymbol{x}) = \sum_{\alpha=1}^{N} \varphi_\alpha(\boldsymbol{x}) \sum_{i \in \mathcal{I}(\alpha)} a_{i\alpha} L_{i\alpha}(\boldsymbol{x}) = \sum_{\alpha=1}^{N} \sum_{i \in \mathcal{I}(\alpha)} a_{i\alpha} \phi_{i\alpha}(\boldsymbol{x}) \quad (10)$$

where

$$\phi_{i\alpha}(\boldsymbol{x}) := \varphi_\alpha(\boldsymbol{x}) L_{i\alpha}(\boldsymbol{x}) \qquad \text{(no sum on } \alpha) \quad (11)$$

are denoted partition of unity, cloud or generalized finite element shape functions.

In the generalized finite element method the partition of unity is in general provided by linear Lagrangian finite element shape functions. The support ω_α of φ_α is then given by the union of the finite elements sharing a vertex node $\boldsymbol{x}_\alpha$. Figure 3 shows a one-dimensional finite element discretization. The partition of unity functions φ_α are the usual global finite element shape functions, the classical "hat-functions", associated with node $\boldsymbol{x}_\alpha$. The support ω_α is thus the union of the elements $\tau_{\alpha-1}$ and τ_α. The resulting shape functions are called generalized finite element shape functions. Figure 1 illustrates the construction of these functions in two dimensions. The number of partition of unity functions, N, is given by the number of vertex nodes, $\boldsymbol{x}_\alpha$, in the finite element mesh. The enrichment, $L_{i\alpha}$, and corresponding shape functions, $\phi_{i\alpha}$, are defined on a node-by-node or cloud-by-cloud basis. Each node may have a different set of enrichment functions [15, 31].

Using property (7), we can write a partition of unity and, in particular, a generalized finite element approximation as

$$\begin{aligned} u_h(\boldsymbol{x}) &= \sum_{\alpha=1}^{N} \sum_{i\in\mathcal{I}(\alpha)} a_{i\alpha}\varphi_\alpha(\boldsymbol{x})L_{i\alpha}(\boldsymbol{x}) \\ &= \sum_{\alpha=1}^{N} \varphi_\alpha(\boldsymbol{x}) \left[a_{1\alpha} + \sum_{i\in\mathcal{I}(\alpha), i\neq 1} a_{i\alpha}L_{i\alpha}(\boldsymbol{x}) \right] \\ &= \underbrace{\sum_{\alpha=1}^{N} a_{1\alpha}\varphi_\alpha(\boldsymbol{x})}_{\text{regular interpolation}} + \underbrace{\sum_{\alpha=1}^{N} \sum_{i\in\mathcal{I}(\alpha), i\neq 1} a_{i\alpha}\varphi_\alpha(\boldsymbol{x})L_{i\alpha}(\boldsymbol{x})}_{\text{POU-based enrichment}} \end{aligned} \tag{12}$$

The above decomposition of $u_h(\boldsymbol{x})$ is used in the extended finite element method [8, 26, 42]. In this case, the enrichment functions are Heaviside or Westergaard functions.

An a-priori error estimate for partition of unity approximations and, in particular, for the generalized finite element method, was proved by Babuška and Melenk [24]. The estimate says that if the partition of unity $\mathsf{PoU}_N = \{\varphi_\alpha\}_{\alpha=1}^N$ satisfies some mild requirements and the error of the local approximations, $u_{h\alpha} \in \chi_\alpha(\omega_\alpha)$, $\alpha = 1, \ldots, N$, are bounded by

$$\|u - u_{h\alpha}\|_{E(\omega_\alpha)} < \epsilon(\alpha, u), \qquad \alpha = 1, \ldots, N, \tag{13}$$

then the error of a partition of unity approximation, u_h, given by (3) is bounded by

$$\|u - u_h\|_{E(\Omega)} < C \left(\sum_{\alpha=1}^{N} \epsilon^2(\alpha, u) \right)^{1/2} \tag{14}$$

where $\|.\|_E$ denotes the energy norm and C is a constant. Details and proofs can be found in [12, 19, 24].

3 A Global-Local Approach to Build Enrichment Functions

In this section, we present the formulation of the proposed global-local approach to build enrichment functions for the generalized FEM. The focus is on 3-D fracture mechanics problems but the formulation can also be applied to other classes of problems. Sections 3.1 and 3.2 present the approach in a generic, abstract, setting. Section 3.3 illustrates the application of the procedure using a more computationally oriented setting.

3.1 Formulation of Global Problem

Consider a domain $\bar{\Omega}_G = \Omega_G \cup \partial\Omega_G$ in $\mathbb{R}^3$. The boundary is decomposed as $\partial\Omega_G = \partial\Omega^u_G \cup \partial\Omega^\sigma_G$ with $\partial\Omega^u_G \cap \partial\Omega^\sigma_G = \emptyset$. For simplicity of notation, let us consider the case of a single crack surface Γ_c with front Γ_{front} in the domain Ω_G.

The strong form of the equilibrium equations is given by

$$\nabla \cdot \boldsymbol{\sigma} = \mathbf{0} \qquad \boldsymbol{\sigma} = \boldsymbol{C} : \boldsymbol{\varepsilon} \qquad \text{in } \Omega_G, \tag{15}$$

where $\boldsymbol{C}$ is Hooke's tensor. The following boundary conditions are prescribed on $\partial\Omega$

$$\boldsymbol{u} = \bar{\boldsymbol{u}} \text{ on } \partial\Omega^u_G \qquad \boldsymbol{\sigma} \cdot \boldsymbol{n} = \bar{\boldsymbol{t}} \text{ on } \partial\Omega^\sigma_G, \tag{16}$$

where $\boldsymbol{n}$ is the outward unit normal vector to $\partial\Omega^\sigma_G$ and $\bar{\boldsymbol{t}}$ and $\bar{\boldsymbol{u}}$ are prescribed tractions and displacements, respectively.

The solution of the following problem provides a GFEM approximation of $\boldsymbol{u}$:

Find $\boldsymbol{u}^0_G \in \boldsymbol{X}^{hp}_G(\Omega_G) \subset H^1(\Omega_G)$ such that, $\forall\, \boldsymbol{v}^0_G \in \boldsymbol{X}^{hp}_G(\Omega_G)$,

$$\int_{\Omega_G} \boldsymbol{\sigma}(\boldsymbol{u}^0_G) : \boldsymbol{\varepsilon}(\boldsymbol{v}^0_G) d\boldsymbol{x} + \eta \int_{\partial\Omega^u_G} \boldsymbol{u}^0_G \cdot \boldsymbol{v}^0_G d\boldsymbol{s} = \int_{\partial\Omega^\sigma_G} \bar{\boldsymbol{t}} \cdot \boldsymbol{v}^0_G d\boldsymbol{s} + \eta \int_{\partial\Omega^u_G} \bar{\boldsymbol{u}} \cdot \boldsymbol{v}^0_G d\boldsymbol{s} \tag{17}$$

where $\boldsymbol{X}^{hp}_G(\Omega_G)$ is a discretization of $H^1(\Omega_G)$ built with the GFEM shape functions defined in Section 2 and η is a penalty parameter. In the computations of Section 4, we adopt $\eta = 10^8 * E * J$, where E and J are the Young modulus of the material and the Jacobian on a volume element with a face on Ω^u_G, respectively. Problem (17) leads to a system of linear equations for the unknown degrees of freedom of $\boldsymbol{u}^0_G$. The global problem is typically solved on a coarse mesh, without refinements around the crack front.

3.2 Local Problems

Let $\boldsymbol{x}_\alpha$ denote a point along the crack front Γ_{front} and $\Omega_{loc,\alpha}$ a neighborhood of $\boldsymbol{x}_\alpha$. This *local domain*, $\Omega_{loc,\alpha}$, is composed of a set of finite elements containing point $\boldsymbol{x}_\alpha$ enlarged with neighboring elements. Examples of local domains are given in Section 3.3.

The following *local problem* is solved on $\Omega_{loc,\alpha}$ after the global solution $\boldsymbol{u}^0_G$ is computed as described above:

Find $\boldsymbol{u}_{loc,\alpha} \in \boldsymbol{X}^{hp}_{loc,\alpha}(\Omega_{loc,\alpha}) \subset H^1(\Omega_{loc,\alpha})$ such that, $\forall\, \boldsymbol{v}_{loc,\alpha} \in \boldsymbol{X}^{hp}_{loc,\alpha}(\Omega_{loc,\alpha})$

$$\begin{aligned}
&\int_{\Omega_{loc,\alpha}} \boldsymbol{\sigma}(\boldsymbol{u}_{loc,\alpha}) : \boldsymbol{\varepsilon}(\boldsymbol{v}_{loc,\alpha}) d\boldsymbol{x} + \eta \int_{\partial\Omega_{loc,\alpha} \setminus (\partial\Omega_{loc,\alpha} \cap \partial\Omega^\sigma_G)} \boldsymbol{u}_{loc,\alpha} \cdot \boldsymbol{v}_{loc,\alpha} d\boldsymbol{s} \\
&= \eta \int_{\partial\Omega_{loc,\alpha} \setminus (\partial\Omega_{loc,\alpha} \cap \partial\Omega_G)} \boldsymbol{u}^0_G \cdot \boldsymbol{v}_{loc,\alpha} d\boldsymbol{s} \\
&\quad + \eta \int_{\partial\Omega_{loc,\alpha} \cap \partial\Omega^u_G} \bar{\boldsymbol{u}} \cdot \boldsymbol{v}_{loc,\alpha} d\boldsymbol{s} + \int_{\partial\Omega_{loc,\alpha} \cap \partial\Omega^\sigma_G} \bar{\boldsymbol{t}} \cdot \boldsymbol{v}_{loc,\alpha} d\boldsymbol{s}
\end{aligned} \tag{18}$$

where, $\boldsymbol{X}^{hp}_{loc,\alpha}(\Omega_{loc,\alpha})$ is a discretization of $H^1(\Omega_{loc,\alpha})$ using GFEM shape functions.

A key aspect of problem (18) is the use the generalized FEM solution of the (crude) global problem, $\boldsymbol{u}^0_G$, as boundary condition on $\partial\Omega_{loc,\alpha}\backslash(\partial\Omega_{loc,\alpha} \cap \partial\Omega_G)$. The explanation why this is admissible is given in Section 1.

The GFEM solution $\boldsymbol{u}_{loc,\alpha}$ can now be used as an enrichment function for nodes in the global mesh whose support, ω_α, is contained in the local domain $\Omega_{loc,\alpha}$. The corresponding global GFEM shape function is given by

$$\phi_\alpha = \varphi_\alpha \boldsymbol{u}_{loc,\alpha} \tag{19}$$

where φ_α denotes a global partition of unity. The global problem (17) is then solved using these global shape functions. As discussed in Section 1, we expect that $\boldsymbol{u}_{loc,\alpha}$ can approximate well the exact solution $\boldsymbol{u}$ *inside* $\Omega_{loc,\alpha}$ although close to the boundary of $\Omega_{loc,\alpha}$ the approximation can be worse because the global crude boundary condition can be innacurate. Nevertheless, this possibly worse approximation is eliminated since the global partition of unity φ_α is zero at the boundary of $\Omega_{loc,\alpha}$.

The proposed approach is related to global-local techniques developed for the classical finite element method in the 1970s [29] and broadly used in many practical applications of the FEM. A fundamental difference, however, is that the global-local GFEM procedure accounts for possible interactions of local (near crack, for example) and global (structural) behavior. This is in contrast with standard global-local FEM. Our approach is also related to upscaling techniques for microscale problems [22]. However, the proposed approach does not lead to non-conforming approximations like in some upscaling techniques [22]. In our case, the global shape functions build with local solution enrichment functions $\boldsymbol{u}_{loc,\alpha}$ overlap and are conforming.

3.3 Example: Three-Dimensional Through-the-Thickness Crack

The construction of enrichment functions using the proposed global-local approach is illustrated in this section. Consider the edge-cracked rectangular bar shown in Figure 4 and denoted henceforth as the global domain Ω_G. It contains a through-the-thickness edge crack, Γ_c, with front Γ_{front}.

A discretization of the global problem is denoted by $G^{p=(p_x,p_y,p_z)}$ where p_x, p_y and p_z denote the polynomial degree of the shape functions in the x-, y-, and z-direction, respectively. Orthotropic p-enrichment of tetrahedral meshes is discussed in [13]. No mesh refinement is done at the crack front. Therefore, the global GFEM solution, $\boldsymbol{u}_G$, computed on the mesh shown in Figure 4 has a large error in the energy norm since this discretization is not capable of capturing the singular behavior of the solution near the crack front. On the other hand, the computational cost of computing $\boldsymbol{u}_G$ is small if the polynomial order of the approximation is not high.

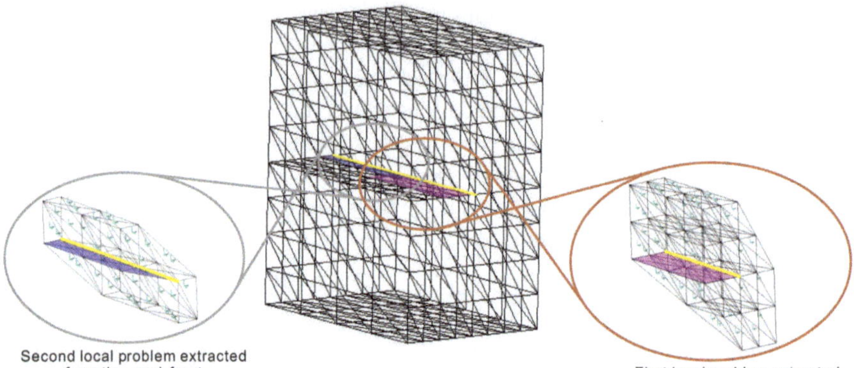

Fig. 4. Global problem with an edge crack and the first step of the global-local approach. Local problems are automatically created along a crack front of a coarse global mesh.

Step 1: Construction of Local Discretizations: Discretizations for local problems defined in a neighborhood of the crack front are constructed as follows. We restrict the definitions to the case of a global discretization with cracks fronts located along element boundaries, like the case shown in Figure 4. A situation like that shown in Figure 2b, in which the crack front is arbitrarily located in the mesh, can be dealt with analogously.

Let $\mathcal{N}_{front}$ denote the indices of a set of nodes from the global mesh located along the crack front Γ_{front}. Local meshes are created using elements extracted from the global mesh around the crack front. A local domain corresponding to a mesh with one layer of elements around the crack front is given by

$$\Omega^{nlay=1}_{loc,N_{front}} := \bigcup_{\alpha \in N_{front}} \omega_\alpha$$

where ω_α is the union of (copy of) global elements sharing vertex node $\boldsymbol{x}_\alpha$. Local domains with additional layers of elements around the crack front are defined analogously.

Figure 4 illustrates two local problems extracted from the coarse global mesh shown in the figure. The number of local problems created can be chosen such that each local problem can be efficiently solved in a single processor using, for example, a direct sparse solver.

The boundary conditions applied to a local problem are displacements computed by solving the global problem on a coarse mesh. These boundary conditions are also illustrated in Figure 4.

Step 2: Solution of Local Problems on *hp* Refined Meshes: In this step, the local meshes are *h*-refined and *p*-enriched in order to capture the singularity at the crack

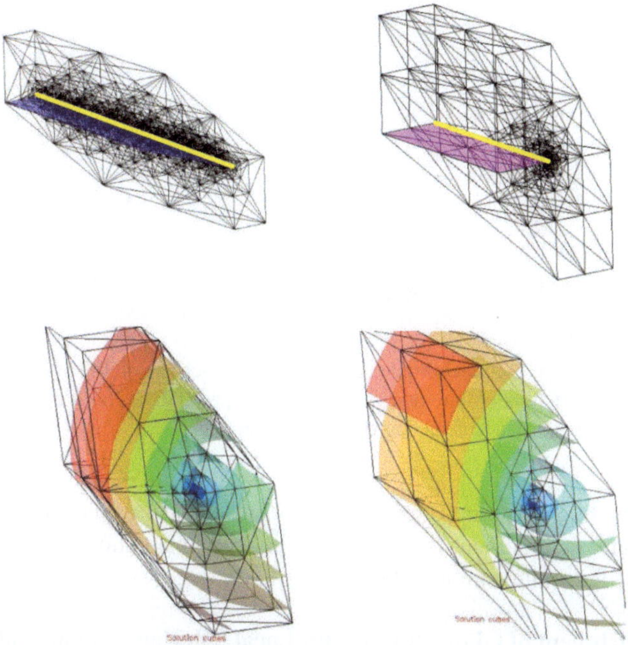

Fig. 5. Refined local meshes and computed local solutions. Each local problem can be automatically refined and solved in parallel and thus a large number of local problems can be efficiently solved.

front. Very refined meshes can be used in each local problem since the local domains are much smaller than the global one. This allows the computation of highly accurate local approximations of the global solution. Figure 5 illustrates this step of the global-local GFEM.

A discretization of a local problem is denoted by $L_{nref,\, nlay}^{p=(p_x,p_y,p_z)}$ where: (i) p_x, p_y and p_z denote the polynomial degree in the x-, y-, and z-direction, respectively; (ii) $nlay$ denotes the number of layers of elements around the crack front in the mesh extracted from the global problem as previously described; (iii) $nref$ indicates the level of the local (uniform or non-uniform) mesh refinement. A non-uniform mesh refinement with $nref = 1$ is performed by first bisecting all elements in the initial mesh with nodes on the crack front and then bisecting additional elements in order to recover a conforming discretization. The marked-edge algorithm [1, 5] is used to select the refinement edges of the elements. This procedure is repeated $n-1$ times for a refinement level $nref = n$. The initial mesh extracted from the global mesh corresponds to $nref = 0$. Figure 5 shows an example of the application of this algorithm on local meshes extracted from the global mesh. Isosurfaces of the computed solutions of the local problems are also shown in the figure. This local

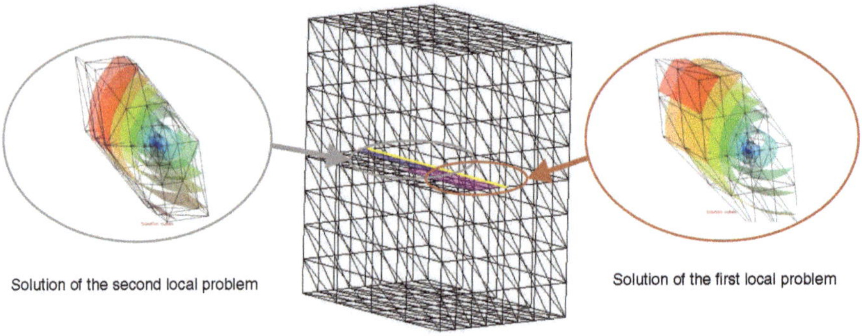

Fig. 6. The solutions of the local problems are used to enrich the global problem. Only a few additional degrees of freedom are added to the global problem in this step even if the local solutions have several thousands of degrees of freedom.

mesh refinement algorithm preserves the nesting of local elements into the global mesh. This greatly facilitates the computational implementation and provides many opportunities for optimization of the code.

Step 3: Enrichment of Global Space with Local Solutions: Figure 6 illustrates the last step in the global-local GFEM. The global problem is enriched with the solutions of the local problems. This is performed using Equation (19). Only a few additional degrees of freedom are added to the global problem even if the local solutions have a large number of degrees of freedom. The enriched global space has very good approximation properties even though it is built on a coarse mesh. This is due to the use of custom-built enrichment functions provided by the local problems and the error estimate given in (14).

4 Numerical Example

As an illustrative example of the proposed global-local GFEM, a Multiple Site Damage(MSD) example is introduced in this section. MSD problems focus on the combined effect of multiple growing cracks where each individual crack can be harmless, but the combined effect caused by several cracks can be disastrous. Hence, it is necessary to consider many possible damage locations and damage sizes in the structure [2]. MSD problems are a good application for the global-local GFEM since they contain a series of localized stress distributions which can be separately modeled in a local problem of the global-local GFEM. In this numerical experiment, we deal with only one combination of possible MSD cracks.

We consider the same MSD example as presented in [44]. In this example, there are two small MSD cracks on the left and right side of the main crack as shown in

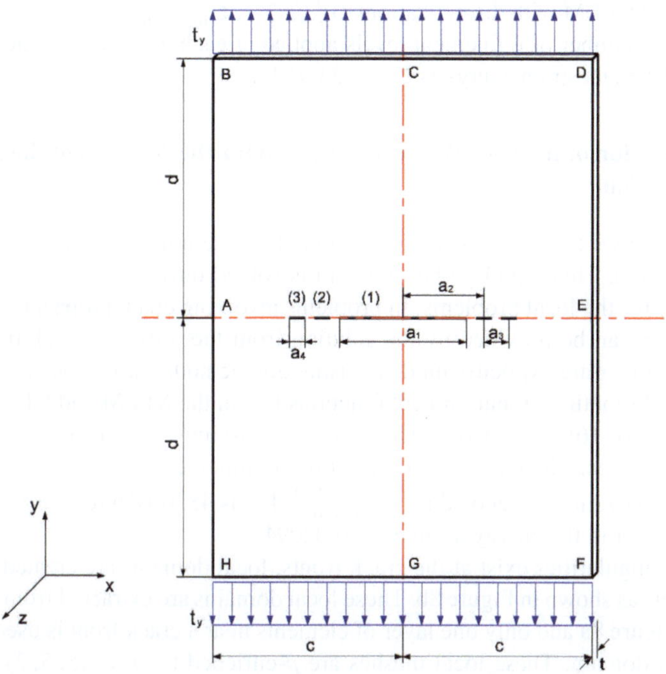

Fig. 7. Multiple Site damage(MSD) crack example [44] (not drawn to scale). The main crack is indicated as (1) and MSD cracks on the left of the main crack as (2) and (3).

Figure 7. The constant far field stress $t_y = 20$ ksi is applied on the top and bottom of the panel and it causes the opening of the main and MSD cracks. In the computations, only half of the entire domain (domain CDFG in Figure 7) is discretized. Symmetry boundary conditions are applied along the vertical plane of symmetry. A quasi uniform mesh with $(20 * 6) \times (14 * 6) \times (1 * 6)$ tetrahedral elements is used to discretize the global problem. No refinement of the global mesh is done at the crack fronts. The following parameters are assumed in this model: Poisson's ratio $\nu = 0.33$; Young's modulus $E = 10,500$ ksi; In-plane dimensions $d = 75.0$ in, $c = 45.0$ in, $a_1 = 20.0$ in, $a_2 = 11.5$ in, $a_3 = 2.0$ in; Size of MSD cracks $a_4 = 1.0$ in; Domain thickness $t = 1.0$ in.

The exact strain energy is required for the computation of the error in the energy norm. We estimated the value of the exact strain energy using the a-posteriori error estimate of Szabo and Babuška [43] since this problem does not have an analytical solution. In this procedure, the exact strain energy is estimated using a sequence of finite element solutions and a-priori error estimates for *h*- or *p*-extensions. Details are described in [43]. The accuracy of the estimate greatly depends on the accuracy of approximate solutions. For this problem, we solved the MSD problem using the standard GFEM with discretizations $G^{p=(2,2,2)}_{nref=6}$, $G^{p=(3,3,3)}_{nref=6}$ and $G^{p=(4,4,4)}_{nref=6}$ where $nref$ is the number of refinement levels applied at each crack front. The estimated value of the exact strain energy is $U = 134.9445$.

4.1 Illustration of the Global-Local Approach for the Analysis of the MSD Crack Problem

Figure 8a shows the mesh of the initial global problem used in the analysis of the MSD problem. This initial global problem is solved only to provide the boundary conditions for the local problems. In problems involving crack propagation, this initial solution can be provided by the solution from the previous crack propagation step. The cracks are explicitly modeled using double nodes along the crack surfaces instead of discontinuous enrichment functions like in the XFEM and GFEM presented in [26] and [16], respectively. The mesh is almost uniform and very coarse for the given crack sizes. The polynomial order used in this mesh was $p = (5, 5, 2)$. This global discretization is denoted by $G^{p=(5,5,2)}_{nref=0}$. It has 48300 degrees of freedom and a relative error in the energy norm $e_r = 0.02094$

Since singularities exist at the crack fronts, local domains are created at all five crack fronts as shown in Figure 8b. These local domains are extracted from the global mesh in Figure 8a and only one layer of elements near a crack front is used to define each local domain. These local meshes are p-enriched to $p = (5, 5, 2)$ and three levels of refinement are applied to the crack fronts. The resulting local discretizations are denoted by $L^{p=(5,5,2)}_{nref=3,nlay=1}$. Westergaard singular enrichment functions could, of course, be used at the crack fronts as described in, for example [15, 16, 30, 32]. However, since our goal here is to demonstrate the effectiveness of the global-local GFEM for problems in which such enrichment functions are not known, we rely *only* on *hp*-refinements to control the discretization error.

Figure 9 displays von Mises stress distribution and displacement for the five local problems created. The level of von Mises stress in this figure is represented by the legend included in Figure 10. Notice that the local domain created at the front of the main crack has the largest von Mises stress and displacement, whereas other local domains have relatively smaller von Mises stress and displacement.

After the solution of local problems, all nodes on the crack fronts of the global mesh are enriched with the local solutions as shown in Figure 8c. The global problem enriched with local solutions uses the same coarse mesh as the initial global problem and no refinement is done in this global problem. The elements near the crack fronts which came from the local problems are included in Figure 8c just for the visualization of the global solution with local solution enrichment. These elements are also used for numerical integration in the global domain. The enriched global discretization has 47880 degrees of freedom and a relative error in the energy norm $e_r = 0.009737$. The ratio between the error of the enriched and the initial global

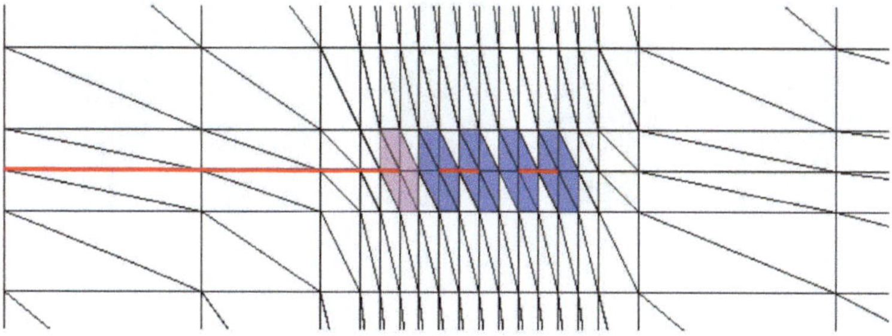

(a) Initial global mesh and corresponding local domains at each one of the five crack fronts.

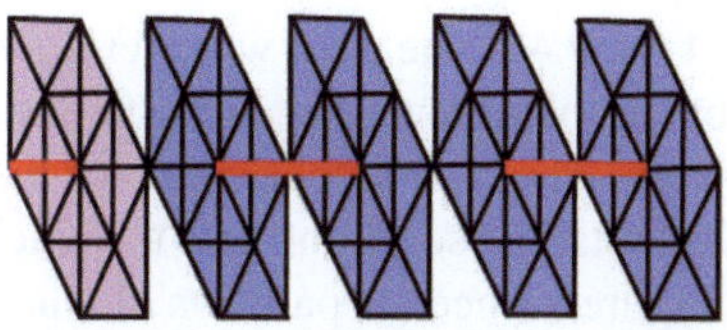

(b) Local meshes(level of refinement: $nref = 3$).

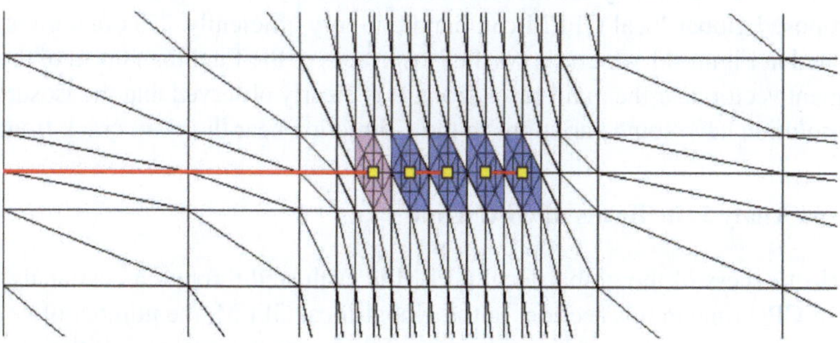

(c) Global mesh with nested local meshes used for visualization of global solution. Yellow squares on the crack fronts represent the nodes enriched by the local solutions.

Fig. 8. Discretization of the MSD problem in the global-local GFEM.

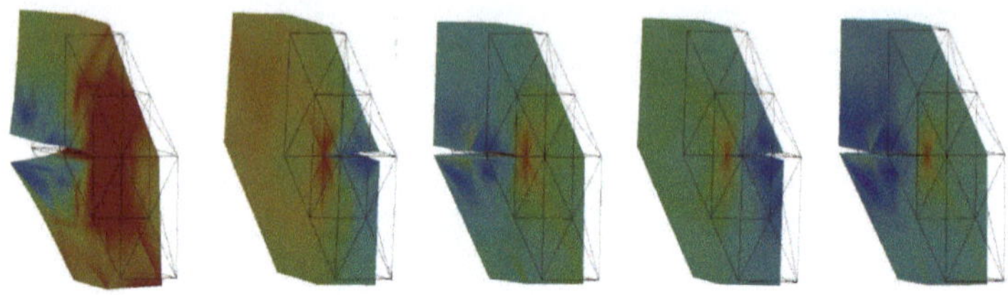

(a) (b) (c) (d) (e)

Fig. 9. Von Mises stress distribution and deformed shape of each local problem. Figure (a) displays the result of the main crack front. Figures (b), (c), (d) and (e) are the results of MSD cracks and ordered according to their relative locations from the main crack front.

solutions is $0.009737/0.02094 = 0.46$. The error was therefore reduced to less than half while the size of the global problem remained almost constant.[1]

Figure 10 displays von Mises stress and displacement of the global problem enriched with local solutions. The obtained solutions near the crack front are at least as good as the local solutions in Figure 9 since the partition of unity not only reproduces the local solution, but also adds one polynomial order to it as indicated in (19).

Although this MSD example has a very small thickness compared to other dimensions, three-dimensional effects near the crack front are developed. Two-dimensional expansions of the elasticity solution have limitations in capturing this effect, whereas the proposed global-local GFEM can handle it very efficiently. This feature is well illustrated in Figure 11 which shows the isosurfaces of the Euclidian norm of the displacement vector near the main crack front. It is clearly observed that the isosurfaces of the solution have some variations in the z-direction near the main crack front.

4.2 Cost Analysis in Terms of CPU Time

The effectiveness of the global-local GFEM is evaluated through a cost analysis in terms of CPU time in this section. In the global-local GFEM, the number of degrees of the freedom cannot be used as a measure of the computational cost of the enriched global problem. This is because the number of degrees of freedom before and after enrichment of the global problem with local solutions is almost the same and they do not account for the computational cost of solving the local problems and the

[1] The size of the enriched global problem (47880) is *smaller* than the initial global problem (48300). This happens because in our implementation, the polynomial order of a node is reduced by one degree when a custom enrichment (a local solution in this case) is assigned to the node. This leads to a net reduction in the number of degrees of freedom at the node. Of course, we can keep the polynomial shape functions at a node by increasing the order assigned to the node by one before adding the custom enrichment.

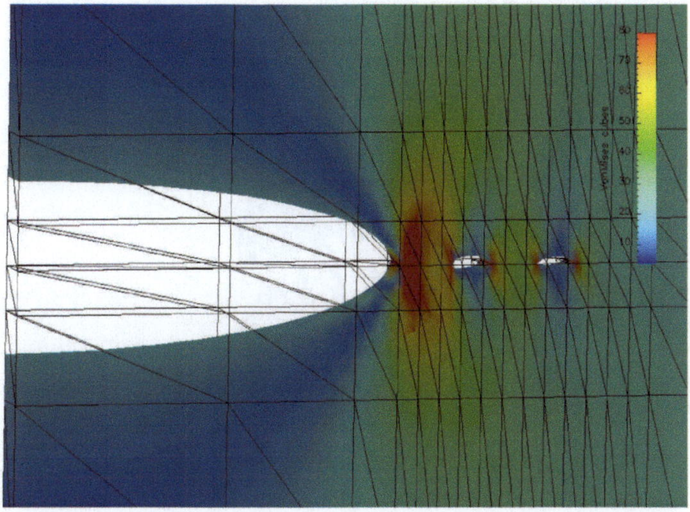

Fig. 10. Von Mises stress distribution and deformed shape of MSD example after enrichment of global problem with the local solutions shown in Figure 9.

additional assemblage procedure used in the enriched global problem. We use the CPU time as a measure of the computational cost of the enriched global problem. The reported CPU time for the enriched global problem includes both the local and global computations. All the computations were performed on a Dell Dimension 4600 PC with a 3.2 MHz Pentium processor and 2 GBs of memory.

The computational performance of the proposed global-local GFEM is compared with the standard GFEM. In the later, the discretization error is controled through *hp* refinements of the global mesh. Two problems are used in the analysis: The MSD problem described in Section 4.1 and the same problem without the MSD cracks indicated as (2) and (3) in Figure 7. In this case, it has only the single main crack shown in Figure 7. The problems with and without MSD cracks use five and one local problem, respectively. Therefore, the analysis presented here can illustrate the effect of the number of local problems on the effectiveness of the proposed global-local GFEM. The details of the analysis are as follows.

The exact strain energy for the problem with the single main crack was also estimated using the a-posteriori error estimate described in the previous section. The standard GFEM with discretizations $G^{p=(2,2,2)}_{nref=10}$, $G^{p=(3,3,3)}_{nref=10}$ and $G^{p=(4,4,4)}_{nref=10}$ were used in this case. The estimated value of the exact strain energy is $U = 134.7412$.

In the analysis using the global-local GFEM, all local problems in the MSD and single main crack problem are solved with the following parameters: $L^{p=(5,5,2)}_{nlay=1,nref}$, $0 \leq nref \leq 13$, and Dirichlet boundary conditions obtained from the solution of the global problem $G^{p=(5,5,2)}_{nref=0}$. A relatively low polynomial order, $p_z = 2$, is used

Fig. 11. Isosurface of the solution near the main crack front. 3D effects near the crack front are captured.

in the z-direction since the thickness of the panel is much smaller than the other dimensions (cf. Figure 7). The polynomial shape functions of degree $p = (5, 5, 2)$ can approximate well the smooth part of the solution. In all local discretizations, non-uniform h-extensions are performed on the elements near the crack fronts using the marked-edge algorithm mentioned in Section 3.3. The level of refinement along the crack fronts is indicated through the parameter $nref$. The solutions of the local problems are used as enrichment functions for the global problem $G_{nref=0}^{p=(5,5,2)}$.

The same polynomial order, $p = (5, 5, 2)$, as in the global-local GFEM analysis is used in the standard GFEM. The only difference is that some elements in the neighborhood of the local domains are slightly more refined in the standard GFEM analysis (compare Figures 8c and 12). This is a consequence of the mesh refinement required to keep the global mesh conforming. As a result, it is expected that this additional refinement can reduce errors in those elements slightly more than in the global-local GFEM analysis. The global mesh used in standard GFEM is illustrated in Figure 12.

In this numerical example, the maximum level of h-refinement is limited by the maximum amount of memory a 32 bit machine can allocate. Therefore, the examples analyzed by the global-local approach are expected to have a higher level of maximum h-refinement since the local problems are much smaller than the global problem solved with standard GFEM and we can release the dynamic memory assigned to local problems after computing their solutions.

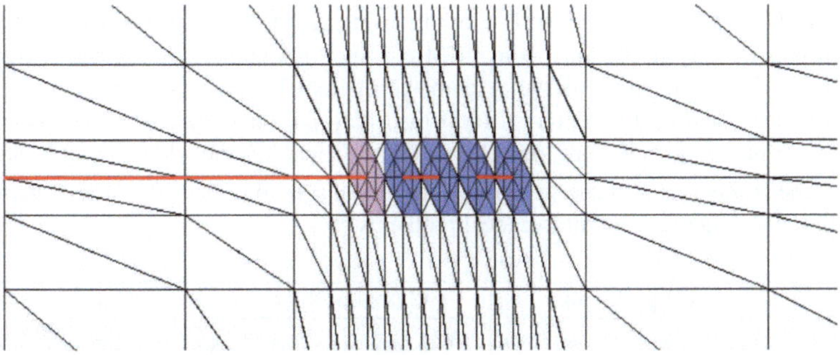

Fig. 12. Discretization of the MSD problem in the standard GFEM.

The results of non-uniform h-extensions done on the domain of the MSD and single main crack example using both the global-local and standard GFEM are presented in Tables 1, 2, 3 and 4. The tables list the number of degrees of freedom, the relative error in the energy norm in the global domain and the CPU time spent on the assemblage (t_{assem}), factorization (t_{fact}) and backward and forward substitution (t_{sub}) of the stiffness matrix in both local and global problems. t_{assem}, t_{fact} and t_{sub} are the three essential components which comprise most of the computational cost of the analysis. The local t_{assem}, t_{fact} and t_{sub} reported in Table 1 are the total CPU time spent in all five local problems. Tables 1, 3 and 2, 4 also report t_{tot} which measures the total CPU time spent in a global-local or in a standard GFEM analysis. In case of a global-local analysis, t_{tot} also includes the CPU time spent in the local problems.

From the results reported in the tables, some unique features of the global-local GFEM in the composition of the CPU time can be identified. As the number of degrees of freedom in the local problem increases (as the local domain is h-refined), all t_{assem}, t_{fact} and t_{sub} of the local problem increase, whereas only t_{assem} increases in the global problem. This is because the enrichment of the global problem with local solutions changes little the number of degrees of freedom of the global problem. In addition, the number of degrees of freedom in the global problem does not depend on the size of the local problems. However, t_{assem} in the global problem increases since mesh refinement in the local problem requires more integration points in the global elements enriched with local solutions. It can also be noted that t_{assem} spent on the global problem of the MSD example is larger than t_{assem} of the single main crack example when the same level of h-refinemnet is done in each local problem (compare t_{assem} from Tables 1 and 3).

Figure 13 plots the values of t_{tot} versus e_r from Tables 1, 2, 3 and 4. Several observations can be made based on the plots. First, we observe a reduction in the convergence rate with mesh refinement in both the standard and the global-local GFEM. This happens because the mesh refinement is applied *only* to elements at

Table 1. CPU time spent on the MSD example using global-local GFEM. Non-uniform h-extension is done only in the local domains. Here, N is the number of degrees of freedom, t_{assem} is the CPU time spent on the assemblage of the stiffness matrix, t_{fact} is the CPU time spent on the factorization of the stiffness matrix, t_{sub} is the CPU time spent on back and forward substitution, t_{tot} measures the total CPU time spent in a global-local GFEM analysis and e_r^{gl} is the relative error in energy norm of the global solution. The CPU times reported for the local problems account for all five local problems.

	Local problems				Global problem					
$nref$	N	t_{assem}	t_{fact}	t_{sub}	N	t_{assem}	t_{fact}	t_{sub}	t_{tot}	e_r^{gl}
0	1200	6.40	5.59	0.14	47880	93.97	329.41	8.16	443.67	0.014657
3	3300	32.00	35.00	0.49	47880	109.19	329.36	8.15	514.19	0.009737
7	8025	113.47	182.79	1.77	47880	157.90	328.36	8.21	792.50	0.007013
10	15600	250.01	841.53	4.06	47880	232.07	328.80	8.41	1664.88	0.004585
13	30075	524.82	3129.59	9.28	47880	381.94	329.35	8.15	4383.13	0.002265
∞	∞				47880					0

Table 2. CPU time spent on the MSD crack example using standard GFEM. Non-uniform h-extension is done in the global domain. Here, t_{tot} is the CPU time spent on a standard GFEM analysis.

$nref$	N	t_{assem}	t_{fact}	t_{sub}	t_{tot}	e_r
0	48300	96.20	378.03	8.55	482.78	0.020942
2	54675	111.33	420.72	9.55	541.60	0.013783
5	70050	156.21	485.56	12.25	654.02	0.008448
∞	∞					0

Table 3. CPU time spent on the single main crack example using global-local GFEM. Non-uniform h-extension is done only in the local domain.

	Local problem				Global problem					
$nref$	N	t_{assem}	t_{fact}	t_{sub}	N	t_{assem}	t_{fact}	t_{sub}	t_{tot}	e_r^{gl}
0	1200	1.31	1.13	0.04	47916	95.42	332.04	8.37	438.31	0.011340
3	3300	6.35	6.88	0.09	47916	98.97	332.28	8.35	452.92	0.007784
7	8025	22.74	36.49	0.33	47916	108.91	332.49	8.39	509.35	0.005910
10	15600	50.18	166.80	1.85	47916	123.91	332.82	8.33	682.87	0.004368
13	30075	104.46	600.13	1.85	47916	152.79	332.97	8.31	1200.51	0.003183
∞	∞				47916					0

Table 4. CPU time spent on the single main crack example using standard GFEM. Non-uniform h-extension is done in the global domain.

$nref$	N	t_{assem}	t_{fact}	t_{sub}	t_{tot}	e_r
0	48000	95.58	348.81	8.43	452.82	0.015684
3	50100	101.04	369.76	8.85	479.65	0.009370
7	54825	118.42	402.65	9.78	530.85	0.005848
10	62400	144.16	580.74	12.10	737.00	0.003607
∞	∞					0

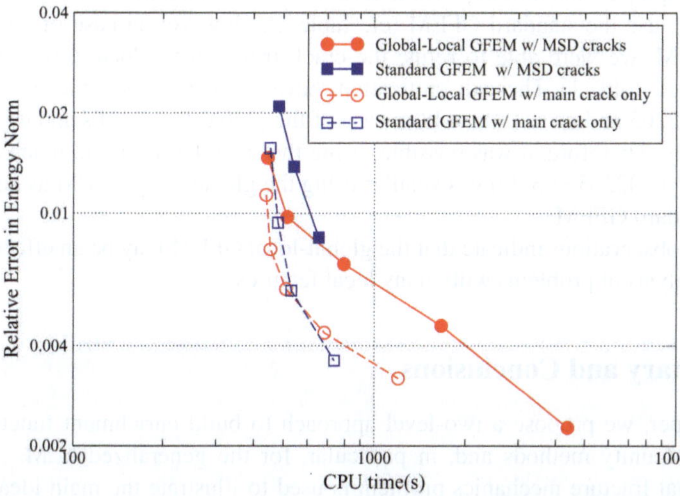

Fig. 13. Computational cost for the MSD and single main crack problems using the proposed local-global GFEM and the standard GFEM. The discreatization error in the local-global GFEM is controled through mesh refinements in the local problems while global mesh refinements are used only in the case of the standard GFEM.

the crack fronts followed by additional refinements required to recover a conforming mesh. Therefore, the mesh refinement is only controlling the error in a small neighborhood of the crack front. The decrease in convergence rate is less pronounced in the case of the standard GFEM since, in this case, the mesh refinement is applied to the global mesh and the refinements used to remove hanging nodes propagates in the global domain. The convergence rate in the global-local GFEM may also be significantly affected by the cost of the local problems. As the local problems are h-refined, the computational cost of the local problems becomes the major cost in the global-local analysis (cf. Tables 1 and 3). This, of course, depends on the size of the original global problem. Therefore, there is a limitation for the size of the local problems where the global-local GFEM is more computationally efficient than the standard GFEM.

Second, the global-local GFEM is more efficent than the standard GFEM over a wider rage of CPU time when solving the MSD problem than when solving the single main crack problem. Therefore, the efficiency of the global-local GFEM as compared to the standard GFEM can be extended by increasing the number of local problems. The number of local problems and consequently the effectiveness of the global-local GFEM can be increased by using more than one local problem at each crack front. The solution of the local problems is also amenable to parallel processing.

Third, the global-local GFEM is less affected by hardware limitations. For the MSD example, the maximum possible level of h-refinement near the crack fronts is 5 when we use the standard GFEM (cf. Table 2). However, in case of the global-local GFEM, we were able to refine the crack fronts in the local problems up to 13 levels (cf. Table 1). These discretizations have relative errors in the energy norm $e_r = 0.002265$ and $e_r = 0.008448$ in case of the global-local and standard GFEM, respectively. Therefore, it was possible, using the same hardware, to reach an error $0.008448/0.002265 = 3.7$ times smaller using the global-local GFEM as compared to the standard GFEM.

These observations indicate that the global-local GFEM may be an effective tool for the analysis of problems with many local features.

5 Summary and Conclusions

In this paper, we propose a two-level approach to build enrichment functions for partition of unity methods and, in particular, for the generalized FEM. A three-dimensional fracture mechanics problem is used to illustrate the main ideas of the procedure. The approach is not, however, limited to this particular application.

The main features of the proposed generalized FEM with global-local enrichment functions are as follows:

- The procedure accounts for possible interactions of local (near crack for example) and global (structural) behavior. This is in contrast with standard global-local FEM [29] which is broadly used in many engineering applications of the FEM.
- Local features, like crack fronts, are several orders of magnitude smaller than the size of the domain of interest (a complex structure, for example). The mesh density required in the neighborhood of a crack front is several orders of magnitude larger than in parts of the domain with smooth solutions. As a result, mesh refinements at the crack front usually propagate beyond the region of interest in order to create a conforming global mesh. In the proposed global-local GFEM, this problem does not exist since the refinement is done only at the local domains. Therefore, no matter how refined are the local domains, no unnecessary refinement is performed in the, usually large, global problem.

- The computation of all local problems can be parallelized without difficulty allowing the solution of large problems very efficiently. This feature of the methodology is related to the various domain decomposition techniques available in the literature [6, 9, 10, 33, 35].
- The size of the global problems solved with the proposed approach is almost independent of the resolution used to resolve the local features. This happens because the number of degrees of freedom added to the global problem does *not* depend on the number of degrees of freedom used in the local problems. It depends only on the number of global nodes enriched with local solutions.
- The enrichment of the global mesh with local solutions does not lead to linear dependencies of the resulting global shape functions since the global partition of unity itself cannot reproduce the solution of the local problem unless it is exactly in the space spanned by the partition of unity itself.

The generalized FEM has had a tremendous impact in the solution of problems in which there is an a-priori knowledge about the behavior of the solution (Helmholtz equation, linear elastic fracture mechanics, domains with many voids or inclusions, etc.). The proposed methodology, by numerically constructing the enrichment functions, brings the benefits of existing generalized FEM to a broader class of problems.

This paper presents the main ideas and investigate the computational performance of the proposed generalized FEM with global-local enrichment functions. There are, however, many open issues that need further investigation. Among them, we have:

- The effect of using the solution of the global problem, instead of the unknown exact solution, as boundary conditions for the local problems;
- The effect of the size of the local domains on the performance of the proposed global-local GFEM. Larger domains are in principle desirable since they would reduce the effect of not using exact boundary conditions for the local problems. However, large domains are more computationally demanding;
- The effectiveness of the proposed technique in terms of convergence rates of the global problem and its relations with the convergence rates of the local problems;
- How to efficiently implement the procedure.

We are currently investigating these topics.

Acknowledgments

The partial support of this work by The University of Illinois (C.A. Duarte and D.-J. Kim), the National Science Foundation under grant DMS-0341982 (C.A. Duarte and I. Babuška) and the Korea Science and Engineering Foundation (D.-J. Kim) Project Number M06-2003-000-10312-0, are gratefully acknowledged.

References

1. D. N. Arnold, A. Mukherjee, and L. Pouly. Locally adapted tetrahedral meshes using bisection. *SIAM Journal of Scientific Computing*, 22(2):431–448, 2000.
2. I. Babuška and B. Andersson. The splitting method as a tool for multiple damage analysis. *SIAM journal on scientific computing*, 26:1114–1145, 2005.
3. I. Babuška, G. Caloz, and J. E. Osborn. Special finite element methods for a class of second order elliptic problems with rough coefficients. *SIAM J. Numerical Analysis*, 31 (4):745–981, 1994.
4. I. Babuška and J. M. Melenk. The partition of unity finite element method. *International Journal for Numerical Methods in Engineering*, 40:727–758, 1997.
5. E. Bansch. Local mesh refinement in 2 and 3 dimensions. *Impact of Computing in Science and Engineering*, 3:181–191, 1991.
6. R. Becker and P. Hansbo. A finite element method for domain decomposition with non-matching grids. Technical Report RR-3613, INRIA, 1999.
7. T. Belytschko and T. Black. Elastic crack growth in finite elements with minimal remeshing. *International Journal for Numerical Methods in Engineering*, 45:601–620, 1999.
8. T. Belytschko, N. Moes, S. Usui, and C. Parimi. Arbitrary discontinuities in finite elements. *International Journal for Numerical Methods in Engineering*, 50:993–1013, 2001.
9. C. Bernadi, Y. Maday, and A. Patera. A new non-conforming approach to domain decomposition: The mortar element method. In H. Brezis and J. L. Lions (Eds), *Nonlinear Partial Differential Equations and Their Applications*. Pitman, 1994, pp. 13–51.
10. E. G. D. Carmo and A. V. C. Duarte. A discontinuous finite element-base domain decomposition method. *Computer Methods in Applied Mechanics and Engineering*, 190: 825–843, 2000.
11. J. Dolbow, N. Moes, and T. Belytschko. Discontinuous enrichment in finite elements with a partition of unity method. *Finite Elements in Analysis and Design*, 36:235–260, 2000.
12. C. A. Duarte. *The hp Cloud Method*. PhD Dissertation, The University of Texas at Austin, December 1996. Austin, TX, USA.
13. C. A. Duarte and I. Babuška. Mesh-independent directional *p*-enrichment using the generalized finite element method. *International Journal for Numerical Methods in Engineering*, 55(12):1477–1492, 2002.
14. C. A. Duarte, I. Babuška, and J. T. Oden. Generalized finite element methods for three dimensional structural mechanics problems. In S. N. Atluri and P. E. O'Donoghue (Eds), *Modeling and Simulation Based Engineering*, Volume I, Proceedings of the International Conference on Computational Engineering Science, Atlanta, GA, October 5–9, 1998. Tech Science Press, 1998, pp. 53–58.
15. C. A. Duarte, I. Babuška, and J. T. Oden. Generalized finite element methods for three dimensional structural mechanics problems. *Computers and Structures*, 77:215–232, 2000.
16. C. A. Duarte, O. N. Hamzeh, T. J. Liszka, and W. W. Tworzydlo. A generalized finite element method for the simulation of three-dimensional dynamic crack propagation. *Computer Methods in Applied Mechanics and Engineering*, 190:2227–2262, 2001.
17. C. A. M. Duarte and J. T. Oden. *Hp* clouds – A meshless method to solve boundary-value problems. Technical Report 95-05, TICAM, The University of Texas at Austin, May 1995.
18. C. A. M. Duarte and J. T. Oden. An *hp* adaptive method using clouds. *Computer Methods in Applied Mechanics and Engineering*, 139:237–262, 1996.

19. C. A. M. Duarte and J. T. Oden. *Hp* clouds – An *hp* meshless method. *Numerical Methods for Partial Differential Equations*, 12:673–705, 1996.
20. M. Griebel and M. A. Schweitzer. A particle-partition of unity method for the solution of elliptic, parabolic and hyperbolic PDEs. *SIAM Journal Scientific Computing*, 22(3): 853–890, 2000.
21. P. Grisvard. *Singularities in Boundary Value Problems*. Research Notes in Appl. Math. Springer-Verlag, New York, 1992.
22. T. Y. Hou and X.-H. Wu. A multiscale finite element method for elliptic problems in composite materials and porous media. *Journal of Computational Physics*, 134:169–189, 1997.
23. P. Lancaster and K. Salkauskas. Surfaces generated by moving least squares methods. *Mathematics of Computation*, 37(155):141–158, 1981.
24. J. M. Melenk and I. Babuška. The partition of unity finite element method: Basic theory and applications. *Computer Methods in Applied Mechanics and Engineering*, 139:289–314, 1996.
25. N. Moes, M. Cloirec, P. Cartraud, and J.-F. Remacle. A computational approach to handle complex microstructure geometries. *Computer Methods in Applied Mechanics and Engineering*, 192:3163–3177, 2003.
26. N. Moes, J. Dolbow, and T. Belytschko. A finite element method for crack growth without remeshing. *International Journal for Numerical Methods in Engineering*, 46:131–150, 1999.
27. N. Moës, A. Gravouil, and T. Belytschko. Non-planar 3D crack growth by the extended finite element and level sets – Part I: Mechanical model. *International Journal for Numerical Methods in Engineering*, 53:2549–2568, 2002.
28. S. A. Nazarov and B. A. Plamenevsky. *Elliptic Problems in Domains with Piecewise Smooth Boundaries*, De Gruyter Expositions in Mathematics, Vol. 13, Walter de Gruyter, Berlin, 1994.
29. A. K. Noor. Global-local methodologies and their applications to nonlinear analysis. *Finite Elements in Analysis and Design*, 2:333–346, 1986.
30. J. T. Oden and C. A. Duarte. Clouds, Cracks and FEM's. In B. D. Reddy (Ed.), *Recent Developments in Computational and Applied Mechanics*. International Center for Numerical Methods in Engineering, CIMNE, Barcelona, Spain, 1997, pp. 302–321.
31. J. T. Oden, C. A. Duarte, and O. C. Zienkiewicz. A new cloud-based *hp* finite element method. *Computer Methods in Applied Mechanics and Engineering*, 153:117–126, 1998.
32. J. T. Oden and C. A. M. Duarte. Solution of singular problems using *hp* clouds. In J. R. Whiteman (Ed.), *The Mathematics of Finite Elements and Applications – Highlights 1996*. John Wiley & Sons, New York, 1997, pp. 35–54.
33. A. Quarteroni and A. Valli. *Domain Decomposition Methods for Partial Differential Equations*. Oxford University Press, 1999.
34. A. Simone, C. A. Duarte, and E. van der Giessen. A generalized finite element method for polycrystals with discontinuous grain boundaries. *International Journal for Numerical Methods in Engineering*, in press, 2006.
35. B. Smith, P. Bjorstad, and W. Gropp. *Domain Decomposition: Parallel Multilevel Methods for Elliptic Partial Differential Equations*. Cambridge University Press, 2004.
36. T. Strouboulis, I. Babuška, and K. Copps. The design and analysis of the generalized finite element mehtod. *Computer Methods in Applied Mechanics and Engineering*, 81 (1–3):43–69, 2000.

37. T. Strouboulis, K. Copps, and I. Babuška. The generalized finite element method: An example of its implementation and illustration of its performance. *International Journal for Numerical Methods in Engineering*, 47(8):1401–1417, 2000.
38. T. Strouboulis, K. Copps, and I. Babuška. The generalized finite element method. *Computer Methods in Applied Mechanics and Engineering*, 190:4081–4193, 2001.
39. T. Strouboulis, L. Zhang, and I. Babuška. Generalized finite element method using mesh-based handbooks: Application to problems in domains with many voids. *Computer Methods in Applied Mechanics and Engineering*, 192:3109–3161, 2003.
40. T. Strouboulis, L. Zhang, and I. Babuška. *p*-version of the generalized FEM using mesh-based handbooks with applications to multiscale problems. *International Journal for Numerical Methods in Engineering*, 60:1639–1672, 2004.
41. N. Sukumar, D. Chopp, N. Moes, and T. Belytschko. Modeling holes and inclusions by level sets in the extended finite element method. *Computer Methods in Applied Mechanics and Engineering*, 190:6183–6200, 2001.
42. N. Sukumar, N. Moes, B. Moran, and T. Belytschko. Extended finite element method for three-dimensional crack modelling. *International Journal for Numerical Methods in Engineering*, 48(11):1549–1570, 2000.
43. B. Szabo and I. Babuška. *Finite Element Analysis*. John Wiley and Sons, New York, 1991.
44. L. Wang, F. W. Brust, and S. N. Atluri. The elastic-plastic finite element alternating method (EPFEAM) and the prediction of fracture under WFD conditions in aircraft structures. *Computational Mechanics*, 19:356–369, 1997.

Study of Some Optimal *XFEM* Type Methods

Elie Chahine[1], Patrick Laborde[2], Julien Pommier[1], Yves Renard[1] and Michel Salaün[3]

[1] *INSA, Laboratoire MIP, CNRS UMR 5640, Complexe scientifique de Rangueil, 31077 Toulouse, France; E-mail: {elie.chahine, pommier, renard}@insa-toulouse.fr*
[2] *Université P. Sabatier, Laboratoire MIP, CNRS UMR 5640, 118 route de Narbonne, 31062 Toulouse Cedex 4, France; E-mail: laborde@math.ups-tlse.fr*
[3] *ENSICA, 1 pl. Émile Blouin, 31056 Toulouse Cedex 5, France; E-mail: msalaun@ensica.fr*

Abstract. The *XFEM* method in fracture mechanics is revisited. A first improvement is considered using an enlarged fixed enrichment subdomain around the crack tip and a bonding condition for the corresponding degrees of freedom. An efficient numerical integration rule is introduced for the nonsmooth enrichment functions. The lack of accuracy due to the transition layer between the enrichment aera and the rest of the domain leads to consider a pointwise matching condition at the boundary of the subdomain. An optimal numerical rate of convergence is then obtained using such a nonconformal method.

Key words: Fracture, finite elements, XFEM method, optimal rate of convergence, pointwise matching.

1 Introduction

In computational fracture mechanics, the *eXtended Finite Element Method* was introduced in order to use a finite element mesh independent of the crack geometry [2, 8, 15, 16, 19]. A better accuracy was obtained for a lower computational cost thanks to *XFEM* instead of considering a classical finite element method. However, numerical experiments show that the rate of convergence is not improved, when the mesh parameter h goes to zero, for the elasticity problem on a cracked body [18]. So, we are interested in the abilities of the methodology *XFEM* to achieve an *optimal* accuracy for such non-smooth problems. Optimality refers here to an error of the same order than the one given by a classical finite element method for a *smooth* problem.

The principle of the extended finite element method consists in incorporating some *enrichment functions* into the finite element basis. Singular enrichement functions are used to take into account the nonsmooth behavior of the displacement field near the crack tip. In the standard *XFEM*, the size of the enrichment area at the crack

V.M.A. Leitão et al. (eds), Advances in Meshfree Techniques, 27–38.

tip vanishes when h goes to zero (the enrichment area is the union of the supports of these new singular basis functions). So the influence in the global error of the enrichment decreases with h, which explains the above-mentioned unsatisfactory numerical behavior. To overcome the difficulty, a first variant of *XFEM* was considered in which a whole fixed area (independent of h) around the crack tip is enriched [3, 11].

In the present paper, some improvements of the previous approach are studied in order to obtain better computational performances (in terms of numerical rate of convergence, number of degrees of freedom or well-conditioned system).

The *outline* of the paper is the following. In Section 2, the model problem of a cracked body in linear plane elasticity is considered. Section 3 is devoted to a new *XFEM* type method where the crack tip enrichment functions are localized by using a smooth *cut-off* function. A mathematical result of *optimal* error estimate is stated and confirmed by numerical tests for linear finite elements. In Section 4, a piecewise linear cut-off function is considered for the singular enrichment. The method comes to introduce some *bonding condition* between the enrichment degrees of freedom in *XFEM* with a fixed enrichment area. The numerical rate of convergence is improved for high order finite elements (of degree two or three) with respect to the classical *XFEM* method. However, optimality is not achieved because of the lack of accuracy coming from the elements in the *transition layer* (the finite elements between the enrichment area and the rest of the body). An efficient numerical integration rule for the nonsmooth enrichment functions is presented in Section 5. In the last section, we study a nonconformal method where a *pointwise matching condition* at the boundary of the enriched area takes the place of the transition layer. On a computational test, we then obtain the expected optimality.

2 The Elasticity Problem on a Cracked Domain

Consider the model problem of the equilibrium of a cracked body in plane elasticity. Let Ω be the bounded cracked domain in $\boldsymbol{R}^2$; the crack Γ_C is assumed to be straight. The boundary $\partial\Omega$ of the body is partitioned into Γ_C, Γ_D and Γ_N; a traction free condition is considered on Γ_C, on Γ_D the displacement is prescribed and the surface forces are known on Γ_N (Figure 1).

The weak formulation of the elasticity problem on the cracked domain Ω consists in finding a displacement field $u = (u_1, u_2)$ such that

$$u \in V, \;\; a(u, v) = L(v) \;\; \forall v \in V, \tag{1}$$

in the space of admissible displacements:

$$V = \{v \; : \; v \in \boldsymbol{H}^1(\Omega), \; v = 0 \; \text{ on } \Gamma_D\}.$$

We have denoted

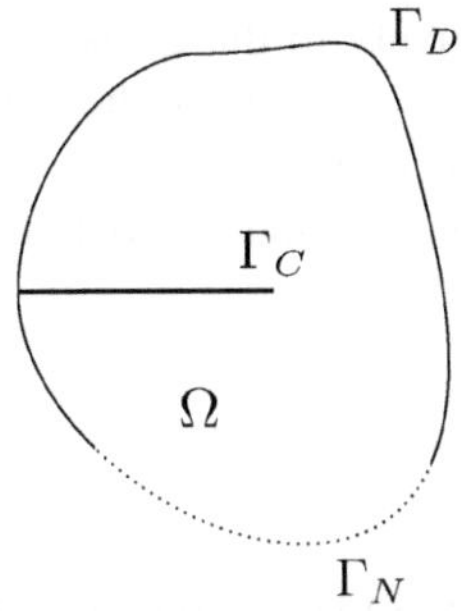

Fig. 1. The cracked domain Ω.

$$a(u,v) = \int_\Omega \sigma(u) : \varepsilon(u)\, d\Omega,$$

$$\sigma(u) = \lambda\, tr\varepsilon(u)\, I + 2\mu\, \varepsilon(u),$$

$$L(v) = \int_\Omega g.v\, d\Omega + \int_{\Gamma_N} f.v\, d\Gamma.$$

The notation $\varepsilon(u)$ stands for the linearized strain tensor, the Lamé coefficients satisfy $\lambda > 0$, $\mu > 0$, the force densities f and g are given on Γ_N and Ω respectively. The inner product of vectors in $\boldsymbol{R}^2$ is written $u.v = \sum_i u_i\, v_i$ and the associated norm $|.|$; for tensors, the inner product is denoted as usually $\sigma : \varepsilon = \sum_{ij} \sigma_{ij}\, \varepsilon_{ij}$. Finally, the functional spaces for vector-valued functions are distinguished by bold characters, e.g. $\boldsymbol{H}^s(\Omega) = H^s(\Omega; \boldsymbol{R}^2)$ equipped with its canonical norm $\|.\|_{s,\Omega}$.

There exists a unique displacement solution u to (1) under standard assumptions, i.e. mes $\Gamma_D > 0$, f and g defining a continuous linear form $L(.)$ on V. Assuming smoothness conditions on the data, the solution u can be written as a sum of a singular part u^* and regular one $u - u^*$ satisfying the following properties:

$$u^* = K_I\, u_I + K_{II}\, u_{II}, \tag{2}$$

$$u - u^* \in \boldsymbol{H}^{2+m}(\Omega), \tag{3}$$

for some integer $m \geq 0$ such that (in particular) $g \in \boldsymbol{H}^m(\Omega)$. In the definition of u^*, the constants K_I, K_{II} are the so-called *stress intensity factors* and the displacements u_I, u_{II} denote the *opening modes* of the crack. For a bi-dimensional crack [12, 13]:

$$u_I = \frac{1}{E}\sqrt{\frac{r}{2\pi}}(1+\nu)\begin{pmatrix} \cos\frac{\theta}{2}(3 - 4\nu - \cos\theta) \\ \sin\frac{\theta}{2}(3 - 4\nu - \cos\theta) \end{pmatrix}, \tag{4}$$

$$u_{II} = \frac{1}{E}\sqrt{\frac{r}{2\pi}}(1+\nu)\begin{pmatrix} \sin\frac{\theta}{2}(\gamma + 2 + \cos\theta) \\ \sin\frac{\theta}{2}(\gamma - 2 + \cos\theta) \end{pmatrix}, \tag{5}$$

in polar coordinates (r, θ) relatively to the crack tip, where ν is the Poisson ratio and the constant γ is equal to $\gamma = 3 - 4\nu$ for the plane stress problem. The normal (respectively tangential) component of function u_I (resp. u_{II}) is discontinuous along the crack. Note that the functions u_I and u_{II} belong to $\boldsymbol{H}^{3/2-\epsilon}(\Omega)$ for any $\epsilon > 0$ (see [9, 10]).

3 The Cut-Off Method

Assume that the *uncracked* body $\overline{\Omega}$ is a polyhedric domain and consider a *regular* triangulation $\mathcal{T}_h$ of $\overline{\Omega}$. The mesh parameter h corresponds to the maximum of the diameters of the triangles in $\mathcal{T}_h$. Denote $\varphi_1 \ldots \varphi_N$ the P_k finite element basis functions on the triangulation, where P_k represents the polynomials of degree $k \geq 1$.

Let H be the function defined on Ω, equal to $+1$ on the one side of the crack Γ_C and equal to -1 on the other one:

$$H(x) = \begin{cases} +1 & \text{if } (x - x^*).n > 0, \\ -1 & \text{elsewhere.} \end{cases}$$

In this definition, n is a given normal vector to the crack line. The asymptotic displacement (2) at the crack tip x^* can be written as a linear relation between the following singular functions $F_1, \ldots, F_4$ (see (4) and (5)):

$$F_1 = \sqrt{r} \sin\frac{\theta}{2}, \; F_2 = \sqrt{r}\cos\frac{\theta}{2}, \; F_3 = \sqrt{r}\sin\frac{\theta}{2}\cos\theta, \; F_4 = \sqrt{r}\cos\frac{\theta}{2}\cos\theta.$$

Let us introduce a C^2-function χ satisfying

$$\begin{cases} \chi(r) = 1 & \text{if } r \leq R_0, \\ 0 < \chi_0(r) < 1 & \text{if } R_0 < r < R_1, \\ \chi_0(r) = 0 & \text{if } R_1 \geq r. \end{cases} \tag{6}$$

Parameters R_0 and R_1 are given such that $0 < R_0 < R_1$.

We seek an approximate displacement field of the following form:

$$u_h = \sum_{1 \leq i \leq N} a_i \, \varphi_i + \sum_{i \in I_H} b_i \, H\varphi_i + \sum_{1 \leq j \leq 4} c_j \, F_j \chi. \tag{7}$$

The degrees of freedom are vector-valued: $a_i, b_i, c_j \in \boldsymbol{R}^2$. The corresponding discrete problem is the following: find u_h such that

$$u_h \in V_h, \;\; a(u_h, v_h) = L(v_h) \;\; \forall v_h \in V_h, \tag{8}$$

where V_h is the vector space of the displacements of the form (7).

The present method differs from the standard *XFEM* by the definition of the singular enrichment term. Namely, for the classical extended finite element, the last term in expression (7) of the approximate displacement u_h is changed into:

$$\sum_{i \in I_F} \sum_{1 \le j \le 4} c_{ij} \, F_j \psi_i, \quad c_{ij} \in \boldsymbol{R}^2, \tag{9}$$

where the local *partition of unity* ψ_i $(i \in I_F)$ is equal to the *linear* finite element basis functions associated to the vertices of the element containing the crack tip x^*. A variant consists in enriching all the finite elements nodes in a fixed area around x^*, say the disk $B(x^*, R)$ of radius $R > 0$ independent of h. The crack tip enrichment term then becomes:

$$\sum_{i \in I_F(R)} \sum_{1 \le j \le 4} c_{ij} \, F_j \psi_i, \tag{10}$$

where $I_F(R)$ corresponds now to the nodes in $B(x^*, R)$ [3, 11].

In the following result of convergence, the exact solution u satisfies the smoothness condition:

$$u - u^* \in \boldsymbol{H}^{2+\epsilon}(\Omega), \tag{11}$$

for some $\epsilon > 0$ (see (3)). In the statement below, only the case $k = 1$ is considered. To our knowledge, this is the first mathematical result about the accuracy of *XFEM* type methods [4, 5].

Theorem 1. *Let u be the displacement field solution to the model problem (1) on the cracked domain, and u_h the discrete solution defined from the enriched linear finite element method (7), (8). Under assumption (11), the following error estimate holds:*

$$\|u - u_h\|_{1,\Omega} \le Ch \|u - \chi u^*\|_{2+\epsilon,\Omega}, \tag{12}$$

where u^* is the asymptotic displacement (2) at the crack tip x^*, χ the cut-off function for the singular enrichment and $C > 0$ a constant only depending on Ω.

Remark 1. For a classical affine finite element method over a cracked domain, the error is of order $\sqrt{h}$, since the displacement field belongs to $H^{3/2-\epsilon}$ for any $\epsilon \ge 0$. The error estimate obtained in Theorem 1 is *optimal* in the sense that the rate of convergence is the same than using a classical P_1 finite element method for a *smooth* problem (the presence of $\epsilon > 0$ in the assumption (11) only corresponds to a technical difficulty).

The numerical tests are relative to the model problem (1) on the square domain $\overline{\Omega} = [-0.5, 0.5] \times [-0.5, 0.5]$ where the crack is the line segment $\Gamma_C = [-0.5, 0] \times \{0\}$. The exact solution u is the mode I crack displacement (4) prescribed as a Dirichlet condition on the whole domain boundary. The parameters of the cut-off function χ in the definition (6) are equal to $R_0 = 0.01$, $R_1 = 0.49$ and $\chi(x)$ is identical to a

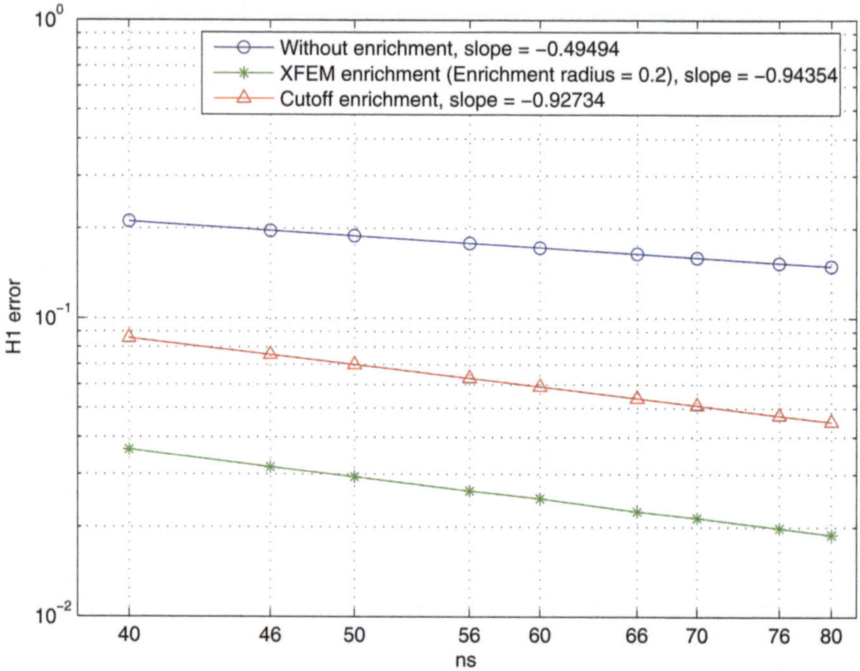

Fig. 2. Energy norm error for classical or enriched P1 elements with respect to the number of cells $ns = 1/h$ in each direction (logarithmic scales).

fifth degree polynomial if $R_0 \leq |x - x^*| \leq R_1$. The triangulation of the domain is defined from a grid of square cells (independently of the crack); let ns be the number of cells of the subdivision in each direction. A linear Lagrange finite element method is considered on $\mathcal{T}_h$.

Figure 2 shows a comparison between the convergence rates of:

- the classical finite element method (without enrichment),
- the *XFEM* method specified by (10), where the radius of the singular enrichement area is equal to $R = 0.2$,
- the previous cut-off enrichment strategy.

The energy norm error $\|u - u_h\|_{1,\Omega}$ is computed by running the test problem for different values of the mesh parameter $h = 1/ns$. It may be seen that the numerical error is of order h^α where the slope α on the figure differs according to the method. With respect to the classical finite element method, the cut-off enrichment reduces the error for a given mesh and presents a convergence rate α almost equal to 1 instead of $1/2$. Compared to the *XFEM* method with a fixed enrichment area, the convergence rate is very close, but the computational cost is better in the case

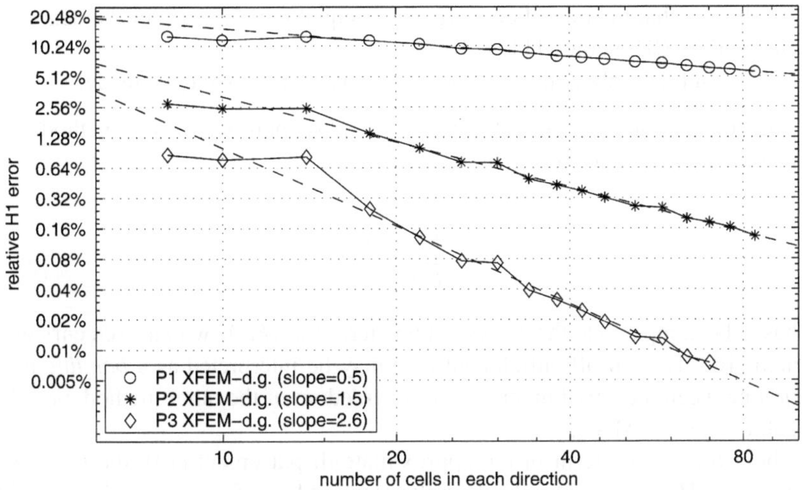

Fig. 3. Energy norm error with respect to $1/h$ in the case of the modified *XFEM* method using a finite element cut-off function (*XFEM*–d.g.).

of the cut-off method. In fact, the number of degrees of freedom increases significantly when h goes to zero in the term (10) specific to *XFEM* with a fixed enrichment area. Another advantage of the cut-off method lies in its significantly better condition number. Further details can be found in [5].

4 Piecewise Linear Cut-Off Function

The cut-off function is now defined as the continuous piecewise linear function on the finite element mesh, which is equal to 1 at the vertices in $B(x^*, R)$ and 0 at the other ones. It is denoted:

$$\chi_h = \sum_{i \in I_F(R)} \psi_i \tag{13}$$

with the previous notations. The approximate displacement considered is now written:

$$u_h = \sum_{1 \leq i \leq N} a_i\, \varphi_i + \sum_{i \in I_H} b_i\, H\varphi_i + \sum_{1 \leq j \leq 4} c_j\, F_j \chi_h \tag{14}$$

and the degree k of the finite element basis functions φ_i here is equal to 1, 2 or 3.

For the Mode I test problem considered in the previous section, the convergence curves are given on Figure 3. The radius R of the enrichment area is fixed to $1/10th$ of the domain size. The figure shows that the convergence rate is equal to 0.5, 1.5, 2.6 according to the different choices of the polynomials degree $k = 1, 2, 3$ respectively.

Let us observe that the rate of convergence is equal to 0.5 whatever the degree k in the classical *XFEM* method [11, 18].

Some comments about the method under consideration may be made.

(i) The crack tip term in the approximate displacement u_h for the method of the piecewise linear cut-off function:

$$\sum_{j=1}^{4} c_j \, F_j \chi_h \tag{15}$$

has to be compared to the corresponding term for *XFEM* with a fixed enrichment area (10). The cut-off enrichment (15) may be interpreted as a *bonding condition* between the enrichment degrees of freedom of the other method, or a d.o.f. gathering (XFEM–d.g.).

(ii) About the second term in the approximate displacement (14), the partition of unity for H is defined by the P_k finite element basis functions φ_i instead of the P_1 partition of unity ψ_i in the standard *XFEM*. So the approximation of the jump $[u_h]$ of displacement along Γ_C:

$$[u_h] = 2 \sum_{i \in I_H} b_i \, \varphi_i \quad \text{on } \Gamma_C$$

is compatible with the finite element method (i.e. of the same order).

(iii)In the chosen approximation strategy (14), the number of degrees of freedom for the F_j enrichment is minimal for a given enrichment aera $B(x^*, R)$. Moreover, the condition number is significantly better than using a classical *XFEM* enrichment on a fixed subdomain (when h decreases and for high degree k) [11]. An explanation may be found in the fact that the enrichment functions are not linearly independent. For instance, in the case of a P_1 partition of unity, we observe that

$$p_2(F_1 - F_4) + p_1 F_3 = 0, \quad p_2(F_3 - F_2) + p_1 F_4 = 0,$$

where p_i are the linear shape functions on the reference triangle:

$$p_1(x, y) = x, \;\; p_2(x, y) = y \;\text{ and }\; p_3(x, y) = 1 - x - y.$$

If a P_2 partition of unity is used, there are six relations of that kind.

5 The Polar Numerical Integration

Special care has to be taken in the numerical integration of the elementary stiffness matrix for the triangle containing the crack tip. First, expressing the integral

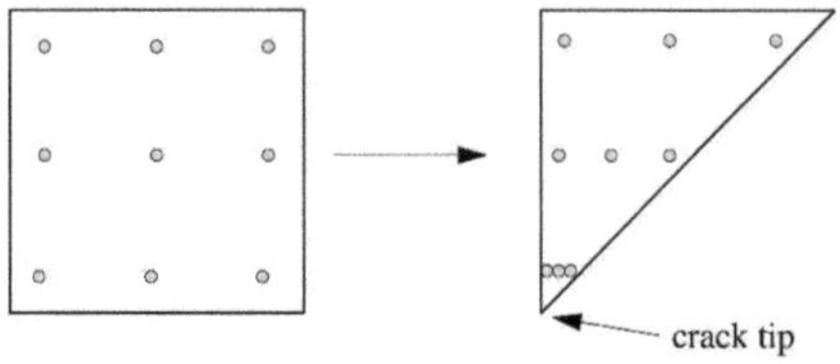

Fig. 4. Transformation of an integration method on a square into an integration method on a triangle for crack-tip functions.

$$\int_T \nabla(F_i\varphi_j).\nabla(F_k\varphi_l)\, dx$$

in polar coordinates, the $r^{-1/2}$ singularity of $\nabla F_i(x)$ is canceled. The finite element is then divided in (a few number of) subtriangles such that the crack tip is a vertex of some of them. For such subtriangles, the following integration method gives excellent results with a low number of integration points (keeping a classical Gaussian curvature formulae on the other subtriangles).

The geometric transformation $\tau : (x_1, x_2) \to (x_1x_2, x_2)$ maps the unit square onto a triangle (Figure 4). Using this transformation, it is possible to build a curvature formulae on the triangle from each one defined on the unit square. The new integration points $\bar{\xi}$ and their weights $\bar{\eta}$ are obtained from those of the original curvature formulae by

$$\bar{\xi} = \tau(\xi), \quad \bar{\eta} = \eta \det(\nabla\tau).$$

This curvature formulae will be called in the following the *polar integration* method.

The performances of the classical refined numerical integration and the polar integration curvature formulae are compared computing a *XFEM* elementary matrix. The reference elementary matrix is computed on a very refined subdivision near the singularity point x^*. Figure 5 presents the relative error in infinity norm between this reference elementary matrix and a computation of the elementary matrix by the following different strategies:

- using a regular refinement of the triangle and a fixed Gaussian formulae on each refined triangle (of order 3 and 10),
- using the polar integration method without any refinement, but for Gaussian curvatures on the square of increasing order.

This figure shows that the polar integration approach offers an important gain. Practically, 25 Gauss points were enough for the most accurate convergence test we have done.

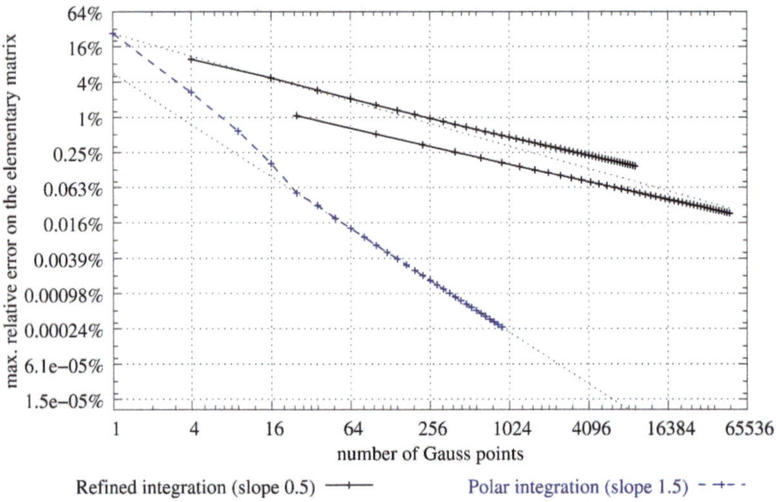

Fig. 5. Comparison of numerical integration methods for nonsmooth functions: uniform refinement with order 3 or 10 Gaussian method, and polar integration.

6 Pointwise Matching at the Boundary of Enriched Zone

The rates of convergence obtained in Figure 3 are not optimal. The only potential problem comes from the *transition layer* between the enrichment area and the rest of the domain, i.e. the triangles partially enriched. An analysis of interpolation error shows a lack of accuracy of *XFEM* methods due to the transition layer [11]. Let us note that a different analysis of this kind of problem is done in [6] where a specific *reproducing condition* is introduced. But this analysis cannot be straightforwardly applied to the present problem.

Let Ω^1 and Ω^2 be a partition of Ω where Ω^i is a union of mesh triangles, the crack tip belonging to Ω^2. The interface between Ω^1 and Ω^2 is denoted Γ_{12}. The approximate displacement u_h is such that $u_h = u_h^1$ on Ω^1 *without* a F_j enrichment:

$$u_h^1 = \sum_{i \in I(\Omega^1)} a_i \, \varphi_i + \sum_{i \in I_H(\Omega^1)} b_i \, H\varphi_i$$

and $u_h = u_h^2$ on Ω^2 *with* a F_j enrichment:

$$u_h^2 = \sum_{i \in I(\Omega^2)} a_i \, \varphi_i + \sum_{i \in I_H(\Omega^2)} b_i \, H\varphi_i + \sum_{j=1}^{4} c_j \, F_j.$$

Finally, $u_h^1 = u_h^2$ at the nodes on Γ_{12}. Naturally, this approximation procedure is no longer a conformal method. The matching condition at the interface may be seen

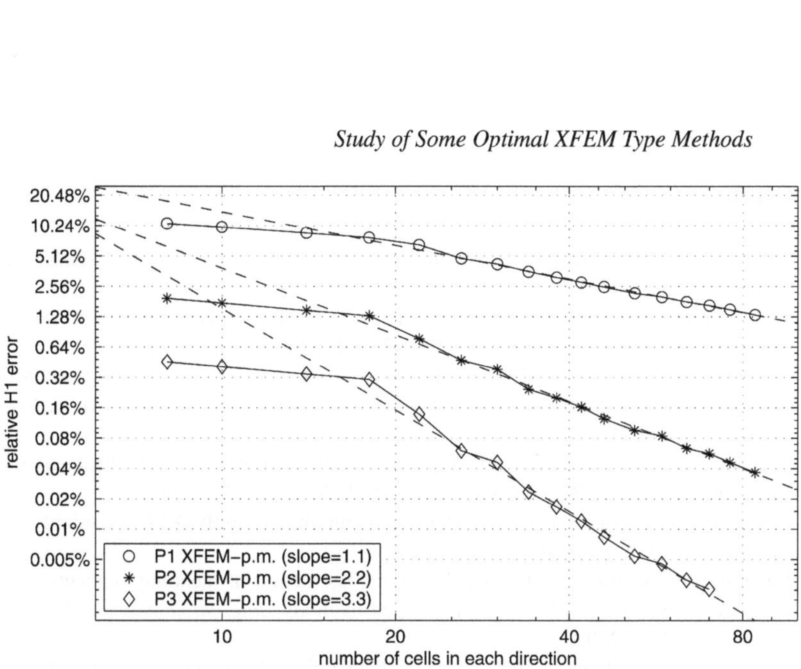

Fig. 6. Convergence of *XFEM* with pointwise matching condition (*XFEM*–p.m.) for the mode I problem.

as a linear relation between the concerned degrees of freedom from the one or the other side of Γ_{12}. The convergence curves in Figure 6 show that optimality is reached (actually, with a slight superconvergence) without any increasing of the number of degrees of freedom or making the condition number worse. For more details, see [11].

Remark 2. The *XFEM* method is based on a partition of unity principle. The *Partition of Unity Finite Element Method* does not exhibit such a lack of accuracy [14] [1] [7]. Thus, an idea is to be closer to *PUFEM* original principle. So let consider a *PUFEM* method using two overlapping subdomains Ω^1 and Ω^2 such that the crack tip x^* belongs to Ω^1 but not to $\bar{\Omega}^2$; a *XFEM* enrichment is defined on Ω^1 with a standard finite element approximation on Ω^2. This method is different from the classical *XFEM* only on the transition layer. It may be seen that the size of the transition layer does not influence the interpolation error estimate and consequently the convergence rate. So a transition layer with a vanishing width should be convenient (i.e. when Ω^1 and Ω^2 define a partition of Ω): this is the motivation of the previous *XFEM* method with pointwise matching.

The numerical experiments were performed with the finite element library *Getfem++* [17].

References

1. I. Babuška and J. M. Melenk. The Partition of Unity Method. *Int. J. Numer. Meth. Engng.*, 40:27–758, 1997.
2. T. Belytschko, N. Moës, S. Usui and C. Parimi. Arbitrary discontinuities in finite elements. *Int. J. Numer. Meth. Eng.*, 50(4): 993–1013, 2000.
3. E. Béchet, H. Minnebo, N. Moës and B. Burgardt. Improved implementation and robustness study of the X-FEM for stress analysis around cracks. *Int. J. Numer. Meth. Engng.*, 64:1033–1056, 2005.
4. E. Chahine, P. Laborde, Y. Renard. A quasi-optimal convergence result for fracture mechanics with XFEM. *C. R. Acad. Sci., Paris*, Ser. I, 342:527–532, 2006.
5. E. Chahine, P. Laborde, Y. Renard. Crack tip enrichment in the XFEM method using a cut-off function. *Submitted.*
6. J. Chessa, H. W. Wang, T. Belytschko. On construction of blending elements for locally partition of unity enriched finite element methods. *Internat. J. Numer. Meth. Engr.*, 57(7):1015–1038, 2003.
7. C.A.M. Duarte, I. Babuška and J.T. Oden. Generalized finite element method for three-dimensional structural mechanics problems. *Computers and Structures*, 77:219–232, 2000.
8. A. Gravouil, N. Moës and T. Belytschko. Non-planar 3D crack growth by the extended finite element and level sets, Part II: Level set update. *Int. J. Numer. Meth. Eng.*, 53(11): 2569–2586, 2002.
9. P. Grisvard. *Singularities in Boundary Value Problems.* Masson, 1992.
10. P. Grisvard. Problèmes aux limites dans les polygones – Mode d'emploi *EDF Bull. Direction Etudes Rech. Série C Math. Inform.*, 1:21–59, 1986.
11. P. Laborde, Y. Renard, J. Pommier and M. Salaün. High order extended finite element method for cracked domains. *Int. J. Numer. Meth. Engng.*, 64:354–381, 2005.
12. J. B. Leblond. *Mécanique de la Rupture Fragile et Ductile.* Hermes, Lavoisier, 2003.
13. J. Lemaitre and J.-L. Chaboche *Mechanics of Solid Materials.* Cambridge University Press, 1994.
14. J.M. Melenk and I. Babuška. The partition of unity finite element method: Basic theory and applications. *Comput. Meth. Appl. Mech. Engng.*, 139:289–314, 1996.
15. N. Moës, J. Dolbow, and T. Belytschko. A finite element method for crack growth without remeshing. *Int. J. Numer. Meth. Engng.*, 46:131–150, 1999.
16. N. Moës, A. Gravouil and T. Belytschko. Non-planar 3D crack growth by the extended finite element and level sets, Part I: Mechanical model. *Int. J. Numer. Meth. Engng.*, 53(11):2549–2568, 2002.
17. J. Pommier and Y. Renard. Getfem++. *An Open Source generic C++ library for finite element methods*, http://www.gmm.insa-toulouse.fr/getfem.
18. F.L. Stazi, E. Budyn, J. Chessa and T. Belytschko. An extended finite element method with higher-order elements for curved cracks. *Comput. Mech.*, 31:38–48, 2003.
19. N. Sukumar, N. Moës, B. Moran and T. Belytschko. Extended finite element method for three dimensional crack modelling. *Int. J. Numer. Meth. Engng.*, 48:1549–1570, 2000.

Generalized Finite Element Method in Mixed Variational Formulation: A Study of Convergence and Solvability

W. Góis and S.P.B. Proença

Structural Engineering Department, São Carlos School of Engineering, University of São Paulo, 400 Trabalhador SanCarlense Avenue, São Carlos, SP 13500, Brazil; E-mail: persival@sc.usp.br

Abstract. The Generalized Finite Element Method (GFEM) is first applied to hybrid-mixed stress formulations (HMSF). Generalized shape approximation functions are generated by means of polynomials of three independent approximation fields: stresses and displacements in the domain and displacements field on the static boundary. Firstly, the enrichment can independently be conducted over each of the three approximation fields. However, solvability and convergence problems are induced mainly due to spurious modes generated when enrichment is arbitrarily applied. With the aim of efficiently exploring enrichments in HMSF, an extension of the patch-test is proposed as a necessary condition to ensure enrichment, thus preserving convergence and solvability. In the present work, the inf-sup test based on Babuška–Brezzi condition was used to demonstrate the effectiveness of the Patch-Test. In particular, the inf-sup test was applied over selectively enriched quadrilateral bilinear and triangular finite element meshes. Numerical examples confirm the Patch-Test as a necessary but not sufficient condition for convergence and solvability.

Key words: Generalized finite element method, hybrid-mixed stress formulation, Babuška–Brezzi condition, inf-sup test.

1 Introduction

Boundary value problems (BVPs) can be variationally expressed using different principles. Depending on the variables involved in the variational principle, a specific weak form results and the FEM can then be applied to generate approximated solutions.

Among the non-conventional weak forms, three variants of the Hybrid formulation can be emphasized: the Hybrid-Mixed, Hybrid and Hybrid-Trefftz forms, all of which are detailed in [6]. In the present work, the Hybrid-Mixed Stress Model Formulation (HMSF) is addressed.

V.M.A. Leitão et al. (eds), Advances in Meshfree Techniques, 39–53.

This non-conventional form is called mixed for the reason that two incompatible stress and displacements fields are approximated in the domain. Since a second displacement field is approximated independently on the static boundary (where surface forces are imposed), the formulation is also typified as a hybrid one. Finally, the model is referred to as a stress model in view of the fact that continuity is primarily imposed over the stress approximation field.

Concerning numerical tools to solve BVPs, meshless methods provide approximations which are totally or relatively independent of the finite element mesh concept. The hp-Clouds Method [5], is distinguished among the meshless methods by the enrichment alternative of a basic approximation (partition of unity) without the definition of any additional nodal points in the domain. The Generalized Finite Element Method (GFEM) [8], allows combining the enrichment scheme of the hp-cloud method and advantageous features such as the strong imposition of boundary conditions of the conventional FEM.

Pimenta et al. [9] and Góis [7] applied the nodal enrichment technique to HMSF, resulting in a new application of the GFEM. Apparently, since displacements and stresses can be independently approximated, the enrichment could also be unconditionally imposed to each of those fields. However, that is not true because not all the combinations of enrichment supply stable and convergent solutions. Actually, in spite of noticeably good convergent responses, sometimes bad combinations of enrichments furnish unsatisfactory values of strain energy or stress and displacement fields as well.

Based on the above, in the present work, two numerical tests are proposed to verify the stability of HMSF solutions provided by GFEM. The first, a kind of Patch Test is linked to solvability, while the second is used to verify whether the (inf-sup) condition is satisfied by GFEM approximation functions obtained from clouds composed by four nodes quadrilaterals.

2 Hybrid-Mixed Stress Formulation in Plane Elasticity

The basis of the Hybrid-Mixed stress model in isotropic linear elasticity employed in the present work is founded on the variational principle of Reissner–Hellinger expressed as:

$$\begin{aligned}\Pi(u,\sigma,u_\Gamma) = &-\int_\Omega \frac{1}{2}\sigma^T f\sigma \,\mathrm{d}\Omega - \int_\Omega u^T(L\sigma + b)\,\mathrm{d}\Omega \\ &+ \int_{\Gamma_t} u_\Gamma^T(N\sigma - t)\,\mathrm{d}\Gamma + \int_{\Gamma_u} u^T(N\sigma)\,\mathrm{d}\Gamma. \qquad (1)\end{aligned}$$

The above expression comprises stress (σ) and displacement (u) fields defined in the domain (Ω) and a displacement field (u_Γ) defined on the static part (Γ_T) of the boundary. In addition, L is the differential divergent operator; b the vector of body forces; N the matrix formed by the components of the unit vector normal to the boundary; u the vector of prescribed displacements on Γ_u and t the vector of applied superficial forces on Γ_t. The treatment will be restricted to plane elasticity. Thus, f represents the flexibility matrix for linear and isotropic elastic materials.

Concerning numerical approximations and assuming the domain (Ω) and boundary (Γ) of the solid discretized by nodal points, interpolations of nodal values can be used to approximate the three independent fields as indicated below:

$$\hat{\sigma} = S_\Omega\, s_\Omega; \quad \hat{u} = U_\Omega\, q_\Omega; \quad \hat{u}_\Gamma = U_\Gamma\, q_\Gamma. \tag{2}$$

In the preceding relations, s_Ω represents the vector of nodal stress variables while q_Ω and q_Γ are vectors of displacement nodal degrees of freedom. S_Ω, U_Ω and q_Γ are respectively the matrices collecting the approximation functions for stress and displacement fields. Taking into consideration the approximated fields and imposing the stationary condition to the functional given in Equation (1), the following system of linear equations can be generated:

$$\begin{bmatrix} F & A_\Omega & -A_\Gamma \\ A_\Omega^T & 0 & 0 \\ -A_\Gamma^T & 0 & 0 \end{bmatrix} \begin{bmatrix} s_\Omega \\ q_\Omega \\ q_\Gamma \end{bmatrix} = \begin{bmatrix} e_\Gamma \\ -Q_\Omega \\ -Q_\Gamma \end{bmatrix}. \tag{3}$$

The system matrix can be rearranged using the following sub-matrices:

$$F = \int_\Omega S_\Omega^T f S_\Omega \,\mathrm{d}\Omega, \tag{4a}$$

$$A_\Omega = \int_\Omega (L S_\Omega)^T U_\Omega \,\mathrm{d}\Omega, \tag{4b}$$

$$A_\Gamma = \int_{\Gamma_t} (N S_\Omega)^T U_\Gamma \,\mathrm{d}\Gamma, \tag{4c}$$

$$e_\Gamma = \int_{\Gamma_u} (N S_\Omega)^T u \,\mathrm{d}\Gamma, \tag{4d}$$

$$Q_\Omega = \int_\Omega U_\Omega^T b \,\mathrm{d}\Omega, \tag{4e}$$

$$Q_\Gamma = \int_{\Gamma_t} U_\Gamma^T (t) \,\mathrm{d}\Gamma. \tag{4f}$$

In this work body forces (b) are not considered and the vector of displacements u is prescribed as null on the boundary (Γ_u) (then $e_\Gamma = Q_\Omega = 0$).

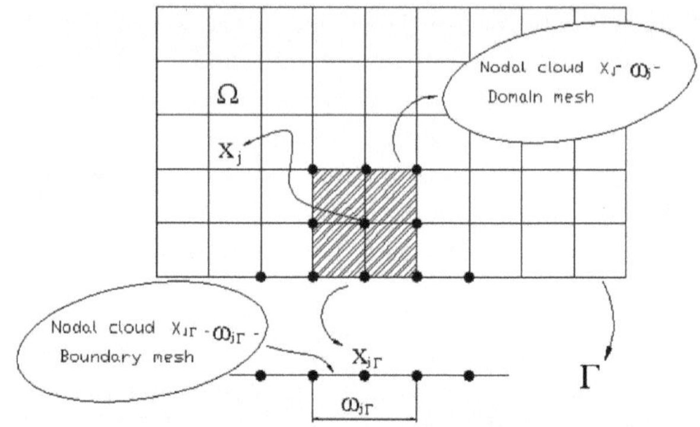

Fig. 1. Nodal clouds on domain and boundary nodes.

3 Bilinear Quadrilateral Elements with Nodal Enrichment

Consider a plain domain covered by a mesh of four nodes quadrilateral elements. The covering mesh is employed to define (ω) clouds in the domain, Figure 1. The boundary mesh is composed of linear elements connecting the nodes positioned on the boundary. Each cloud on both the domain and boundary is then formed by the elements sharing a common node. For those meshes composed of quadrangular elements, conventional bilinear Lagrangian functions are used as interpolating partition of unity for both stresses and displacements fields in the domain. Linear partitions of unit are used to interpolate displacements on boundary.

In order to provide enrichment to any approximation field attached to the nodes of the covering mesh, polynomial functions h_{kj}, $k = 1, \ldots, I(j)$ can be adopted in the domain Ω. Here $I(j)$ is the number of chosen functions at each node of index j. The following GFEM approximation family then results:

$$\Im_N^2 = \{\{S_{\Omega_j}\}_{j=1}^N \cup \{S_{\Omega_j} h_{kj}\}_{j=1}^N : j = 1, \ldots, N; k = 1, \ldots, I(j)\}. \quad (5)$$

For instance the enriched stress field can be constructed as:

$$\hat{\sigma} = \sum_{j=1}^{N} S_{\Omega_j} \left\{ s_{\Omega_j} + \sum_{i=1}^{I(j)} h_{ij}\, b_{ij} \right\}, \quad (6)$$

where N is the total number of nodes in the domain, s_{Ω_j} are the stress degrees of freedom associated with the original shape functions and b_{ij} are new nodal parameters introduced by each of the enrichment portions. With respect to the fields of displacements in domain and boundary, analogous procedure can be used to generate enriched approximations.

Thus, the interpolation matrices of the HSMF stress and displacement fields in domain can be represented as:

$$S_{\Omega_e} = [\phi_1\Delta_1 \;\; \phi_2\Delta_2 \;\; \phi_3\Delta_3 \;\; \phi_4\Delta_4], \tag{7}$$

$$U_{\Omega_e} = [\phi_1\Delta_1 \;\; \phi_2\Delta_2 \;\; \phi_3\Delta_3 \;\; \phi_4\Delta_4], \tag{8}$$

where $\phi_j, j = 1, \ldots, 4$ represents the Lagrangian bilinear functions connected to node j. The interpolation matrix for the two nodes element employed on the boundary mesh is given by:

$$U_\Pi = [\psi_1\Delta_{\Gamma_1} \;\; \psi_2\Delta_{\Gamma_2}], \tag{9}$$

where $\psi_{j_\Gamma}, j_\Gamma = 1, 2$ being the Lagrangian linear functions connected to node j_Γ of the element.

In Equation (7) through Equation (9), Δ_j and $\Delta_{\Gamma_{j_\Gamma}}$ are named the polynomial enrichment matrices attached respectively to node j of the domain and node j_Γ of the boundary element. Such matrices are given by:

$$\Delta_j = [I_3 \;\; h_{1j}I_3 \quad \ldots \quad h_{kj}I_3 \quad \ldots \quad h_{i(j)j}I_3] \tag{10}$$

when enrichment is over domain stress fields;

$$\Delta_j = [I_2 \;\; h_{1j}I_2 \quad \ldots \quad h_{kj}I_2 \quad \ldots \quad h_{I(j)j}I_2] \tag{11}$$

when enrichment is over domain displacements fields; and

$$\Delta_{\Gamma_{j_\Gamma}} = [I_2 \;\; h_{1j_\Gamma}I_2 \quad \ldots \quad h_{kj_\Gamma}I_2 \quad \ldots \quad h_{I(j)j_\Gamma}I_2] \tag{12}$$

when enrichment is over boundary displacements fields.

Clearly, if the functions h_{kj} and h_{kj_Γ} are null the conventional structure of the FEM is recovered. In this work the functions and are such that they are null in the enriched nodes ("bubble like functions"). The advantage of this procedure is that the original physical meaning of nodal degrees of freedom s_Ω, q_Ω and q_Γ is preserved.

The forms selected for the bubbles functions h_{kj} (later referred to as levels 2, 3 and 4 enrichment) and h_{kj_Γ} (later referred to as level 2 enrichment) are:

$$h_{kj} = (Y - Y_j)^2, (X - X_j)^2, (X - X_j)(Y - Y_j)^2, \ldots \text{ and } h_{kj_\Gamma} = (\xi - \xi_{j_\Gamma})^2, \ldots, \tag{13}$$

where X_j, Y_j are dimensionless coordinates to the finite element in the domain covering mesh and ξ_{j_Γ} is a dimensionless coordinate for the one-dimensional finite element of the boundary covering mesh.

Obviously the set of approximating functions involved in the HMSF can be limited not only to polynomials but also extended to include customized functions. On the other hand, with respect to the number of Gauss points necessary to carry out a numerical integration of the enriched matrices of each quadrilateral element, the following cases are distinguished:

- for the domain elements where the maximum degree of the h_{kj} polynomials in a direction is g_{ap}, $g_{ap} + 2$ Gauss points are necessary in each direction for integrating while;
- for the boundary elements where the maximum degree of the h_{kj_Γ} polynomials is $g_{ap\Gamma}$, the number of Gauss points necessary for integrating is $g_{ap\Gamma} + 2$;

4 On the Conditions of Convergence of HMSF with GFEM Spaces

A first step towards the study of the conditions of convergence of GFEM solutions for the case of HMSF presented here is founded in the work of Zienkiewicz et al. [11]. Initially a simple algebraic condition called Patch Test is suggested to ensure solvability (non-singularity condition) of the discrete linear system. Then a numerical study is proposed following the Babuška–Brezzi (*inf-sup*) sufficient condition to solvability and convergence.

4.1 The Patch Test Applied to HMSF with GFEM

Taking into account the system of equations given by Equation (3), limited to the case without enrichment and based on the proposal of Zienkiewicz et al. [11], the following algebraic conditions are necessary for the existence of a solution:

$$s_\Omega \geq q_\Omega, \tag{14}$$

expressed in terms of degrees of freedom attached to domain patches and

$$s_\Omega \geq q_\Gamma, \tag{15}$$

expressed in terms of degrees of freedom of stress and displacements on boundary patches. Both conditions are essentially important to assure good properties to the sub-matrices, A_Ω and A_Γ composed of the system given by Equation (3).

When enrichment is considered, the patch test is extended by including the new set of degrees of freedom introduced at each node. Then, Equation (14) and Equation (15) become:

$$a_\Omega + b_{ij} \geq q_\Omega + c_{mn}, \tag{16}$$

$$s_\Omega + b_{ij} \geq q_\Gamma + d_{kl}, \tag{17}$$

where c_{mn} and d_{kl} are the additional degrees of freedom introduced by displacement enrichments in the domain and boundary respectively.

Now, analyzing the enrichment possibilities and considering the conditions expressed in formuae (16) and (17), it can be concluded initially that:

- Only stress field enrichment in the domain is unrestricted since both conditions would always be satisfied;
- The simultaneous enrichment of stress and displacement fields in the domain is also effective;
- Simultaneous enrichment of stress and displacement fields (in domain and boundary) is allowed provided enrichment applies to those nodes in the domain coincident with nodes of the enriched mesh on the boundary;

- The previous conclusion can be extended to enrichment including stress fields in the domain and displacements on the boundary;
- Enrichment restricted to displacement fields is ineffective since Equations (16) and (17) would not be satisfied.

It is important to note once more that the patch test conditions are necessary but not sufficient for solvability and consequently for the stability of the linear system.

4.2 Study of the Babuška–Brezzi (inf-sup) Condition Apllied to the HMSF with GFEM

The Babuška–Brezzi or *inf-sup* condition [1, 3] is a necessary and sufficient condition to ensure stability and convergence of any linear numerical approach supplied by the MEF. Based on Babuška [2] and Chapelle and Bathe [4], a numerical verification of this condition extended to GFEM in HMSF is presented here.

4.2.1 On the Well-Posedness of a Boundary Value Problem: The Babuška–Brezzi (inf-sup) Condition

A boundary value problem (BVP) expressed in its variational form consists in finding a solution $u_0 \in H_1$ such that:

$$B(u_0, v) = F(v) \quad \forall v \in H_2 \tag{18}$$

for every continuous linear functional F in H_2. Well-posedness of the problem is assured if the following conditions are verified: Continuity of the bilinear form $B(\ldots)$ and the Babuška–Brezzi (inf-sup) condition.

Continuity of the bilinear form

Let H_1 and H_2 be Hilbert spaces endowed with norms $\|\cdot\|_{H_1}$ and $\|\cdot\|_{H_2}$ respectively. The bilinear form, $B(\cdot) : H_1 x H_2 \to R$ is continuous if for a given positive scalar C_B, the inequality below is valid:

$$|B(u, v)| < C_B \|u\|_{H_1} \|v\|_{H_2} \quad \forall\, u \in H_1 \quad \text{and} \quad v \in H_2. \tag{19}$$

The Babuška–Brezzi (inf-sup) condition

The class of problems governed by the addressed continuous bilinear form is well-posed if:

$$\inf_{u \neq 0 \in H_1} \sup_{v \neq 0 \in H_2} \frac{B(u, v)}{\|u\|_{H_1} \ \|v\|_{H_2}} \geq \lambda > 0, \tag{20}$$

$$\sup_{u \in H_1} B(u, v) > 0 \quad \forall\, v \neq 0 \in H_2. \tag{21}$$

In the above equations, v represents the space of test functions. Furthermore, by definition:

$$H_1 = H_2 = \left\{ v | v \in L^2(\Omega); \frac{\partial v_i}{\partial x_j} \in L^2(\Omega), \right.$$

$$\left. (i, j = 1, 2, 3); \; v_{i|\Gamma_u} = 0, \; (i = 1, 2, 3) \right\}. \tag{22}$$

Accordingly, any element v in H_2 must satisfy boundary conditions which are homogeneous on Γ_u. Moreover,

$$\int_{\Omega} (v)^2 \, d\Omega < \infty, \tag{23}$$

$$\int_{\Omega} \left(\frac{\partial v_i}{\partial x_j} \right)^2 d\Omega < \infty, \quad i, j = 1, 2, 3. \tag{24}$$

In fact, the preceding conditions are related to the existence of a solution to the problem in question. Essentially, it can be demonstrated that the conditions given by Equation (20) and Equation (21) are necessary and sufficient to ensure the existence and uniqueness of the BVP solution.

Once the original problem is well-posed, the performance of the formulation using finite elements is dependent on the choice of n-dimensional linear subspaces $S_1^n \subset H_1$ and $S_2^n \subset H_2$ defined as:

$$s_1^n = s_2^n \tag{25}$$

$$= \left\{ v_n | v_n \in L^2(\Omega); \frac{\partial v_{n_i}}{\partial x_j} \in L^2(\Omega), \; (i, j = 1, 2, 3); \; v_{n_i}|_{\Gamma_u} = 0, \; (i = 1, 2, 3) \right\}.$$

Usually, v_u is a polynomial function of degree n. As a rule, it is assumed that there are $v_k \in S^k$ such that the sequence $v_k (k = 1, 2, \ldots)$ converges to u_0.

The partition of unity features used in the approximation spaces, S_1^n and S_2^n, ensures continuity of the bilinear form involved in HMSF and also the verification of Equation (21). Hence, the Babuška–Brezzi (inf-sup) condition is the main focus of what follows.

Assuming that $B(u, v)$ satisfies Equation (19), the discrete form of Equation (21) can be written as:

$$\inf_{u \neq 0 \in s_1^n} \sup_{v \neq 0 \in S_2^n} \frac{B(u_n, v_n)}{\|u_n\|_{H_1} \|v_n\|_{H_2}} \geq \lambda(n) > 0. \tag{26}$$

It can be perceived that $\lambda(n)$ now depends on the dimension of the approximation spaces. However, if the problem is well-posed, convergence to a value λ_0 should be verified provided mesh refinement or enrichment of the approximation space is carried out.

Let then $u_0 \in H_1$ be the exact solution derived from the strong formulation, for instance, and $u_n \in S_1^n$ a numerical estimate. Then, since the exact solution is also a solution for the weak form, both conditions below are valid:

$$B(u_0, v) = F(v), \quad \forall v \in H_2, \tag{27}$$

$$B(u_n, v_n) = F(v_n), \quad \forall v_n \in S_2^n. \tag{28}$$

Subtracting Equation (27) from Equation (28) and using the continuity property of the bilinear form, we have:

$$\|u_0 - u_n\|_{h_1} \leq \left(1 + \frac{C_B}{\lambda(n)}\right) \inf_{\chi \in S_1^n} \|u_0 - \chi\|_{H_1}. \tag{29}$$

Hence, if the inequality Equation (26) is satisfied one can conclude that convergence and unicity of the numerical solution is also ensured. In practical terms, in a first stage, the solution exists if the Babuška–Brezzi condition gives ($\lambda(n) > 0$) for a certain approach. In a second stage, by considering successively more refined approaches, the result $\lambda(n) \to \lambda_0$ confirms convergence and unicity.

Numerical determination of $\lambda(n)$

Babuška [2] presents a mathematical development showing that the determination of $\lambda(n)$ is equivalent to finding the square root of the nonzero smallest eigenvalue of the following generalized eigenvalue problem:

$$B^T A_2^{-1} Bx = \mu A_1 x. \tag{30}$$

In the above relation, B can be obtained from

$$B(u_n, v_n) = v^{\mathrm{T}} Bu, \tag{31}$$

while v and u are vectors with components $u_n \in S_1^n$ and $v_n \in S_2^n$. Finally, A_1 and A_2 are symmetric and positive-definite matrices associated to the norms:

$$\|u\|_{H_1}^2 = \mathrm{u}^{\mathrm{T}} A_1 \mathrm{u}; \; \|v\|_{H_2}^2 = \mathrm{v}^{\mathrm{T}} A_2 \mathrm{v}. \tag{32}$$

The inf-sup test applied to HMSF with GFEM

In the present work, we extend the numerical inf-sup test to a hybrid mixed formulation. An important aspect of this is the identification of all matrices involved in Equation (30).

Consider then the governing equations of the HMSF derived from the stationary condition of the variational principle expressed by Equation (1):

$$\int_{\Omega} \delta\sigma^T f\sigma \, d\Omega + \int_{\Omega} u^T (L\delta\sigma) \, d\Omega - \int_{\Gamma_t} u_{\Gamma}^T (N\delta\sigma) \, d\Gamma = \int_{\Gamma_u} u^T (N\delta\sigma) \, d\Gamma, \quad (33)$$

$$\int_{\Omega} \delta u^T (L\sigma) \, d\Omega = -\int_{\Omega} \delta u^T \, b \, d\Omega, \quad (34)$$

$$\int_{\Gamma_t} \delta u^T (N\sigma) \, d\Gamma = \int_{\Gamma_t} \delta u^t \, t \, d\Gamma. \quad (35)$$

It follows that the bilinear form $B(\ldots)$ and the linear form $F(\cdot)$ can be written as:

$$B(U, V) = \int_{\Omega} \delta\sigma^T f\sigma \, d\Omega + \int_{\Omega} u^T (L\delta\sigma) \, d\Omega - \int_{\Gamma_t} u_{\Gamma}^T (N\delta\sigma) \, d\Gamma$$
$$+ \int_{\Omega} \delta u^T (L\sigma) \, d\Omega + \int_{\Gamma_t} \delta u^T (N\sigma) \, d\Gamma, \quad (36)$$

$$F(V) = \int_{\Gamma_u} u^T (N\delta\sigma) \, d\Gamma - \int_{\Omega} \delta u^T \, b \, d\Omega + \int_{\Gamma_t} \delta u^T t \, d\Gamma. \quad (37)$$

It can be assumed [10] that the spaces $U = (\sigma, u, u_\Gamma)$ and $V = (\delta\sigma, \delta u, \delta u_\Gamma)$ can be defined in XxY as:

$$X = \{(\sigma, u, u_\Gamma) : \sigma \in H_1; \sigma, u, u_\Gamma \in L^2(\Omega)\}, \quad (38)$$

$$Y = \{(\delta\sigma, \delta u, \delta u_\Gamma) : \delta\sigma \in H_1; \delta\sigma, \delta u, \delta u_\Gamma \in L^2(\Omega)\}. \quad (39)$$

Furthermore, the space U is endowed with the norm:

$$\|U\|_x^2 = \|(\sigma, u, u_\Gamma)\|_x^2$$
$$= \int_{\Omega} \sigma^2 \, d\Omega + \int_{\Omega} (L\sigma)^2 \, d\Omega + \int_{\Omega} u^2 \, d\Omega + \int_{\Gamma_t} u_{\Gamma}^2 \, d\Gamma. \quad (40)$$

Finally, from Equations (36) and (37), all matrices in Equation (30) can be derived.

5 Inf-sup Test: Numerical Results

Here the methodology of the inf-sup test is now applied considering the problem depicted in Figure 2. For reasons of simplification, no units for the elastic parameters and dimensions are adopted in both problems. Furthermore, the Young's modulus and Poisson's coefficient are assumed to be respectively: $E = 1000$ and $\nu = 0.3$.

The problem considers a 5 units side square plate and stressed by a uniformly distributed force of $p = 10$ units applied along its length. Essential boundary conditions are imposed over the left vertical side by constraining the plane displacement components ($u_x = 0$ and $u_y = 0$). Essential boundary conditions are also imposed over

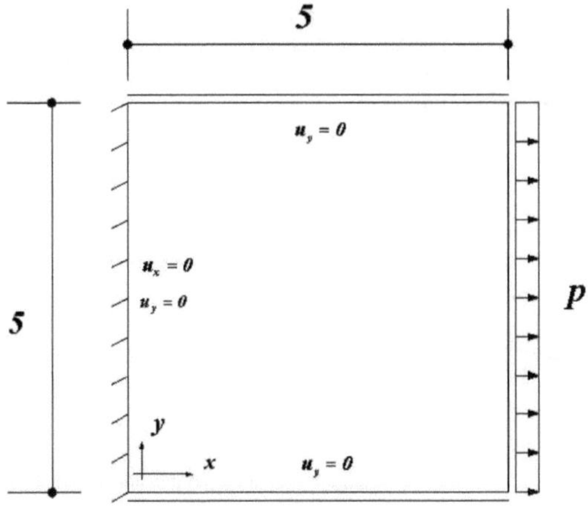

Fig. 2. Uniformly tensioned plate.

the upper and lower horizontal sides by constraining the displacements in direction $y(u_y = 0)$. The reference value of the strain energy in this problem is: 1.14.

The inf-sup test was applied to the HMSF by considering sequences of regular quadrilateral element meshes indicated as (1× 1), (2× 2), (4× 4), (8×8) and (16× 16). The numbers in the preceding nomenclature represent the amount of regular divisions in the x ey directions respectively. Polynomials as indicated by Equation (13) were often adopted as enrichment functions for the stress and displacement fields in the domain and displacement field on the boundary.

For each mesh, the value of $\lambda(n)$ was computed from the square root of the smallest eigenvalue determined on solving Equation (30). The results were then plotted on a $\log(\lambda(n)) - \log(1/N)$, ($N$ is the total number of degrees of freedom) scale. The inf-sup test was considered to be satisfied if the $(\log(1/N) - \log(\lambda(n)))$ curve showed an asymptotic behavior towards positive values.

5.1 Numerical Results

The results of the inf-sup test applied to the sets of quadrilateral element meshes without enrichment are presented in Figure 3.

Even though the inf-sup condition is not satisfied, we note that the patch test is verified and there is convergence in terms of strain energy.

With the enrichment restricted to the stress field in all nodes in domain, one can conclude that the quadrilateral element satisfies the *inf-sup test*, see Figure 4. This test confirms the results predicted by the patch-test. Another conclusion derived from the control of 'spurious modes' is that enrichment of the stress field eliminates such modes, improving the solvability of the problem.

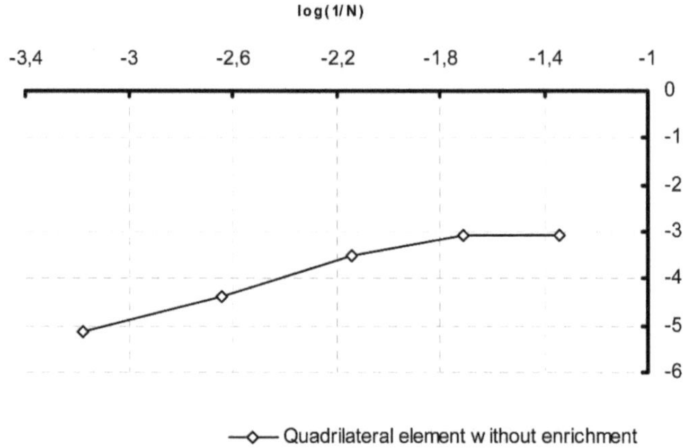

Fig. 3. Results for regular meshes without enrichment.

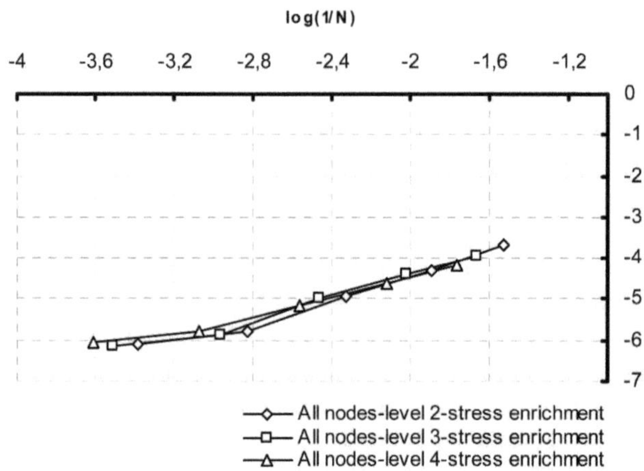

Fig. 4. Results for regular meshes with enrichment of the stress field in the domain.

Figure 5 shows results obtained by adopting selective polynomial enrichments over the stress field. It can be concluded that the number of enriched nodes, compared to the total number of nodes, affects the inf-sup test results. In fact, the responses for few enriched nodes are more comparable to the situation without enrichment.

The quadrilateral element with simultaneous enrichment of the domain fields satisfies the *inf-sup test* as illustrated in Figure 6. Spurious displacement modes are still present but not affecting solvability and convergence aspects.

In Figure 7 two other possibilities of enrichment satisfying the patch test are presented. Enrichment limited to displacement field at all nodes of the domain is not recommended (patch test fails). The enrichment of the boundary displacement field

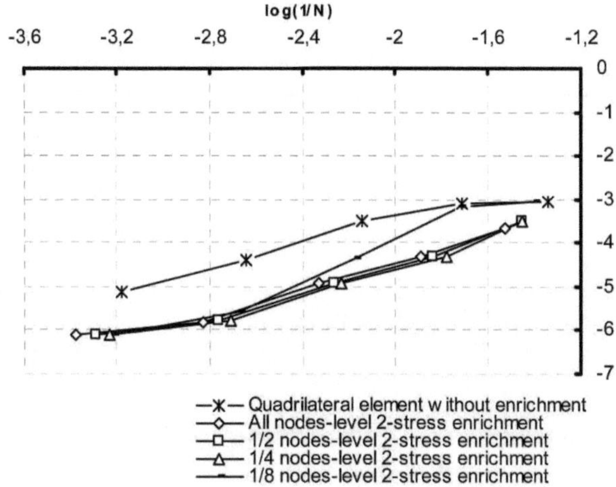

Fig. 5. Results for regular meshes with selective stress polynomial enrichment.

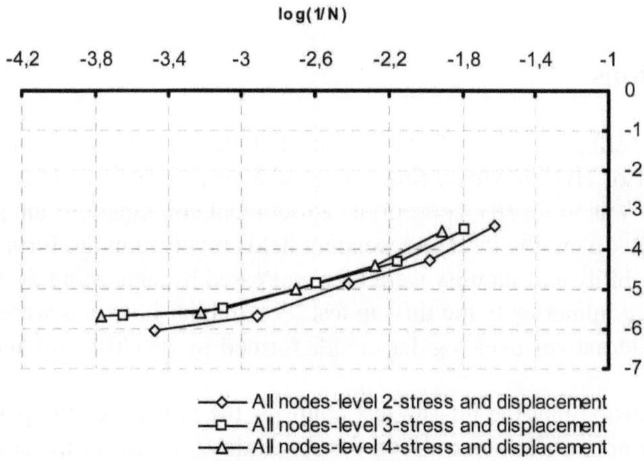

Fig. 6. Results for enrichment of the stress and displacement fields in the domain.

is effective if supplemented by enrichment of the stress fields. In such a case the patch test is verified.

Although the inf-sup test fails, the convergence strain energy is verified when the displacement field in the domain is enriched.

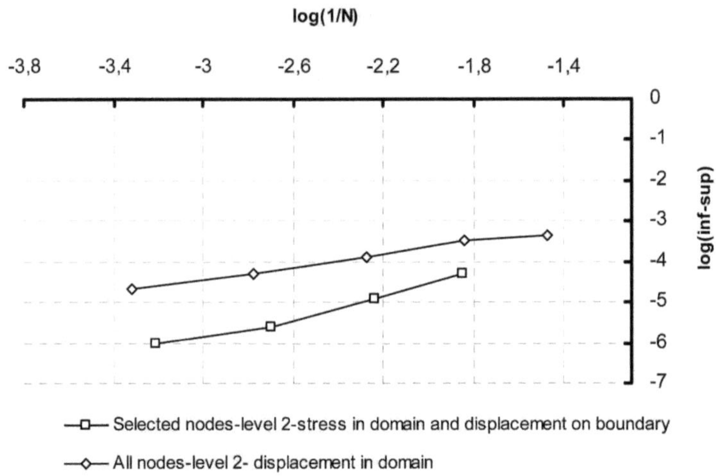

Fig. 7. Results for enrichments of displacements field in the domain or displacements on the boundary accompanied by enrichment over the stress field.

6 Conclusions

A study on the effectiveness of the Generalized Finite Element approximation spaces in the context of Hybrid Mixed Stress Formulation problems was conducted. The basic purpose was to discern appropriate enrichment combinations among multiple possibilities involving the three independent fields involved in the formulation. The issues of solvability and stability were then addressed by suggesting a patch test and by carrying out numerically the inf-sup test. Polynomial functions were selected as enrichment alternatives over regular clouds formed by quadrilateral and triangular elements.

The conclusion that the inf-sup test confirms the efficacy of the patch test as a necessary but not sufficient condition for solvability, at least in the sort of HMSF problems analyzed, is highlighted. For instance, in spite of the fact that the patch test was verified considering the basic case of partition of unity without enrichment, the inf-sup test was able to identify cinematic spurious modes and for this reason was not satisfied. Furthermore, in all situations of enrichment imposed over the basic fields wherein the patch test was not verified, the inf-sup test was not satisfied as well.

In essence, in the framework of HMSF, the enrichment over the stress field is always effective. However, enrichments over approximations for displacements in the domain and displacements on boundary always introduce spurious modes. This can be effective only if accompanied by stress field enrichment.

Acknowledgement

The authors would like to thank CAPES for the granted support.

References

1. Babuška I., Error bounds for finite element methods. *Numerische Mathematik* 16:322–333, 1971.
2. Babuška I., On the inf-sup (Babuška–Brezzi) condition. The University of Texas at Austin. Technical Report #5, TICAM, 1996.
3. Brezzi F., On the existence, uniqueness and approximation of saddle point problems arising from Lagrange multipliers. *RAIRD* 8 (r-2):127–151, 1974.
4. Chapelle D. and Bathe K.J., The inf-sup test. *Computers & Structures*, 47:537–545, 1993.
5. Duarte C.A., A review of some meshless methods to solve partial differential equations. The University of Texas at Austin. Technical Report, TICAM, 1995.
6. Freitas J.A.T., Almeida J.P.B.M. and Pereira E.M.B.R., Non-conventional formulations for the finite element method. *Structural Engineering and Mechanics*, 4:655–678, 1996.
7. Góis W., Generalized finite element method in hibryd mixed stress formulation, Master Dissertation. São Carlos School of Engineering. University of São Paulo, 2004 [in Portuguese]
8. Oden J.T., Duarte C.A. and Zienkiewicz O.C., A new cloud-based hp finite element method. *Computer Methods in Applied Mechanics and Engineering*, 153:117–126, 1998.
9. Pimenta P.M., Proença S.P.B. and Freitas J.A.T., Elementos finitos híbridos mistos com enriquecimento nodal. In: *Proceedings of Métodos Numéricos em Ingeniería V*, J.M. Gaicolea, C. Mota Soares, M. Pastor and G. Bugeda (Eds), SEMNI, 2002.
10. Schwab Ch., *P- and hp-Finite Element Methods: Theory and Applications in Solid and Fluid Mechanics*, Oxford Science Publications, New York, 1998.
11. Zienkiewicz O.C. et al., The patch test for mixed formulation. *International Journal for Numerical Methods in Engineering*, 23:1873–1882, 1986.

Stability in Lagrangian and Semi-Lagrangian Reproducing Kernel Discretizations Using Nodal Integration in Nonlinear Solid Mechanics

Jiun-Shyan Chen[1] and Youcai Wu[2]

[1] *Civil & Environmental Engineering, 5713 Boelter Hall, UCLA, Los Angeles, CA 90095, U.S.A.; E-mail: jschen@seas.ucla.edu*
[2] *Karagozian & Case, 2550 N Hollywood Way, Suite 500, Burbank, CA 91505, U.S.A.*

Abstract. Stability analyses of Lagrangian and Semi-Lagrangian Reproducing Kernel (RK) approximations for nonlinear solid mechanics are performed. It is shown that the semi-Lagrangian RK discretization yields a convective term resulting from the non-conservative coverage of material points under the kernel support. The von Neumann stability analysis shows that the discrete equations of both Lagrangian and semi-Lagrangian discretizations are stable when they are integrated using stabilized conforming nodal integration. On the other hand, integrating the semi-Lagrangian discretization with a direct nodal integration yields an unstable discrete system which resembles the tensile instability in SPH. Under the framework of semi-Lagrangian discretization, it is shown that the inclusion of convective term yields a more stable discrete system compared to the semi-Lagrangian discretization without convective term as was the case in SPH.

Key words: Semi-Lagrangian, reproducing kernel, nodal integration, meshfree method, large deformation.

1 Introduction

Particle and meshfree methods offer considerable advantages over the finite element method in modeling large deformation solid mechanics problems, where mesh entanglement difficulty in the finite element method can be considerably reduced. Smoothed Particle Hydrodynamics (SPH), originally introduced for astrophysics [12, 15] has been extended to model large deformation problems in solids [16, 19]. SPH method, however, exhibits tensile instability [3, 17, 20]. This tension instability has been corrected by the employment of Lagrangian kernel under the framework of reproducing kernel particle method (RKPM) [5, 6]. The Lagrangian kernel based RKPM has been applied to path-independent hyperelasticity [4, 6, 11] and

M.A. Leitão et al. (eds), Advances in Meshfree Techniques, 55–76.

path-dependent plasticity [5–8] large deformation problems. The Lagrangian kernel based meshfree methods in nonlinear solid mechanics, however, have their inherent difficulty due to the regularity requirement of the deformation gradient needed for inverse mapping from deformed configuration to undeformed configuration. A semi-Lagrangian formulation under the framework of reproducing kernel particle method is proposed herein for extremely large deformation solid mechanics problems. Semi-Lagrangian discretization defines kernel function distance measure in the deformed configuration and thus avoids the need for inverse mapping from deformed to undeformed configurations.

The main difference between Lagrangian kernel and semi-Lagrangian kernel is due to the definition of distance measure between point of evaluation and discrete points. The Lagrangian kernel defined at the material discrete points yields vanishing material time derivative of the Lagrangian kernel. On the other hand, material time derivative of the semi-Lagrangian kernel does not vanish due to the advection of materials covered under the kernel support during material deformation. Consequently, nodal mass is not conservative when a continuum is discretized by the semi-Lagrangian kernel, and an additional convective term appears in the conservation laws. Note that SPH is semi-Lagrangian in nature, but it was formulated without considering the above mentioned non-conservative nodal mass and the convective effect in the conservation laws.

Two types of instabilities have been discussed in meshfree literatures: the tensile instability and the rank instability. The tensile instability in SPH has been analyzed by Swegle et al. [20]. Stability analysis of element free Galerkin (EFG) method using Lagrangian and Eulerian kernels has been conducted [1, 2] and the method has been applied to problems with moderate deformations [18]. Belytschko and Xiao [1] showed that tensile instability can be resolved by using a Lagrangian kernel [6]. Other remedies for tensile instability in SPH such as the conservative smoothing approach by Guenther et al. [22] and Swegle et al. [23], the stress point approach by Dyka and Ingel [21], and an artificial stress approach by Monaghan [17] have been proposed. The rank instability due to direct nodal integration (DNI) has been pointed out by Chen et al. [9, 10] and a stabilized conforming nodal integration (SCNI) has been proposed.

In this work, stability analysis of Lagrangian and semi-Lagrangian Galerkin meshfree formulations will be performed. The effects of domain integration using direct nodal integration and stabilized conforming nodal integration on numerical stability will also be analyzed. The remaining of the paper is arranged as follows. An overview of Lagrangian discretization using reproducing kernel (RK) approximation is given in Section 2. In Section 3, the semi-Lagrangian kernel and the corresponding RK shape functions are presented, and the semi-Lagrangian discretizations of equation of motion are derived. The convective effect resulting from the non-conservative nature of semi-Lagrangian kernel is also identified. Von Neumann stability analyses of Lagrangian and semi-Lagrangian discrete equations integrated by SCNI and DNI are carried out in Section 4. Conclusion remarks are given in Section 5.

2 Lagrangian Reproducing Kernel Discretization

2.1 Reproducing Kernel (RK) Approximation

Let the problem domain Ω be discretized into a set of NP points $\{\mathbf{x}_1, \mathbf{x}_2, \ldots, \mathbf{x}_{NP}\}$, where $\mathbf{x}_I$ is the location of node I and NP denotes the total number of points. The variable, for example, displacement $u_i(\mathbf{x})$ in solid mechanics, is approximated by:

$$u_i^h(\mathbf{x}) = \sum_{I=1}^{NP} \Psi_I(\mathbf{x}) d_{iI}, \tag{1}$$

where $u_i^h(\mathbf{x})$ is the approximation of $u_i(\mathbf{x})$, and Ψ_I and d_{iI} are the shape functions and their associated coefficients, respectively. Under the framework of reproducing kernel approximation, the shape function $\Psi_I(\mathbf{x})$ is expressed as:

$$\Psi_I(\mathbf{x}) = \Big(\sum_{i+j=0}^{n} (x_1 - x_{1I})^i (x_2 - x_{2I})^i \ b_{ij}(\mathbf{x}) \Big) \phi_a(\mathbf{x} - \mathbf{x}_I), \tag{2}$$

where $\phi_a(\mathbf{x} - \mathbf{x}_I)$ is a kernel function that defines the smoothness and locality of the approximation with a compact support a, which is the radius of the support. The unknown coefficients $\mathbf{b}(\mathbf{x})$ in Equation (2) are obtained by enforcing the following nth order reproducing conditions:

$$\sum_{I=1}^{NP} \Psi_I(\mathbf{x})\, x_{1I}^{\alpha}\, x_{2I}^{\beta} = x_1^{\alpha} x_2^{\beta}, \quad \alpha + \beta = 0, 1, 2, \ldots, n. \tag{3}$$

Upon solving $b_{ij}(\mathbf{x})$ from Equation (3), the shape functions are obtained:

$$\Psi_I(\mathbf{x}) = \mathbf{H}^T(\mathbf{0})\mathbf{M}^{-1}(\mathbf{x})\mathbf{H}(\mathbf{x} - \mathbf{x}_I)\phi_a(\mathbf{x} - \mathbf{x}_I), \tag{4}$$

where

$$\mathbf{H}^T(\mathbf{x}-\mathbf{x}_I) = [1 \quad x_1-x_{1I} \quad x_2-x_{2I} \quad (x_1-x_{1I})^2 \quad \ldots \quad \ldots \quad (x_2-x_{2I})^n], \tag{5}$$

$$\mathbf{M}(\mathbf{x}) = \sum_{I=1}^{NP} \mathbf{H}(\mathbf{x} - \mathbf{x}_I)\mathbf{H}^T(\mathbf{x} - \mathbf{x}_I)\phi_a(\mathbf{x} - \mathbf{x}_I). \tag{6}$$

2.2 Lagrangian Description and Discretization of Equation of Motion

In solid mechanics large deformation problems, material motion from undeformed (initial) domain Ω_X to deformed (current) domain Ω_x has to be defined. Consider a material particle originally located at $\mathbf{X}$ in the undeformed domain Ω_X . Under certain action this material particle is moved to $\mathbf{x}$ at time t in the deformed domain Ω_x. The motion of the material particle $\mathbf{X}$ is described by a mapping function $\mathbf{x} = \varphi(\mathbf{X}, t)$, and $\mathbf{u}(\mathbf{X}, t) = \mathbf{x}(\mathbf{X}, t) - \mathbf{X}$ is the displacement vector associated with material particle originally positioned at $\mathbf{X}$. In Lagrangian description, the weak form of equation of motion expressed in the undeformed domain Ω_X is expressed as:

$$\int_{\Omega_X} \delta u_i \rho^0 \, \ddot{u}_i \, \mathrm{d}\Omega + \int_{\Omega_X} \delta F_{ij} P_{ji} \, \mathrm{d}\Omega = \int_{\Omega_X} \delta u_i b_i^0 \, \mathrm{d}\Omega + \int_{\Gamma_X^h} \delta u_i h_i^0 \, \mathrm{d}\Gamma, \quad (7)$$

where $F_{ij} = \partial x_i / \partial X_j$ is the deformation gradient, P_{ij} is the first Piola–Kirchhoff stress which in general is a function of F_{ij} and its rate, J is the determinant of the deformation gradient, ρ^0 is the initial density, b_i^0 is the body force defined in the undeformed domain Ω_X, and h_i^0 is the surface traction mapped onto the undeformed traction (natural) boundary Γ_X^h.

In the Lagrangian discretization, the reproducing kernel approximation of material displacements is expressed as [5, 6]:

$$u_i^h(\mathbf{X}, t) = \sum_{I=1}^{NP} \Psi_I^X(\mathbf{X}) d_{iI}(t), \quad (8)$$

where $\mathbf{X}$ is the material coordinate defined in the undeformed domain (configuration). Based on RK approximation, the Lagrangian shape function is expressed as

$$\Psi_I^X(\mathbf{X}) = \mathbf{H}^T(\mathbf{0})\mathbf{M}^{-1}(\mathbf{X})\mathbf{H}(\mathbf{X} - \mathbf{X}_I)\phi_a(\mathbf{X} - \mathbf{X}_I), \quad (9)$$

where the basis functions are defined as

$$\mathbf{H}^T(\mathbf{X} - \mathbf{X}_I) = [1 \quad X_1 - X_{1I} \quad X_2 - X_{2I} \quad \ldots \quad (X_2 - X_{2I})^N] \quad (10)$$

and the moment matrix is computed by

$$\mathbf{M}(\mathbf{X}) = \sum_{I=1}^{NP} \mathbf{H}(\mathbf{X} - \mathbf{X}_I)\mathbf{H}^T(\mathbf{X} - \mathbf{X}_I)\phi_a(\mathbf{X} - \mathbf{X}_I). \quad (11)$$

An example of kernel function is

$$\phi_a(z) = \begin{cases} \frac{2}{3} - 4z^2 + 4z^3 & 0 \le z \le \frac{1}{2} \\ \frac{4}{3} - 4z + 4z^2 - \frac{4}{3}z^3 & \frac{1}{2} < z \le 1 \\ 0 & z > 1. \end{cases} \quad z = \frac{\|\mathbf{X} - \mathbf{X}_I\|}{a} \quad (12)$$

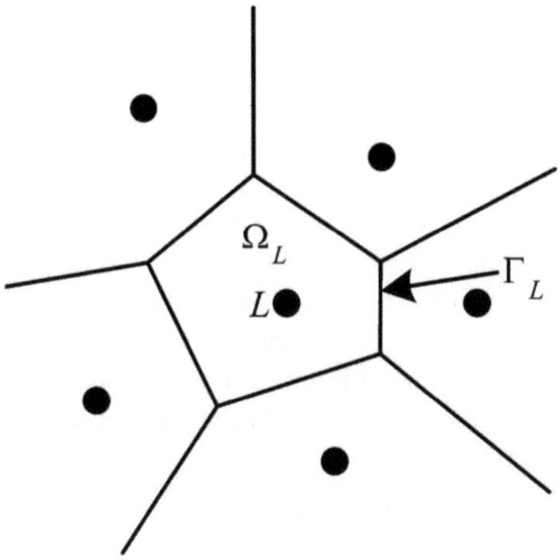

Fig. 1. Representative integration domain for SCNI.

The shape functions $\Psi_I^X(\mathbf{X})$ formulated using material coordinate $\mathbf{X}$ in the undeformed domain Ω_X are called the Lagrangian shape functions.

A stabilized conforming nodal integration (SCNI) [9] has been introduced for the integration of weak form to achieve computational efficiency and stability. Linear exactness in the Galerkin approximation requires (1) first order completeness of the trial and test functions, and (2) a domain integration that satisfies integration constraints. To meet integration constraints for nodal integration of Equation (7) using the deformation gradient as the primary kinematic variable, a smoothing of deformation gradient at a nodal point with material coordinate $\mathbf{X}_L$ has been considered [10]:

$$\bar{F}_{ij}(\mathbf{X}_L) = \frac{1}{A_L}\int_{\Omega_L} F_{ij}\,\mathrm{d}\Omega = \frac{1}{A_L}\int_{\Omega_L} \frac{\partial u_i}{\partial X_j}\,\mathrm{d}\Omega + \delta_{ij}, \quad A_L = \int_{\Omega_L}\mathrm{d}\Omega, \tag{13}$$

where Ω_L is the nodal representative domain of $\mathbf{X}_L$ in the undeformed configuration as shown in Figure 1. Applying the divergence theorem at the undeformed configuration yields

$$\bar{F}_{ij}(\mathbf{X}_L) = \frac{1}{A_L}\int_{\Gamma_L} u_i N_j\,\mathrm{d}\Gamma + \delta_{ij} = \bar{e}_{ij}(\mathbf{X}_L) + \delta_{ij}, \tag{14}$$

where

$$\bar{e}_{ij}(\mathbf{X}_L) = \frac{1}{A_L}\int_{\Gamma_L} u_i N_j\,\mathrm{d}\Gamma. \tag{15}$$

In Equation (15), Γ_L is the boundary of Ω_L as shown in Figure 1, and N_i is the surface normal of Γ_L. Generally, a Voronoi diagram is employed to generate the representative domain. Introducing a Lagrangian shape function into Equation (15) yields

$$\bar{e}_{ij}(\mathbf{X}_L) = \sum_I \bar{b}_{jI}^L d_{iI}, \tag{16}$$

where

$$\bar{b}_{iI}^L = \frac{1}{A_L} \int_{\Gamma_L} \Psi_I^X N_i \, \mathrm{d}\Gamma. \tag{17}$$

It has been shown that by introducing linear basis functions in material coordinate in the reproducing kernel approximation (Equations (9)–(11)), and by employing the above smoothed deformation gradient (Equations (14)–(17)) in the static Lagrangian equilibrium equation integrated by nodal integration, exact solution can be obtained in large deformation problems with homogeneous constant deformation field [10], which is the large deformation version of linear patch test. The corresponding discrete equation of Equation (7) is:

$$\mathbf{M}\ddot{\mathbf{u}} = \mathbf{f}^{\text{ext}} - \mathbf{f}^{\text{int}}, \tag{18}$$

where

$$\mathbf{M}_{IJ} = \sum_L \rho_0 \Psi_I^X(\mathbf{X}_L) \Psi_J^X(\mathbf{X}_L) \mathbf{I} A_L, \tag{19}$$

$$\mathbf{f}_I^{\text{ext}} = \sum_K \Psi_I^X(\hat{\mathbf{X}}_K) \mathbf{h}^0(\hat{\mathbf{X}}_K) W_K + \sum_L \Psi_I^X(\mathbf{X}_L) \mathbf{b}^0(\mathbf{X}_L) A_L, \tag{20}$$

$$\mathbf{f}_I^{\text{int}} = \sum_L \bar{\mathbf{B}}_I^T(\mathbf{X}_L) \mathbf{P}(\mathbf{X}_L) A_L, \tag{21}$$

$$\bar{\mathbf{B}}_I(\mathbf{X}_L) = \begin{bmatrix} \bar{b}_{1I}^L & 0 \\ 0 & \bar{b}_{2I}^L \\ \bar{b}_{2I}^L & 0 \\ 0 & \bar{b}_{1I}^L \end{bmatrix}, \quad \mathbf{P}(\mathbf{X}_L) = \begin{bmatrix} P_{11}(\mathbf{X}_L) \\ P_{22}(\mathbf{X}_L) \\ P_{12}(\mathbf{X}_L) \\ P_{21}(\mathbf{X}_L) \end{bmatrix}, \tag{22}$$

$$\bar{b}_{iI}^L = \frac{1}{A_L} \int_{\Gamma_L} \Psi_I^X N_i \, \mathrm{d}\Gamma, \quad A_L = \int_{\Omega_L} \mathrm{d}\Omega, \tag{23}$$

where $\mathbf{X}_L$ and A_L are the nodal point position and the corresponding weight associated with domain integration in the undeformed domain Ω_X, $\hat{\mathbf{X}}_K$, and W_K are the nodal point position and weight associated with boundary integration on the undeformed natural boundary Γ_X^h, and Γ_L and N_i are the boundary and surface normal of the nodal representative domain Ω_L in the undeformed configuration. Note that the area of nodal representative domain A_L is used as the weight of nodal integration.

3 Semi-Lagrangian Reproducing Kernel Discretization

3.1 Semi-Lagrangian RK Shape Function

For problems involving path-dependent materials such as materials deformed in plastic deformation, the internal energy is expressed as

$$\int_{\Omega_X} \frac{\partial \delta u_i}{\partial X_k} P_{ki} \, \mathrm{d}\Omega = \int_{\Omega_X} \frac{\partial \delta u_i}{\partial X_k} F_{kj}^{-1} \sigma_{ij} \, J \, \mathrm{d}\Omega, \tag{24}$$

where σ_{ij} is Cauchy stress. Cauchy stress calculation requires the spatial derivative of displacements approximated by Lagrangian shape functions, and thus the following chain rule is employed:

$$\frac{\partial \Psi_I^X(\mathbf{X})}{\partial x_i} = \frac{\partial \Psi_I^X(\mathbf{X})}{\partial X_j} \frac{\partial X}{\partial x_i} = \frac{\partial \Psi_I^X(\mathbf{X})}{\partial X_j} F_{ji}^{-1}, \tag{25}$$

where $\mathbf{F}^{-1}$ is obtained by taking a direct inverse of $\mathbf{F}$.

It is clear that this Lagrangian formulation in Equations (24) and (25) breaks down when the mapping $\mathbf{x} = \varphi(\mathbf{X}, t)$ or the inverse mapping $\mathbf{X} = \varphi^{-1}(\mathbf{x}, t)$ is no longer regular (one-to-one). This happens in problems involve situations such as new free surface formation in damage evolution or free surface closure that commonly exists in materials processing, earth moving, and penetration processes. To circumvent this difficulty, a semi-Lagrangian discretization is proposed.

As has been mentioned, the distance measure, $z = \|\mathbf{X} - \mathbf{X}_I\|/a$ for Lagrangian kernel $\phi_a(z)$ is defined in the undeformed configuration. Therefore, the kernel support covers the same set of material particles before and after deformation in the Lagrangian discretization. Similar to the Lagrangian discretization, in semi-Lagrangian discretization the discrete meshfree points follow the material motion, however, the distance measure $z = \|\mathbf{x} - \mathbf{x}(\mathbf{X}_I, t)\|/a$ in the semi-Lagrangian kernel $\phi_a(z)$ is defined in the deformed configuration. Under this definition, the material particles covered under the kernel support vary during material deformation. These two different kernel functions are compared as follows:

$$\phi_a(z) : \begin{cases} \text{Lagrangian kernel:} \quad z = \|\mathbf{X} - \mathbf{X}_I\|/a, \\ \text{Semi-Lagrangian kernel:} \quad z = \|\mathbf{x} - \mathbf{x}(\mathbf{X}_I, t)\|/a. \end{cases} \tag{26}$$

Figure 2 schematically compares kernel supports in Lagrangian and semi-Lagrangian kernels.

As shown in Figure 2, Lagrangian kernel deforms with the material, and the kernel support covers the same group of material particles at all time. For the semi-Lagrangian kernel, in general, material particles can move in and out of the kernel support during the deformation processes.

In semi-Lagrangian discretization, the discrete points follow the material motion, i.e., $\mathbf{x}_I = \mathbf{x}(\mathbf{X}_I, t)$, whereas the approximation is formulated in the current configuration in the following form

$$u_i(\mathbf{x}, t) = \sum_{I=1}^{NP} \Psi_I(\mathbf{x}) d_{iI}(t), \tag{27}$$

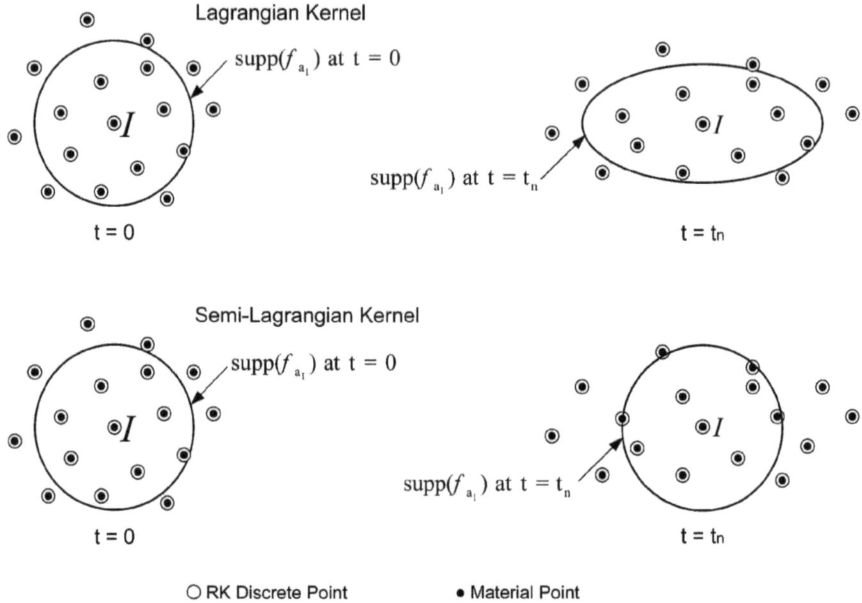

Fig. 2. Comparison of Lagrangian and semi-Lagrangian kernels.

where the semi-Lagrangian shape function $\psi_I(\mathbf{x})$ is constructed based on (i) the current position of material points and (ii) basis functions defined in the current configuration as follows

$$\Psi_I(\mathbf{x}) = \sum_{i+j=0}^{n} [(x_1 - x_1(\mathbf{X}_I, t))^i (x_2 - x_2(\mathbf{X}_I, t))^j \, b_{ij}(\mathbf{x})]\phi_a(\mathbf{x} - \mathbf{x}(\mathbf{X}_I, t)). \quad (28)$$

The coefficients $b_{ij}(\mathbf{x})$ are constructed by imposing the following semi-Lagrangian reproducing conditions

$$\sum_{I=1}^{NP} \Psi_I(\mathbf{x}) x_1(\mathbf{X}_I, t)^i \, x_2(\mathbf{X}_I, t)^j = x_1^i x_2^j, \quad i + j = 0, 1, 2, \ldots, n\,. \quad (29)$$

Solving $b_{ij}(\mathbf{x})$ from Equation (29) yields the following semi-Lagrangian shape function:

$$\Psi_I(\mathbf{x}) = C(\mathbf{x}; \mathbf{x} - \mathbf{x}(\mathbf{X}_I, t))\phi_a(\mathbf{x} - \mathbf{x}(\mathbf{X}_I, t)), \quad (30)$$

where

$$C(\mathbf{x}; \mathbf{x} - \mathbf{x}(\mathbf{X}_I, t)) = \mathbf{H}^T(\mathbf{0})\mathbf{M}^{-1}(\mathbf{x})\mathbf{H}(\mathbf{x} - \mathbf{x}(\mathbf{X}_I, t)), \quad (31)$$

$$\mathbf{M}(\mathbf{x}) = \sum_{I=1}^{NP} \mathbf{H}(\mathbf{x} - \mathbf{x}(\mathbf{X}_I, t))\mathbf{H}^T(\mathbf{x} - \mathbf{x}(\mathbf{X}_I, t))\phi_a(\mathbf{x} - \mathbf{x}(\mathbf{X}_I, t)), \tag{32}$$

$$\mathbf{H}^T(\mathbf{x} - \mathbf{x}(\mathbf{X}_I, t)) = [1 \quad x_1 - x_1(\mathbf{X}_I, t) \quad x_2 - x_2(\mathbf{X}_I, t) \quad \ldots \quad (x_2 - x_2(\mathbf{X}_I, t))^n]. \tag{33}$$

3.2 Material Time Derivatives

The difference between Lagrangian and semi-Lagrangian approximations can be observed in the material time derivative of Lagrangian and semi-Lagrangian kernels:

1. Lagrangian kernel: $\dot{\phi}_a\left(\frac{\|\mathbf{X} - \mathbf{X}_I\|}{a}\right) = 0$ (34)

2. Semi-Lagrangian kernel: $\dot{\phi}_a\left(\frac{\|\mathbf{x} - \mathbf{x}(\mathbf{X}_I, t)\|}{a}\right)$ (35)

$$= \phi_a'\left(\frac{\|\mathbf{x} - \mathbf{x}(\mathbf{X}_I, t)\|}{a}\right)\frac{\mathbf{n}\cdot(\mathbf{v} - \mathbf{v}_I)}{a},$$

where

$$(\dot{\ }) = \left.\frac{\partial(\,)}{\partial t}\right|_{[\mathbf{X}]} \quad \text{is the material time derivative, and}$$

$$\mathbf{n} = \frac{\mathbf{x} - \mathbf{x}(\mathbf{X}_I, t)}{\|\mathbf{x} - \mathbf{x}(\mathbf{X}_I, t)\|}. \tag{36}$$

The term $\mathbf{n}\cdot(\mathbf{v} - \mathbf{v}_I)$ in Equation (35) is the projection of relative velocity between point of evaluation ($\mathbf{x}$) and grid point ($\mathbf{x}(\mathbf{X}_I, t)$) onto the kernel support radial direction centered at $\mathbf{x}(\mathbf{X}_I, t)$, and it represents the relative motion between the neighboring material $\mathbf{x}$ and the kernel support center point $\mathbf{x}(\mathbf{X}_I, t)$ as shown in Figure 3. Positive value of $\mathbf{n}\cdot(\mathbf{v} - \mathbf{v}_I)$ indicates $\mathbf{x}$ moving away from $\mathbf{x}(\mathbf{X}_I, t)$, and negative value of $\mathbf{n}\cdot(\mathbf{v} - \mathbf{v}_I)$ indicates $\mathbf{x}$ moving towards $\mathbf{x}(\mathbf{X}_I, t)$.

3.3 Semi-Lagrangian Discrete Equation of Motion

The semi-Lagrangian discrete equation of motion is obtained by employing the following weak form of equation of motion expressed in the deformed domain Ω_x:

$$\int_{\Omega_x} \delta u_{(i,j)}\sigma_{ij}\,\mathrm{d}\Omega + \int_{\Omega_x} \delta u_i \rho \ddot{u}_i\,\mathrm{d}\Omega = \int_{\Omega_x} \delta u_i b_i\,\mathrm{d}\Omega + \int_{\Gamma_x^h} \delta u_i h_i\,\mathrm{d}\Gamma, \tag{37}$$

where σ_{ij} is the Cauchy stress, $u_{(i,j)} = (\partial u_i/\partial x_j + \partial u_j/\partial x_i)/2$, ρ is the density at the current state, b_i is the body force defined in the deformed domain Ω_x, and h_i is

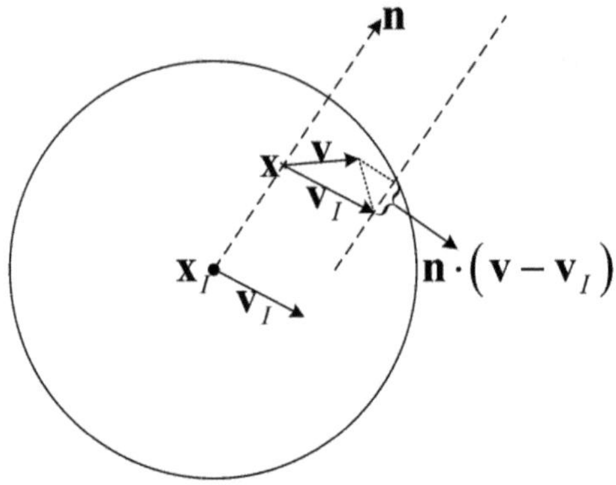

Fig. 3. Relative motion between neighboring material $\mathbf{x}$ and kernel support center point $\mathbf{x}_I = \mathbf{x}(\mathbf{X}_I, t)$.

the surface traction defined on the deformed traction boundary Γ_x^h. Let the velocity v_i be approximated by semi-Lagrangian shape functions,

$$v_i^h(\mathbf{x}, t) = \sum_{I=1}^{NP} \Psi_I(\mathbf{x}) v_{iI}(t). \tag{38}$$

The corresponding semi-Lagrangian approximation of acceleration is given as

$$\ddot{u}_i^h(\mathbf{x}, t) = \dot{v}_i^h(\mathbf{x}, t) = \sum_{I=1}^{NP} (\Psi_I(\mathbf{x}) \dot{v}_{iI}(t) + \Psi_I^*(\mathbf{x}) v_{iI}(t)), \tag{39}$$

where $\Psi_I^*(\mathbf{x})$ is the correction due to time rate of the semi-Lagrangian kernel $\dot{\phi}_a$

$$\Psi_I^*(\mathbf{x}) = C(\mathbf{x}; \mathbf{x} - \mathbf{x}(\mathbf{X}_I, t)) \dot{\phi}_a(\mathbf{x} - \mathbf{x}(\mathbf{X}_I, t)). \tag{40}$$

Substituting Equations (38) and (39) into formula (37), and considering strain smoothing of $u_{(i,j)}$ with similar procedures discussed in Section 2, the following discrete equation is obtained

$$\mathbf{M}\dot{\mathbf{v}} + \mathbf{N}\mathbf{v} = \mathbf{f}^{\text{ext}} - \mathbf{f}^{\text{int}}, \tag{41}$$

where

$$\mathbf{M_{IJ}} = \sum_L \rho(\mathbf{x}(\mathbf{X}_L, t)) \Psi_I(\mathbf{x}(\mathbf{X}_L, t)) \Psi_J(\mathbf{x}(\mathbf{X}_L, t)) \mathbf{I} a_L, \tag{42}$$

$$\mathbf{N}_{IJ} = \sum_L \rho(\mathbf{x}(\mathbf{X}_L, t)) \Psi_I(\mathbf{x}(\mathbf{X}_L, t)) \Psi_J^*(\mathbf{x}(\mathbf{X}_L, t)) \mathbf{I} a_L, \tag{43}$$

$$\mathbf{f}_I^{\text{ext}} = \sum_K \Psi_I(\mathbf{x}(\hat{\mathbf{X}}_K, t))\mathbf{h}(\mathbf{x}(\hat{\mathbf{X}}_K, t))w_k + \sum_L \Psi_I(\mathbf{x}(\mathbf{X}_L, t))\mathbf{b}(\mathbf{x}(\mathbf{X}_L, t))a_L \quad (44)$$

$$\mathbf{f}_I^{\text{int}} = \sum_L \bar{\mathbf{B}}_I^T(\mathbf{x}(\mathbf{X}_L, t))\,\Xi\,(\mathbf{x}(\mathbf{X}_L, t))a_L, \quad (45)$$

$$\bar{\mathbf{B}}_I(\mathbf{x}(\mathbf{X}_L, t) = \begin{bmatrix} \bar{b}_{1I}^L & 0 \\ 0 & \bar{b}_{2I}^L \\ \bar{b}_{2I}^L & \bar{b}_{1I}^L \end{bmatrix} \qquad \Xi\,(\mathbf{x}(\mathbf{X}_L, t) = \begin{bmatrix} \sigma_{11}(\mathbf{x}(\mathbf{X}_L, t)) \\ \sigma_{22}(\mathbf{x}(\mathbf{X}_L, t)) \\ \sigma_{12}(\mathbf{x}(\mathbf{X}_L, t)) \end{bmatrix}, \quad (46)$$

$$\bar{b}_{iI}^L = \frac{1}{a_L}\int_{\gamma_L} \Psi_I n_i \, \mathrm{d}\Gamma \qquad a_L = \int_{\omega_L} \mathrm{d}\Omega, \quad (47)$$

where $\mathbf{x}(\mathbf{X}_L, t)$ and a_L are the nodal point position and the corresponding weight associated with domain integration in the deformed domain Ω_x, respectively, $\mathbf{x}(\hat{\mathbf{X}}_K, t)$ and w_k are the nodal point position and weight associated with boundary integration on the deformed natural boundary Γ_x^h, respectively, and γ_L and n_i are the boundary and surface normal of the nodal representative domain ω_L in the deformed configuration, respectively. The second term on the left hand side of Equation (41) is a convective term resulting from the material time derivative of the semi-Lagrangian kernel function.

4 Stability Analysis of Lagrangian and Semi-Lagrangian Discrete Equations

Von Neumann stability analyses of Lagrangian and semi-Lagrangian discrete equations are presented in this section. The following notations are used in stability analysis:

$$\begin{aligned} C(h) &= C(x_I; x_{I+1} - x_I) & \phi_a(h) &= \phi_a(x_{I+1} - x_I) \\ C(h/2) &= C(x_I; x_{I+1/2} - x_I) & \phi_a(h/2) &= \phi_a(x_{I+1/2} - x_I) \\ \Psi_I(0) &= \Psi_I(x_I) & \Psi_I(h) &= \Psi_I(x_{I+1}) \\ \Psi_{I,x}(h) &= \Psi_{I,x}(x_{I+1}) & \Psi_{I,xx}(h) &= \Psi_{I,xx}(x_{I+1}). \end{aligned}$$

where $h = \Delta x$ is the constant nodal spacing.

4.1 Lagrangian Discrete Equation of Motion Integrated by SCNI

The Lagrangian discrete equation of motion in one-dimension can be expressed as:

$$\mathbf{M}\ddot{\mathbf{u}} = \mathbf{f}^{\text{ext}} - \mathbf{f}^{\text{int}}, \quad (48)$$

where

$$F_I^{\text{int}} = \int_{\Omega_x} \Psi_I^X(X)_{,X}\ P\ \mathrm{d}\Omega, \tag{49}$$

$$F_I^{\text{ext}} = \int_{\Omega_x} \Psi_I^X(X) b^0\ \mathrm{d}\Omega + \Psi_I^X(X) h^0 \Big|_{\Gamma_X^h}. \tag{50}$$

By employing SCNI, Equation (49) can be written as

$$f_I^{\text{int}} = \sum_{L=I}^{NP} \bar{b}_I^L P(X_L) A_L, \tag{51}$$

where $\bar{b}_I^L$ is the smoothed gradient:

$$\bar{b}_I^L = \frac{1}{A_L}\left[\Psi_I^X(X_L^+) - \Psi_I^X(X_L^-)\right]. \tag{52}$$

In one-dimension, A_L is the length of nodal representative domain $\Omega_L =]X_L^-, X_L^+[$ for node L, and X_L^+ and X_L^- are the two end points of Ω_L. The first Piola–Kirchhoff stress P is defined as:

$$P = SF. \tag{53}$$

Here S is the second Piola–Kirchhoff stress obtained by

$$\Delta S = C^{\text{SE}} \Delta E \tag{54}$$

and the Green strain E is given as

$$E = \frac{1}{2}(F^2 - 1). \tag{55}$$

Further considering lumped mass, the Lagrangian discrete equation of motion at a node I can be written as

$$m_I \ddot{u}_I = f_I^{\text{ext}} - f_I^{\text{int}} \qquad \text{no summation on } I. \tag{56}$$

Assuming small perturbation in displacement, the perturbation equation of Equation (56) reads [2]:

$$m_I \ddot{\tilde{u}}_I = -\tilde{f}_I^{\text{int}}. \tag{57}$$

From Equation (51), we have

$$\tilde{f}_I^{\text{int}} = \sum_{L=1}^{NP} \bar{b}_I^L\ \tilde{P}(X_L) A_L. \tag{58}$$

The perturbation of the first Piola–Kirchhoff stress is

$$\tilde{P}(X_L) = (C^{\text{SE}} F^2 + S)\tilde{f}_L, \tag{59}$$

where

$$\tilde{f}_L = \left.\frac{\partial \tilde{u}}{\partial X}\right|_L = \sum_{I=1}^{NP} \bar{b}_I^L \, \tilde{u}_I. \tag{60}$$

Consider uniform particle distribution in an infinite domain, Equation (57) can be written as

$$\ddot{\tilde{u}}_I = -\frac{C^{\text{SE}} F^2 + S}{\rho_0} \sum_{K=-m}^{m} \left[\bar{b}_I^{I+K} \sum_{J=-m}^{m} \bar{b}_I^{I-J} \, \tilde{u}_{I+J+K} \right], \tag{61}$$

where $m = \text{int}(R + 0.5)$ is an integer determined by the normalized support size R (support size divided by nodal distance).

For plane wave, the perturbed displacement takes the form as

$$\tilde{u}_I = g e^{ikI\Delta x = i\omega t}, \tag{62}$$

where g is the amplitude of the perturbation, k is the wave number, ω is the frequency, and $\Delta x = h$ is the nodal spacing. Accordingly, the stability criterion is ω should be real. Substituting Equation (62) into Equation (61) yields:

$$\omega^2 = \frac{C^{\text{SE}} F^2 + S}{\rho_0} \sum_{K=1}^{2m+1} \left[\hat{b}_K \sum_{L=1}^{2m+1} \hat{b}_L \cos(K - L)kh \right], \tag{63}$$

where

$$\hat{b}_K = \bar{b}_I^{I+K-m-1}. \tag{64}$$

Note that since uniform nodal spacing and infinite domain is considered, the index I is ignored in Equation (63) for notational simplicity. The frequency relations based on Equation (63) are shown in Figure 4 for various normalized support sizes, and $c^2 = (C^{\text{SE}} F^2 + S)/\rho_0$. It is observed that the frequency is very close to zero near the cut-off point ($kh = \pi$) for the case with very large support size ($R = 3.5$). This implies that the stiffness matrix becomes more linearly dependent and ill conditioned when very large support size is used and this should be avoided.

4.2 Stability of Semi-Lagrangian Discrete Equation

In one-dimension, with the employment of lumped mass and nodal integration, the discrete equation of motion by the semi-Lagrangian approximation can be written as:

$$m_I \dot{v}_I + \sum_{J=I}^{NP} N_{IJ} v_J = f_I^{\text{ext}} - f_I^{\text{int}} \qquad \text{no summation on } I, \tag{65}$$

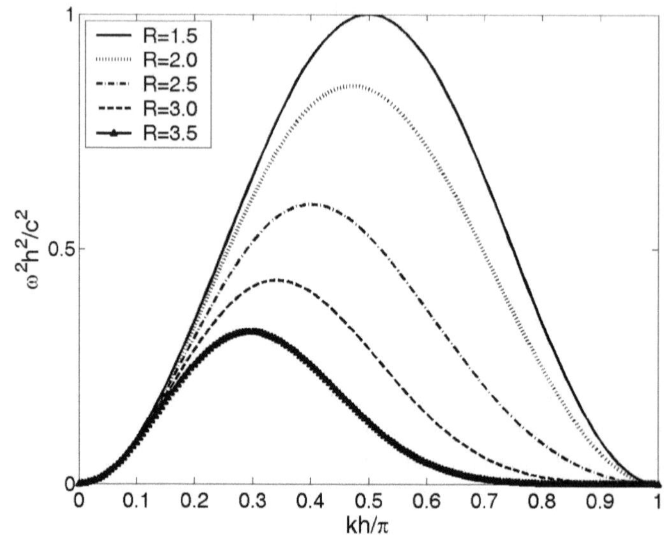

Fig. 4. Frequency characteristic of Lagrangian RK discretization with SCNI.

where

$$N_{IJ} = \sum_{K=1}^{NP} \rho(x_K)\Psi_I(x_K)\Psi_J^*(x_K)a_K, \tag{66}$$

$$f_I^{\text{int}} = \sum_{K=1}^{NP} \bar{B}_I(x_K)\sigma(x_K)a_K, \tag{67}$$

$$\Psi_I^*(x_K) = C(x_K; x_K - x_I)\dot{\phi}_a(x_K - x_I), \tag{68}$$

$$\sigma(x_K) = C^\sigma \varepsilon(x_K) = C^\sigma \sum_{J=1}^{NP} \bar{B}_J(x_K)u_J, \tag{69}$$

$$\dot{\phi}_a(x_K - x_I) = \phi_a'(x_K - x_I)\frac{n_{IK}(v_K - v_I)}{a}, \tag{70}$$

$$n_{IK} = \frac{x_K - x_I}{|x_K - x_I|}. \tag{71}$$

Here a_K is the length of nodal representative domain of node K, $x_K = x(X_K, t)$ is the current position of material point X_K, and C^σ is the material modulus. Note that $\bar{B}_I(x_K)$ is the gradient of shape function which takes the smoothed form (Equation (46)) if SCNI is used, and $\bar{B}_I(x_K) = \Psi_{I,x}(x_K)$ if a direct nodal integration (DNI) is employed.

4.2.1 Semi-Lagrangian Weak Form Integrated by Stabilized Conforming Nodal Integration (SCNI)

When SCNI is employed for domain integration, the smoothed shape function gradient following Equation (46) is

$$\bar{B}_I(x_K) = \bar{B}_I^K = \frac{1}{a_K}\left[\Psi_I(x_K^+) - \Psi_I(x_K^-)\right]. \tag{72}$$

In one-dimension, a_K is the length of nodal representative domain $\Omega_K =]x_K^-, x_K^+[$ for node K, and x_K^+ and x_K^- are the two end points of Ω_K. Under the assumption of small perturbation in displacement and velocity, the perturbation in nodal mass can be neglected, therefore, the perturbed equation corresponding to Equation (65) can be written as

$$m_I\dot{\tilde{v}}_I + \sum_{J=1}^{NP} \tilde{N}_{IJ}\, v_J + \sum_{J=1}^{NP} N_{IJ}\, \tilde{v}_J = -\tilde{f}_I^{\text{int}}. \tag{73}$$

It can be shown that Equation (73) can be expressed explicitly as

$$m_I\dot{\tilde{v}}_I + \sum_{J=1}^{NP}\left(\sum_{K=1}^{NP} m_K \tilde{\Psi}_I(x_K)\Psi_J^*(x_K)\right) v_J + \sum_{J=1}^{NP}\left(\sum_{K=1}^{NP} m_K \Psi_I(x_K)\Psi_J^*(x_K)\right)\tilde{v}_J$$

$$+ \sum_{K=1}^{NP} \tilde{\bar{B}}_I(x_K)\sigma(x_K)a_K + \sum_{K=1}^{NP} \bar{B}_I(x_K)\tilde{\sigma}(x_K)a_K + \sum_{K=1}^{NP} \bar{B}_I(x_K)\sigma_K\, \tilde{a}_K = 0. \tag{74}$$

The perturbed quantities such as $\tilde{\Psi}_I(x_K)$, $\tilde{\sigma}(x_K)$, and $\tilde{\bar{B}}_I(x_K)$ can be obtained according to their definitions. For moderate normalized support size $1 < R < 2$, by approximating $v_{I+1} = v_I + Q\Delta x$, where

$$Q = \left.\frac{\partial v}{\partial x}\right|_{x=x_I},$$

and $\tilde{v} = \dot{\tilde{u}}_I$, Equation (74) can be rewritten as

$$\ddot{\tilde{u}}_I - \frac{2Q^2[C(h)\phi_a'(h)]^2}{R^2}(\tilde{u}_{I+1} - 2\tilde{u}_I + \tilde{u}_{I-1})$$

$$+ \frac{QC(h)\phi_a'(h)}{R}\left[\Psi_I(0)(-\dot{\tilde{u}}_{I+1} + \dot{\tilde{u}}_{I-1}) + \Psi_I(h)(-\dot{\tilde{u}}_{I+2} + \dot{\tilde{u}}_{I-2})\right]$$

$$+ \frac{(C^\sigma + \sigma)}{\rho h^2}\left[C\left(\frac{h}{2}\right)\phi_a\left(\frac{h}{2}\right)\right]^2 (-\tilde{u}_{I+2} + 2\tilde{u}_I - \tilde{u}_{I-2}) = 0. \tag{75}$$

Substituting Equation (62) into Equation (75) leads to

$$\omega^2 - 4Q^2 \frac{[C(h)\phi_a'(h)]^2}{R^2}(1-\cos kh) - 2Q\omega \frac{C(h)\phi_a'(h)}{R}[\Psi_I(0)\sin kh + \Psi_I(h)\sin 2kh] - \frac{2(C^\sigma+\sigma)}{\rho h^2}\left[C\left(\frac{h}{2}\right)\phi_a\left(\frac{h}{2}\right)\right]^2(1-\cos 2kh) = 0. \tag{76}$$

Rewrite Equation (76) as

$$\omega^2 + B\omega + C = 0, \tag{77}$$

where

$$B = -2Q\frac{C(h)\phi_a'(h)}{R}[\Psi_I(0)\sin kh + \Psi_I(h)\sin 2kh]$$

$$C = -4Q^2\frac{[C(h)\phi_a'(h)]^2}{R^2}(1-\cos kh) - \frac{2(C^\sigma+\sigma)}{\rho h^2}\left[C\left(\frac{h}{2}\right)\phi_a\left(\frac{h}{2}\right)\right]^2(1-\cos 2kh). \tag{78}$$

Define a frequency characteristic parameter $D(\omega)$ as

$$D(\omega) = B^2 - 4C. \tag{79}$$

It is observed that if $D(\omega) \geq 0$, real solution for ω is obtained and the discrete system is stable. Note that $D(\omega) \geq 0$ is satisfied as long as $C^\sigma + \sigma \geq 0$, where $C^\sigma + \sigma$ represents the tangent modulus including geometric nonlinearity effect. Therefore, under semi-Lagrangian discretization with SCNI, the sufficient condition for stability is when the tangent modulus is positive ($C^\sigma + \sigma \geq 0$). This stability condition for the semi-Lagrangian discrete system integrated by SCNI is consistent with the stability condition of a continuum.

Note that the sign of the velocity gradient Q does not affect the stability in semi-Lagrangian discretization with SCNI, hence only a non-negative velocity gradient is considered in the following study of the behavior of $D(\omega)$. Consider a discretization with uniform particle distribution, normalized support size $R = 1.5$, small velocity gradient $0 \leq Q \leq 2$, and C^σ is assumed positive.

Figure 5 demonstrates frequency characteristic parameter $D(\omega) = B^2 - 4C$ for the case where the continuum system is stable ($C^\sigma + \sigma > 0$). It can be observed from Figure 5 that in the cases when ($C^\sigma + \sigma > 0$) the frequency is real ($D(\omega) \geq 0$) for all wavelengths, representing a stable discrete system compatible with the continuum system. Figure 6 displays frequency characteristic parameter $D(\omega) = B^2 - 4C$ for the case where continuum system exhibits instability ($C^\sigma + \sigma \leq 0$). It is shown in Figure 6 that $D(\omega) < 0$ in certain range of wavelengths leading to imaginary angular frequency, and hence the discrete system is unstable. This instability in the discrete system is consistent with the continuum instability [2].

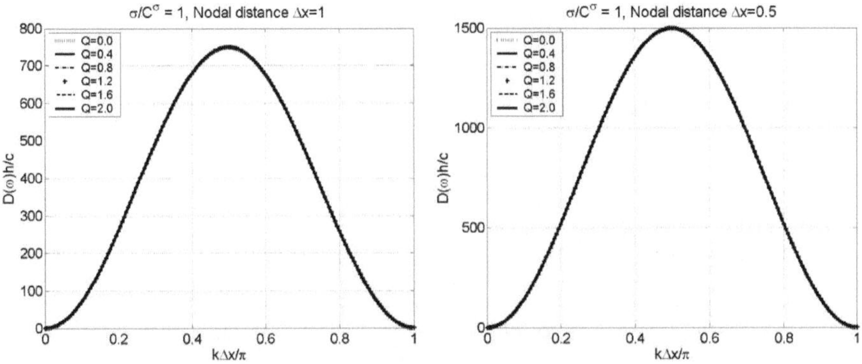

Fig. 5. Frequency characteristic parameter for semi-Lagrangian with SCNI and $C^{\sigma} + \sigma > 0$.

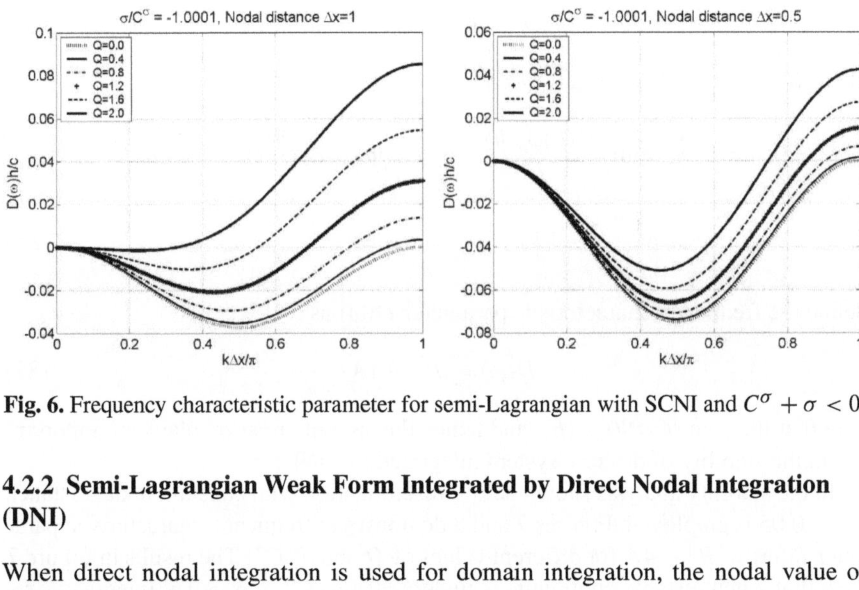

Fig. 6. Frequency characteristic parameter for semi-Lagrangian with SCNI and $C^{\sigma} + \sigma < 0$.

4.2.2 Semi-Lagrangian Weak Form Integrated by Direct Nodal Integration (DNI)

When direct nodal integration is used for domain integration, the nodal value of gradient is calculated as:

$$\bar{B}_I(x_K) = \Psi_{I,x}(x_K). \tag{80}$$

Correspondingly, Equation (74) reduces to

$$\begin{aligned}
&\ddot{\tilde{u}}_I - \frac{2Q^2[C(h)\phi_a'(h)]^2}{R^2}\left(\tilde{u}_{I+1} - 2\tilde{u}_I + \tilde{u}_{I-1}\right) \\
&\quad + \frac{QC(h)\phi_a'(h)}{R}\left[\Psi_I(0)(-\dot{\tilde{u}}_{I+1} + \dot{\tilde{u}}_{I-1}) + \Psi_I(h)(-\dot{\tilde{u}}_{I+2} + \dot{\tilde{u}}_{I_2})\right] \\
&\quad + \frac{\sigma}{\rho}\Psi_{I,xx}(h)\left(\tilde{u}_{I+1} - 2\tilde{u}_I + \tilde{u}_{I-1}\right)
\end{aligned}$$

$$-\frac{(C^\sigma+\sigma)}{\rho}[\Psi_{I,x}(h)]^2(\tilde{u}_{I+2}-2\tilde{u}_I+\tilde{u}_{I-2})=0 \tag{81}$$

Substitution of Equation (62) into Equation (81) yields

$$\begin{aligned}&\omega^2-4Q^2\frac{[C(h)\phi_a'(h)]^2}{R^2}(1-\cos kh)\\&-2Q\omega\frac{C(h)\phi_a'(h)}{R}[\Psi_I(0)\sin kh+\Psi_I(h)\sin 2kh]\\&+2\frac{\sigma}{\rho}\Psi_{I,xx}(h)(1-\cos kh)-4\frac{C^\sigma+\sigma}{\rho}[\Psi_{I,x}(h)]^2\sin^2(kh)=0.\end{aligned} \tag{82}$$

Equation (82) can be written as

$$\omega^2+B\omega+A=0, \tag{83}$$

where B is given in Equation (78), and

$$\begin{aligned}A=&\left\{-4Q^2\frac{[C(h)\phi_a'(h)]^2}{R^2}+2\frac{\sigma}{\rho}\Psi_{I,xx}(h)\right\}(1-\cos kh)\\&-4\frac{C^\sigma+\sigma}{\rho}[\Psi_{I,x}(h)]^2\sin^2(kh).\end{aligned} \tag{84}$$

Define the frequency characteristic parameter $D(\omega)$ as

$$D(\omega)=B^2-4A. \tag{85}$$

Note that the sign of $\sigma\Psi_{I,xx}(h)$, and hence the sign of stress σ, plays an important role in the stability of discrete system integrated by DNI.

In the stability analysis shown below, discretization with constant nodal distance $\Delta x=0.05$ is employed. Figures 7 and 8 demonstrate frequency characteristic parameter $D(\omega)=B^2-4A$ for different values of Q and σ/C^σ. The results in Figure 7 show that when a stable continuum is under tension ($C^\sigma+\sigma>0$), imaginary frequency occurs ($D(\omega)<0$) at certain wavelengths, leading to an unstable discrete system constructed using DNI. This is referred to as the tensile instability observed in SPH [20]. On a separate test as shown in Figure 8, the discrete system constructed using DNI becomes stable when continuum instability occurs ($C^\sigma+\sigma\le 0$), and it is unphysical.

4.3 Effect of Convective Term on Stability

To examine how the convective term in the semi-Lagrangian discretization affects the stability of the discrete equation, consider removing the convective term (second

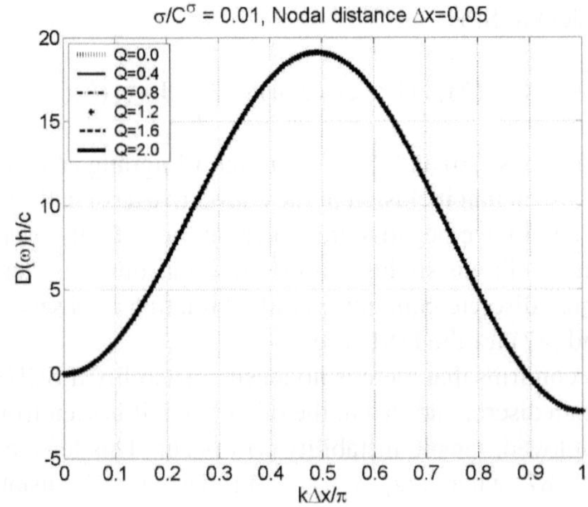

Fig. 7. Frequency characteristic parameter for semi-Lagrangian with DNI for $C^{\sigma} + \sigma > 0$ and $\sigma > 0$.

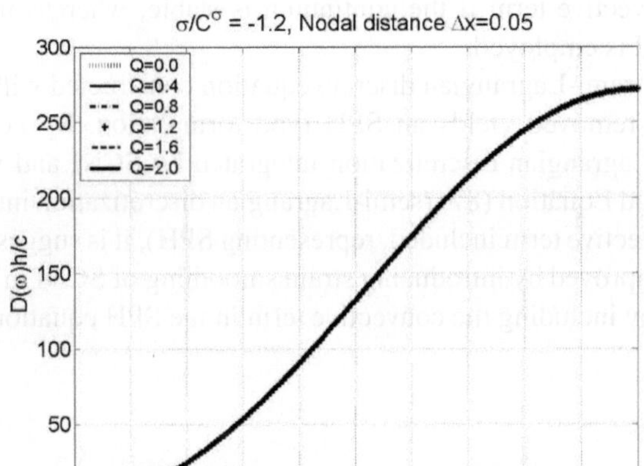

Fig. 8. Frequency characteristic parameter for semi-Lagrangian with DNI for $\sigma + C^{\sigma} < 0$.

term in Equation (65)) to yield the following dispersion equations constructed by SCNI and DNI:

(1) SCNI

$$\omega^2 - \frac{2(C^{\sigma} + \sigma)}{\rho h^2}\left[C\left(\frac{h}{2}\right)\phi_a\left(\frac{h}{2}\right)\right]^2 (1 - \cos 2kh) = 0 \tag{86}$$

(2) DNI (equivalent to SPH)

$$\omega^2 = 2\frac{C^\sigma + \sigma}{\rho}[\Psi_{I,x}(h)]^2\,(1 - \cos 2kh) - 2\frac{\sigma}{\rho}\,\Psi_{I,xx}(h)(1 - \cos kh). \tag{87}$$

By comparing Equations (76) and (86) for the semi-Lagrangian discretization integrated by SCNI, it is seen that inclusion of the convective term in the semi-Lagrangian discretization does not have negative influence on the stability property of discrete system. In fact, the stability is slightly improved when convective term is included in the semi-Lagrangian discrete equation. Similar situation is observed by comparing Equations (82) and (87) for the DNI case.

Equation (86) confirms that there is no tensile instability if SCNI is employed in the semi-Lagrangian discretization. On the other hand, it is seen from Equation (87) that if DNI is employed, tensile instability will occur. This tensile instability condition in Equation (87) when $\sigma\Psi_{I,xx}(h) > 0$ reflects tensile instability in SPH as has been identified by Swegle et al. [20]. Based on the stability analysis of semi-Lagrangian discretization in Sections 4.2.1 and 4.2.2 and in this section, it is shown that SCNI always yields a stable semi-Lagrangian discrete system regardless of the inclusion of convective term if the continuum is stable, whereas instability is observed when DNI is employed.

Note that the semi-Lagrangian discrete equation constructed with DNI and with convective term removed yields an SPH type formulation. By comparing Equation (76) (semi-Lagrangian discretization integrated by SCNI and with convective term included) and Equation (87) (semi-Lagrangian discretization integrated by DNI and without convective term included, representing SPH), it is suggested the stability of SPH can be improved by introducing strain smoothing of SCNI in the nodal strain calculation and by including the convective term in the SPH equation of motion.

5 Conclusions

Lagrangian formulation breaks down when the one-to-one mapping of material point position between undeformed domain and deformed domain cannot be defined. This situation happens in the processes of new free surface formation in damage evolution or free surface closure that commonly exists in materials processing, earth moving, and penetration processes. To circumvent this difficulty, a semi-Lagrangian discretization is proposed. In the Lagrangian discretization the distance measure in the kernel function is defined in the initial undeformed domain, whereas in the semi-Lagrangian discretization the distance measure in the kernel function is defined in the current deformed domain. It has been shown that the semi-Lagrangian discretization yields a convective term resulting from the non-conservative coverage of material points under the kernel support.

Von Neumann stability analyses have been performed for Reproducing Kernel (RK) Lagrangian and semi-Lagrangian discretizations with weak forms integrated using stabilized conforming nodal integration (SCNI). In Lagrangian discretization with SCNI, the stability of the discrete system is always consistent with the stability of the continuum. In the semi-Lagrangian discrete equation, an additional convective term exists due to the change of material particles covered under the semi-Lagrangian kernel function during different states of material deformation. By comparison of stability conditions of Lagrangian and semi-Lagrangian discretizations integrated by SCNI, it has been shown that the inclusion of the convective term in the semi-Lagrangian discretization offers a slightly better stability compared to the semi-Lagrangian discretization without convective term. The stability analysis also showed similar stability conditions between Lagrangian and semi-Lagrangian discretizations if convective term is included in the semi-Lagrangian discretization.

The stability of semi-Lagrangian discretization with domain integration by a direct nodal integration (DNI) has also been analyzed. The analysis results demonstrated that semi-Lagrangian discretization constructed by DNI yields instability under tension, which is referred to as the tensile instability in SPH. Note that semi-Lagrangian discretization integrated by DNI and with removal of convective term resembles SPH. The stability analysis performed in this work suggests that the stability of SPH can be improved with the addition of convective term in the equation of motion and by introducing strain smoothing employed in SCNI in the nodal strain evaluation of SPH.

Acknowledgement

The support of this work by Lawrence Livermore National Laboratory to University of California, Los Angeles, is greatly acknowledged.

References

1. Belytschko T. and Xiao S.P. Stability analysis of particle methods with corrected derivatives. *Computers and Mathematics with Applications*, 43:329–350, 2002.
2. Belytschko T., Guo Y., Liu W.K. and Xiao S.P. A unified stability analysis of meshless particle methods. *International Journal for Numerical Methods in Engineering*, 48:1359–1400, 2000.
3. Bonet J. and Kulasegaram S. Remarks on tension instability of Eulerian and Lagrangian corrected smooth particle hydrodynamics (CSPH) methods. *International Journal for Numerical Methods in Engineering*, 52:1203–1220, 2001.
4. Chen J.S., Pan C. and Wu C.T. Large deformation analysis of rubber based on a reproducing kernel particle method. *Computational Mechanics*, 19:211–217, 1997.
5. Chen J.S., Pan C., Roque C. and Wang H.P. A Lagrangian reproducing kernel particle method for metal forming analysis. *Computational Mechanics*, 22:289–307, 1998.
6. Chen J.S., Pan C., Wu C.T. and Liu W.K. Reproducing kernel particle methods for large deformation analysis of non-linear structures. *Computer Methods in Applied Mechanics and Engineering*, 139:195–227, 1996.

7. Chen J.S., Roque C.M.O.L., Pan C. and Button S.T. Analysis of metal forming process based on meshless method. *Journal of Materials Processing Technology*, 80/81:642–646, 1998.
8. Chen J.S., Wu C.T. and Belytschko T. Regularization of material instability by meshfree approximations with intrinsic length scales. *International Journal for Numerical Methods in Engineering*, 47:1301–1322, 2000.
9. Chen J.S., Wu C.T., Yoon S. and You Y. A stabilized conforming nodal integration for Galerkin mesh-free methods. *International Journal for Numerical Methods in Engineering*, 50:435–446, 2001.
10. Chen J.S., Yoon S. and Wu C.T. Non-linear version of stabilized conforming nodal integration for Galerkin mesh-free methods. *International Journal for Numerical Methods in Engineering*, 53:2587–2615, 2002.
11. Chen J.S., Yoon S., Wang H.P. and Liu W.K. An improved reproducing kernel particle method for nearly incompressible hyperelastic solids. *Computer Methods in Applied Mechanics and Engineering*, 181:117–145, 2000.
12. Gingold R.A. and Monaghan J.J. Smoothed particle hydrodynamics: Theory and application to non-spherical stars. *Monthly Notices of the Royal Astronomical Society*, 181:375–389, 1977.
13. Lancaster P. and Salkauskas K. Surfaces generated by moving least-squares methods. *Mathematics of Computation*, 37:141–158, 1981.
14. Liu W.K., Jun S. and Zhang Y.F. Reproducing kernel particle methods. *International Journal for Numerical Methods in Fluids*, 20:1081–1106, 1995.
15. Lucy L.B. A numerical approach to the testing of the fission hypothesis. *The Astronomical Journal*, 82:1013–1024, 1977.
16. Libersky L.D. and Petschek A.G. Smooth particle hydrodynamics with strength of materials. In: *Advances in the Free Lagrange Method*, Lecture Notes in Physics, Vol. 395, Springer-Verlag, 1990.
17. Monaghan J.J. SPH without a tensile instability. *Journal of Computational Physics*, 159:290–311, 2000.
18. Rabczuk T., Belytschko T. and Xiao S.P. Stable particle methods based on Lagrangian kernels. *Computer Methods in Applied Mechanics and Engineering*, 193:1035–1063, 2004.
19. Randles P.W. and Libersky L.D. Smoothed particle hydrodynamics: Some recent improvements and applications. *Computer Methods in Applied Mechanics and Engineering*, 139:375–408, 1996.
20. Swegle J.W., Hicks D.L. and Attaway S.A. Smoothed particle hydrodynamics stability analysis. *Journal of Computational Physics*, 116:123–134, 1995.
21. Dyka C.T. and Ingel R.P. An approach for tension instability in smoothed particle hydrodynamics. *Computers & Structures*, 57:573–580, 1995.
22. Guenther C., Hicks D.L. and Swegle J.W. Conservative smoothing versus artificial viscosity. Sandia Report SAND94-1853, Sandia National Lab., 1994.
23. Swegle J.W., Attaway S.W., Heinstein M.W., Mello F.J. and Hicks D.L. An analysis of smoothed particle hydrodynamics. Sandia Report SAND94-2513, Sandia National Lab., 1994.

Simulation of Forming Processes by the α-Shapes-Based Natural Element Method

I. Alfaro[1], E. Cueto[1], F. Chinesta[2] and M. Doblaré[1]

[1] *Group of Structural Mechanics and Material Modelling, Aragón Institute of Engineering Research (I3A), University of Zaragoza, Maria de Luna, 3, E-50018 Zaragoza, Spain; E-mail: ecueto@unizar.es*
[2] *LMSP UMR 8106 CNRS-ENSAM-ESEM, 151 Boulevard de l'Hôpital, F-75013 Paris, France; E-mail: francisco.chinesta@paris.ensam.fr*

Abstract. In this paper we review some of the authors' more recent developments in the field of Forming Processes simulation by means of meshless methods. In particular, all simulations are performed by employing the Natural Element Method (NEM), which has shown some particular characteristics that make it appear as an appealing tool for this kind of problems. Applications include forging, aluminum extrusion and other related forming processes. Particularly, the treatment of the free surface deserves some comments, since it is done in this work by means of alpha-shapes, a particular shape constructor.

Key words: Meshless methods, forming processes, natural element method, alpha-shapes.

1 Introduction

Forming processes constitute a field of engineering where very complex phenomena occur and where very often multi-physic and coupled problems arise. In addition, these usually involve large transformations (i.e., large strains or large displacements.) This has traditionally posed some problems to the Finite Element technology, related to mesh distortion, remeshing, etc. Thus, forming processes seem to be a natural candidate for the use of meshless methods.

In this article we review some of the authors' work on meshless simulation of forming processes, particularly those related to aluminium extrusion and other processes where coupled thermo-mechanical problems are of utmost importance.

Although many different meshless methods exist nowadays, our developments have been focused on the so-called Natural Element Method (NEM) or, equivalently, Natural Neighbour Galerkin methods [4, 18]. This choice has been motivated mainly by the unique features of the NEM among meshless methods, that include, for instance, its ability to exactly reproduce essential boundary conditions.

V.M.A. Leitão et al. (eds), Advances in Meshfree Techniques, 77–95.

The NEM also allows for an updated Lagrangian description of the movement, and this is especially interesting when dealing with free surface flows, for instance. This framework can thus be competitive to existing ALE formulations, avoiding advection terms in the equations of conservation and their associated numerical complexity.

This article is organized as follows. Firstly, we describe the NEM and its main properties. In Section 3 we describe the technique used to track the free-surface during extrusion, forging or related processes. We then include some numerical examples showing the performance of the proposed method. The article is finished with some conclusions.

2 The Natural Element Method

The NEM is basically a Galerkin procedure where the essential field is approximated by any of the existing natural neighbour interpolation schemes [9, 15, 20]. Prior to the description of these schemes, it is necessary to review some basic geometrical concepts, such as the Voronoi diagram of a cloud of points or the Delaunay triangulation.

2.1 Natural Neighbour Interpolation

Models are constructed in our version of the NEM based on nodes only. Consider a cloud of irregularly distributed nodes $N = \{n_1, n_2, \dots, n_N\}$. The Delaunay triangulation [5] of the set N, $\mathcal{D}$, is the only triangulation of the cloud that verifies the *empty circumcircle criterion*, i.e., no circle containing the three nodes of a triangle contains an additional node of the cloud, see Figure 1. The dual structure of a Delaunay triangulation is the so-called Voronoi diagram [21], which is the unique decomposition of the space into non-overlapping cells containing the locus of points closer to a given node than to any other. Formally,

$$T_I = \{\boldsymbol{x} \in \mathbb{R}^n : d(\boldsymbol{x}, \boldsymbol{x}_I) < d(\boldsymbol{x}, \boldsymbol{x}_J)\ \forall\ J \neq I\}, \tag{1}$$

where T_I is the Voronoi cell and $d(\cdot, \cdot)$ represents the Euclidean distance. n represents the space dimension (two or three in the examples presented in this article). Two nodes whose Voronoi cells share one edge are called *natural neighbours* and hence the name of these interpolation schemes.

The first, and most obvious, interpolation scheme based on natural neighbours is the so-called *nearest neighbour* or Thiessen interpolation [20]. If we give the nodal value to the whole associated Voronoi cell, we obtain a $\mathcal{C}^{-1}$ interpolation scheme. This interpolation scheme is not suitable for solving second-order partial differential equations, but has been employed in [8] to construct mixed velocity-pressure approximations for the simulation of incompressible media.

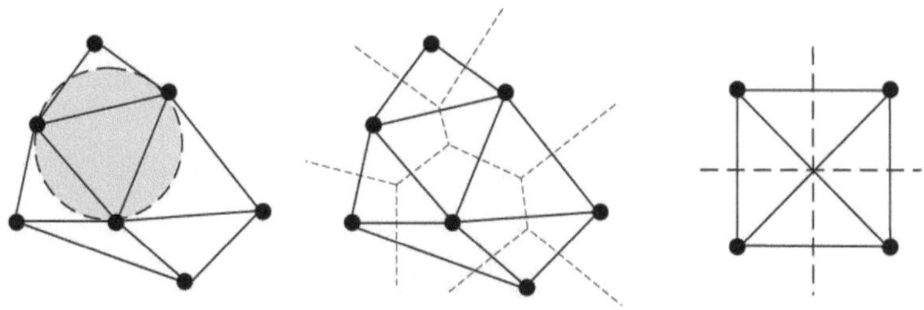

Fig. 1. Delaunay triangulation and Voronoi diagram of a cloud of points. On the right, an example of a degenerate distribution of nodes, with the two possible triangulations depicted. In this last case, four points lie in the same circumcircle and thus no single triangulation exists.

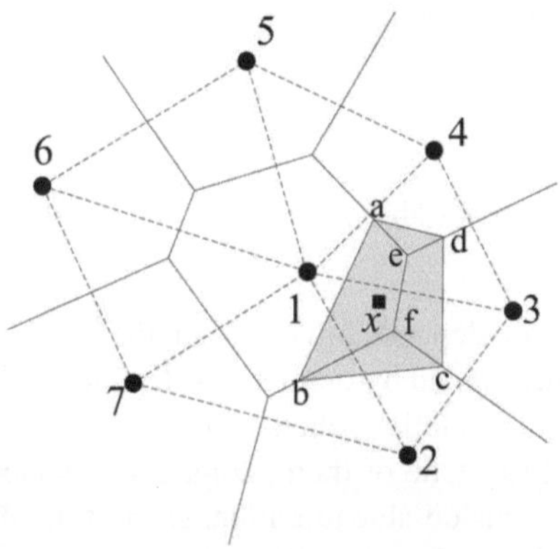

Fig. 2. Definition of the Natural Neighbour coordinates of a point $\boldsymbol{x}$.

The most extended form of natural neighbour-based interpolation is due to Sibson [16]. For the definition of Sibson interpolation it is necessary to previously introduce the concept of *second-order* Voronoi cell. It is defined as the locus of the points that have the node n_I as the closest node and the node n_J as the second closest node:

$$T_{IJ} = \{\boldsymbol{x} \in \mathbb{R}^n : d(\boldsymbol{x}, \boldsymbol{x}_I) < d(\boldsymbol{x}, \boldsymbol{x}_J) < d(\boldsymbol{x}, \boldsymbol{x}_K)\ \forall\ K \neq J;\ \forall\ K \neq I\}. \qquad (2)$$

If a new point is added to a given cloud of points, the Voronoi cells will be modified by its presence. Sibson [15] defined the natural neighbour coordinates of a point $\boldsymbol{x}$ with respect to one of its neighbours I as the ratio of the cell T_I that is transferred to T_x to the initial cloud of points to the total area of T_x. In other words, being $\kappa(\boldsymbol{x})$ and $\kappa_I(\boldsymbol{x})$ the Lebesgue measures of T_x and T_{xI} respectively, the natural neighbour coordinates of $\boldsymbol{x}$ with respect to the node I is defined as

$$\phi_I^{sib}(\boldsymbol{x}) = \frac{\kappa_I(\boldsymbol{x})}{\kappa(\boldsymbol{x})}. \qquad (3)$$

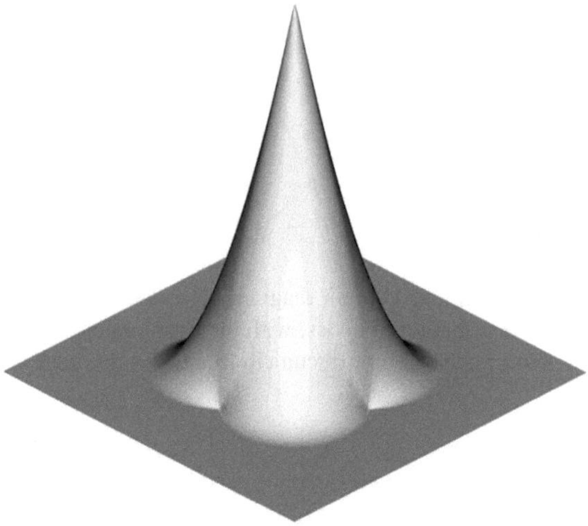

Fig. 3. Natural Element (Sibson) shape function (courtesy N. Sukumar).

The resultant shape function depends obviously on the relative position of the nodes. An example of a node surrounded by other six is depicted in Figure 3.

Recently, some new interpolation schemes based on the concept of natural neighbors have been proposed [9]. One of them, coined as Laplace or *non-Sibsonian* interpolation, has received considerable attention, since it involves magnitudes of one order less of the space dimension (i.e., the calculus of areas in three-dimensional problems, for instance, instead of volumes). If we define the cell intersection $t_{IJ} = \{\boldsymbol{x} \in T_I \bigcap T_J, J \neq I\}$ (note that t_{IJ} may be an empty set) we can define the value

$$\alpha_J(\boldsymbol{x}) = \frac{|t_{IJ}|}{d(\boldsymbol{x}, \boldsymbol{x}_J)}. \tag{4}$$

Thus, the point $\boldsymbol{x}$ shape function value with respect to node 4 in Figure 4 is defined as

$$\phi_4^{ns}(\boldsymbol{x}) = \frac{\alpha_4(\boldsymbol{x})}{\sum_{J=1}^{n} \alpha_J(\boldsymbol{x})} = \frac{s_4(\boldsymbol{x})/h_4(\boldsymbol{x})}{\sum_{J=1}^{n} \left[s_J(\boldsymbol{x})/h_J(\boldsymbol{x})\right]}, \tag{5}$$

where s_J represent the length of the Voronoi segment associated to node J and n represents the number of natural neighbours of the point under consideration, $\boldsymbol{x}$.

Derivatives of the Laplace shape function are not defined along the edges of the Delaunay triangles that lie inside its support (see [19]). For the purposes of the work here presented, Sibson's interpolation has been considered.

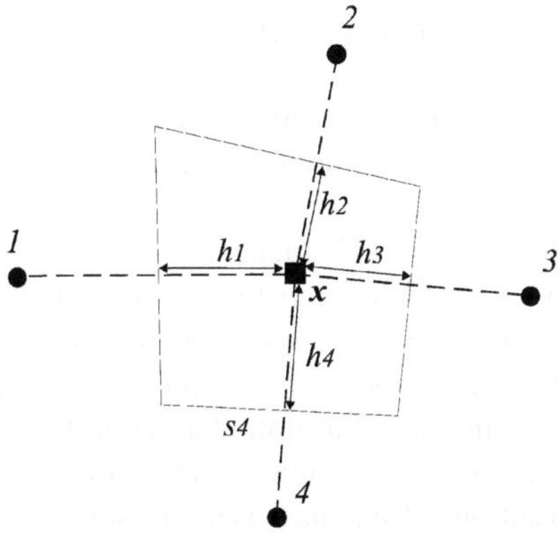

Fig. 4. Definition of non-Sibsonian coordinates.

2.2 Main Properties of the NEM

Firstly, unlike most approximation techniques used in meshless methods, Sibson's interpolation scheme is strictly interpolant, i.e., the approximated surface pass through the data. This can be expressed as

$$\phi_I(\boldsymbol{x}_J) = \delta_{IJ} \tag{6}$$

with δ_{IJ} the Kronecker delta tensor.

The natural neighbour (Sibson) interpolant possesses linear consistency, and thus it is amenable to be used in the approximation of second-order partial differential equations. This can be demonstrated from the before mentioned partition of unity property and its ability to exactly reproduce linear fields (also known as *local coordinate property*, see [16]):

$$\sum_{I=1}^{n} \phi_I(\boldsymbol{x})\boldsymbol{x}_I = \boldsymbol{x} \tag{7}$$

One important property that derives from Equation (6) is that the imposition of essential boundary conditions in NE methods is straightforward. Like in the FEM, it is sufficient to prescribe nodal values to reproduce essential boundary conditions. In addition, natural neighbour interpolants can be strictly interpolant (up to linear precision) on the boundaries (convex or not), under very weak conditions, as demonstrated in [3] by using the theory of α-shapes. This proof was later generalised in [2]. Idelsohn and co-workers later adopted this same approach based on α-shapes in the context of the *Meshless Finite Element Method* [10]. Another method to impose essential boundary conditions in the context of the NEM by employing visibility criteria was developed by Chinesta and co-workers [22].

The *exact* imposition of essential boundary conditions means that no interior point of the domain takes influence on the boundary. This is of first importance in the simulation of processes where friction or other phenomena occurring on the boundary are significant.

In its application to two- and three-dimensional linear elasticity, the natural element method has proved to render more accurate results than linear triangular and tetrahedral finite elements respectively [1, 19].

But, maybe the most important characteristic of the Galerkin scheme obtained by using natural neighbour interpolation is that the accuracy of the approximation is not affected by the distortion of the triangulation [18]. This confers the method the "meshless" character that allows us to implement updated Lagrangian descriptions of the processes. It is also important to note that natural neighbour interpolation is continuous when the data sites move continuously, while other mesh-based methods (i.e., finite elements based on a Delaunay triangulation) are not. This is especially important when dealing when updated Lagrangian descriptions of the movement.

In the following section we describe the technique used to track the free surface appearing in the description of the different forming processes.

3 The α-Shapes-Based Natural Element Method

When dealing with processes involving a free surface – which is commonly the case in classic forming processes simulations, such as forging or extrusion – it is extremely important to accurately track its position. In aluminium extrusion, for instance, non-uniform fluxes of the hot billet through the die can lead to non-straight profiles. Extrusion of hollow profiles, such as tubes, is particularly interesting from a geometrical point of view, since the flow of the hot metal should divide in the interior of the die and re-join just before exiting it.

In many Lagrangian or Lagrangian–Eulerian codes tracking of the free-surface is done by means of markers. The free surface segments (in 2d) or facets (in 3d) are detected and stored at each time step, see [11] for an elegant description of such algorithms in mould filling simulations. Previously, checking of inter-penetration of free-surfaces or development of holes should also be done. This is usually a cumbersome process.

In our approach a different approach has been developed, based on the use of *shape constructors*. Shape constructors are geometrical entities that give a continuous shape to a discrete cloud of nodes. One of such constructors is the family of α-shapes of the cloud [6].

3.1 The Family of α-Shapes of a Cloud of Points

A cloud of points defines a finite set of shapes that can be parameterized by the level of detail up to which we want to represent the geometry. The idea behind α-shapes is simple: the Delaunay triangulation will provide a straightforward connectivity of the nodes. We will eliminate from this triangulation those triangles (or tetrahedra) whose circumradius is greater than the desired level of detail, say α.

Formally, an α-shape is a polytope that is not necessarily convex nor connected, being triangulated by a subset of the Delaunay triangulation of the points. Let N be a finite set of points in $\mathbb{R}^3$ and α a real number, with $0 \leq \alpha < \infty$. A k-simplex σ_T with $0 \leq k \leq 3$ is defined as the convex hull of a subset $T \subseteq N$ of size $| T | = k+1$. Let b be an α-ball, that is, an open ball of radius α. A k-simplex σ_T is said to be α-exposed if there exist an empty α-ball b with $T = \partial b \bigcap N$ where ∂ means the boundary of the ball. In other words, a k-simplex is said to be α-exposed if an α-ball that passes through its defining points contains no other point of the set N.

Thus, we can define the family of sets $F_{k,\alpha}$ as the sets of α-exposed k-simplices for the given set N. This allows us to define an α-shape of the set N as the polytope whose boundary consists on the triangles in $F_{2,\alpha}$, the edges in $F_{1,\alpha}$ and the vertices or nodes in $F_{0,\alpha}$.

A three-dimensional simplicial complex is a collection, $\mathcal{C}$, of closed k-simplices $(0 \leq k \leq 3)$ that satisfies:

(i) If $\sigma_T \in \mathcal{C}$ then $\sigma_{T'} \in \mathcal{C}$ for every $T' \subseteq T$.
(ii) The intersection of two simplexes in $\mathcal{C}$ is empty or is a face of both.

Each k-simplex σ_T included in the Delaunay triangulation, $\mathcal{D}$, defines an open ball b_T whose bounding spherical surface (in the general case) ∂b_T passes through the $k+1$ points of the simplex. Let ϱ_T be the radius of that bounding sphere, then, the family $G_{k,\alpha}$, is formed by all the k-simplexes $\sigma_T \in \mathcal{D}$ whose ball b_T is empty and $\varrho_T < \alpha$. The family $G_{k,\alpha}$ does not necessarily form simplicial complexes, so Edelsbrunner and Mücke [6] defined the α-complex, $\mathcal{C}_\alpha$, as the simplicial complex whose k-simplexes are either in $G_{k,\alpha}$, or else they bound $(k+1)$-simplexes of C_α. If we define the underlying space of $\mathcal{C}_\alpha$, $|\mathcal{C}_\alpha|$, as the union of all simplexes in $\mathcal{C}_\alpha$, the following relationship between α-shapes and α-complexes is found:

$$\mathcal{S}_\alpha = |\mathcal{C}_\alpha| \quad \forall 0 \leq \alpha < \infty \tag{8}$$

In order to clarify the before presented concepts, consider some examples of α-shapes computed from a cloud of points corresponding to one particular simulation of a two-dimensional extrusion process. We restrict ourselves to geometrical concepts only.

Consider the extrusion example shown in Figure 5, where the contour plot of equivalent plastic strain rate is depicted. The key idea of the method here proposed is to extract the shape of the domain at each time step by invoking the concept of α-shape of the cloud of points. The α parameter will be obtained by geometrical considerations. In this case the radius at the inlet of the die, for instance, seems to be the smallest level of detail up to which the domain (i.e., the billet) must be represented. In order to appropriately represent this value, the nodal distance h must be accordingly chosen.

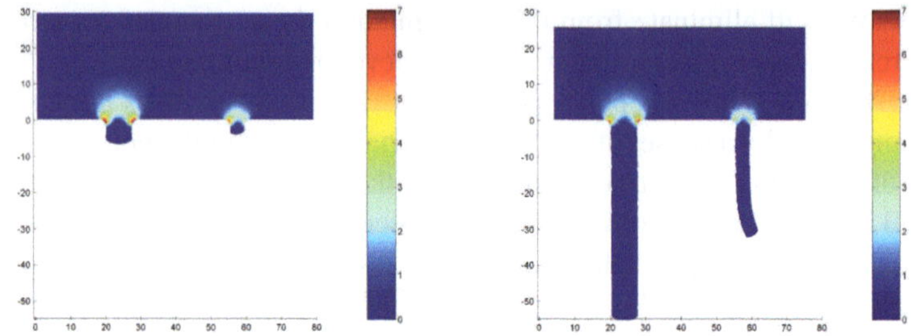

(a) (b)

Fig. 5. Two snapshots of a two-dimensional simulation of an extrusion process. Equivalent plastic strain rate is depicted.

In Figure 6 some members of the family of α-shapes of the cloud of points in its final configuration (corresponding to Figure 5(b)) are depicted. In Figure 6(a) the member for $\alpha = 0$, i.e., the cloud of points itself, is shown. Note how, as α is increased, the number and size of the simplexes (in this case, triangles) that belong to the shape is increasing. For $\alpha = 1.0$ we obtain an appropriate shape for the cloud. Note, however, that this is not an exact value to be determined at each time step. Since the number of α-shapes is finite, there generally exists an interval of valid α values for a single shape. Finally, by increasing the α value, we arrive to the convex hull of the cloud of points (Figure 6(f)).

3.2 Properties of the α-Shapes-Based Natural Element Method

Constructing the model by taking into account the actual shape of the domain has consequences not only in geometrical aspects of the method, but also in the quality of the approximation. As demonstrated in [3], the use of a proper α-shape for the definition of the domain – which means that we reproduce the geometry to the desired level of detail – ensures the linear interpolation of the essential field along the boundary. This is extremely important, since many of the meshless methods lack of the so-called Kronecker delta property, i.e., do not interpolate the field along the boundary, thus complicating the imposition of essential boundary conditions.

In practice, this requirement means that an appropriate cloud of points should be used in order to achieve a correct interpolation on the boundary. Here, the term "appropriate" should be understood in the sense that the cloud of points is dense enough so as to properly reproduce the geometry up to the desired level of detail.

Other methods exist to impose essential boundary conditions in the NEM, such as the so-called C-NEM [22], that uses visibility criteria and renders exactly the same approximation. Both methods can be considered equivalent.

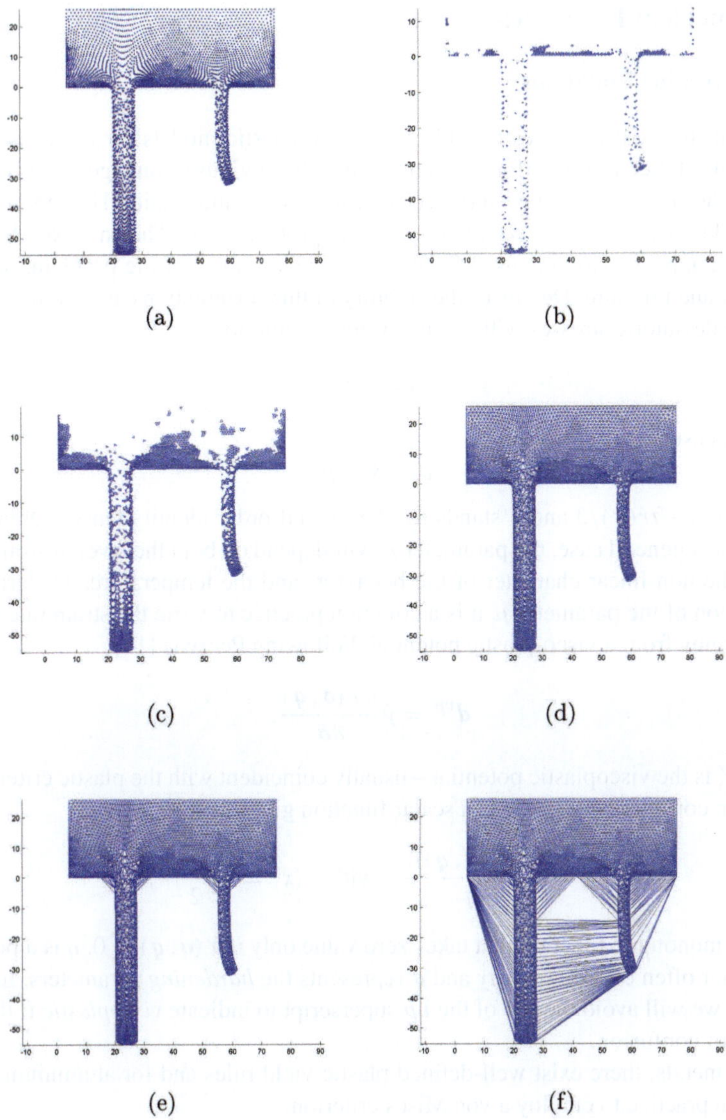

Fig. 6. Some members of the family of α-shapes of the cloud of points used in the extrusion example. (a) $\mathcal{S}_0$ (the cloud of points) (b) $\mathcal{S}_{0.3}$ (c) $\mathcal{S}_{0.5}$ (d) $\mathcal{S}_{1.0}$ (e) $\mathcal{S}_{1.5}$ and (f) $\mathcal{S}_\infty$ (the convex hull of the set).

4 Numerical Examples

4.1 Governing Equations

Many of the existing codes consider rigid-viscoplastic models for the constitutive equations of metals during forming processes. The obvious advantage of this method is that the material can be modelled as a non-Newtonian fluid. This approach is usually known as the *flow formulation* of the problem [23]. Thus a mixed Sibson–Thiessen approximation is used for the essential variables of the problem, namely, velocity and pressure. Details of the stability of this formulation can be found in [8].

The deviatoric stresses will be, under this assumption,

$$\boldsymbol{s} = 2\mu \boldsymbol{d}, \tag{9}$$

being, as usual,

$$\boldsymbol{\sigma} = \boldsymbol{s} - p\boldsymbol{I} \tag{10}$$

where $p = -tr(\boldsymbol{\sigma})/3$ and $\boldsymbol{I}$ stands for the second-order identity tensor. Obviously, in the most general case, the parameter μ will depend on both the level of strain (and hence the non-linear character of the behavior) and the temperature. To derive the expression of the parameter μ it is a common practice to write the strain rate tensor as emerging from a visco-plastic potential. Following Perzyna [13]

$$\boldsymbol{d}^{vp} = \dot{\gamma}\frac{\partial Y(\boldsymbol{\sigma}, \boldsymbol{q})}{\partial \boldsymbol{\sigma}}, \tag{11}$$

where Y is the viscoplastic potential – usually coincident with the plastic criterion as has been considered here – $\dot{\gamma}$ is a scalar function given by

$$\dot{\gamma} = \frac{\langle g(Y(\boldsymbol{\sigma}, \boldsymbol{q}))\rangle}{\eta} \quad \text{with} \quad \langle x\rangle = \frac{x + |x|}{2}, \tag{12}$$

$\langle g\rangle$ is a monotonic function that takes zero value only if $Y(\boldsymbol{\sigma}, \boldsymbol{q}) \leq 0$, η is a positive parameter often called *viscosity* and $\boldsymbol{q}$ represents the *hardening* parameters. In what follows we will avoid the use of the vp superscript to indicate *viscoplastic* if there is no risk of confusion.

For metals, there exist well-defined plastic yield rules and for aluminium it is a common practice to employ a von Mises criterion:

$$Y(\boldsymbol{\sigma}, \boldsymbol{q}) = \overline{\sigma}(\boldsymbol{s}) - \sigma_y(\overline{d}, T), \tag{13}$$

where

$$\overline{\sigma} = \sqrt{\frac{3}{2}\boldsymbol{s} : \boldsymbol{s}} = \sqrt{3J_2}$$

represents the *effective* stress and σ_y represents the uniaxial yield stress. $\overline{d}$ is the only internal variable in this model and is sometimes called *effective strain rate*:

$$\overline{d} = \sqrt{\frac{2}{3}\boldsymbol{d} : \boldsymbol{d}}. \tag{14}$$

Among the most extended laws to account for the aluminium yield stress is the so-called hyperbolic sine or Sellars–Tegart law [14]

$$\sigma_y(\overline{d}) = S_m \operatorname{arcsinh}\left[\left[\left(\frac{\overline{d}}{A}\right) e^{Q/RT}\right]^{1/m}\right]. \tag{15}$$

S_m, m and A are material parameters. A is a factor that depends on the magnesium and silicium matrix solute content (see [12] and references therein), Q represents the activation energy of the deformation process, R is the universal gas constant and, finally, T is the absolute temperature. Note that, as the temperature increases, the yield stress decreases, as expected.

Following this model, a state of null strain rate will give a null yield stress, and this is not in accordance to the well-known behavior of metals in general, and aluminium in particular. So it is a common practice to correct this effect by adding to Equation (15) a initial strain rate, $\overline{d}_0$ so as to render a *modified* Sellars–Tegart law:

$$\sigma_y(\overline{d}) = S_m \operatorname{arcsinh}\left[\left[\left(\frac{\overline{d}_1}{A}\right) e^{Q/RT}\right]^{1/m}\right] \quad \text{with} \quad \overline{d}_1 = \max\{\overline{d}, \overline{d}_0\}. \tag{16}$$

If we combine now the general form of the strain rate tensor given in Equation (11), with Equation (13), we arrive to

$$\boldsymbol{d} = \dot{\gamma}\frac{3\boldsymbol{s}}{2\overline{\sigma}}. \tag{17}$$

It is immediate now, by combining Equation (14) and the definition of effective stress, $\overline{\sigma}$, to prove that $\dot{\gamma}$ is precisely the effective strain rate:

$$\overline{d} = \frac{\dot{\gamma}}{\overline{\sigma}}\sqrt{\frac{3}{2}\boldsymbol{s} : \boldsymbol{s}} = \dot{\gamma}. \tag{18}$$

On the other hand, and by following the Perzyna-like model employed in Equations (11) and (12) and taking $g(f) = f$, we arrive to a relationship between equivalent stress and equivalent strain rate:

$$\overline{d} = \dot{\gamma} = \frac{\langle \overline{\sigma} - \sigma_y \rangle}{\eta} \quad \Rightarrow \quad \overline{\sigma} = \eta\overline{d} + \sigma_y \text{ if } \overline{\sigma} \geq \sigma_y \tag{19}$$

that, introduced in Equation (17), accounting Equation (18), gives the following visco-plastic constitutive equation:

$$\boldsymbol{s} = 2\frac{\eta\overline{d} + \sigma_y(\overline{d})}{3\overline{d}}\boldsymbol{d}. \tag{20}$$

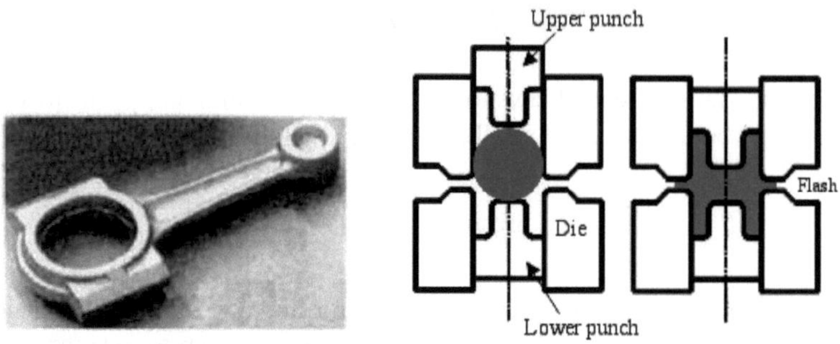

(a) (b)

Fig. 7. The piece to be forged (a) and schematic representation of the forging process (b)

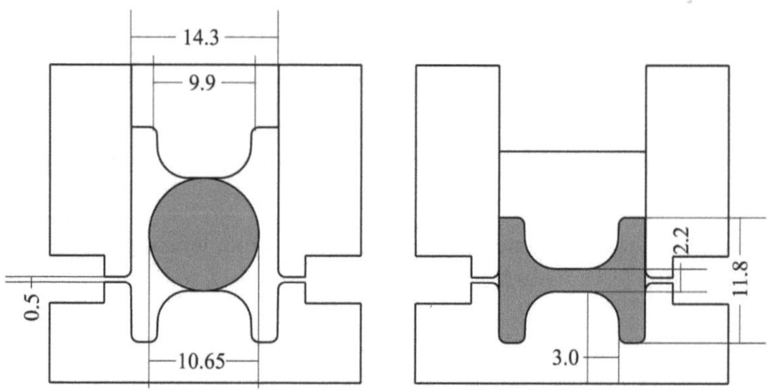

Fig. 8. Geometry and dimensions of the forged piece.

Note that, depending on the η value, the return to the yield surface is done with different velocity. Since it is common to describe aluminium behaviour as rigid-plastic (rather than viscoplastic) we employ null viscosity, so as to enforce $Y = \overline{\sigma} - \sigma_y = 0$, leading to

$$s = \frac{2\sigma_y}{3\overline{d}} \boldsymbol{d}. \tag{21}$$

Finally, the constitutive equation, accounting the incompressibility of plastic flow results:

$$\boldsymbol{\sigma} = 2\mu \boldsymbol{d} - p\boldsymbol{I} \quad \text{with} \quad \mu = \frac{\sigma_y}{3\overline{d}}. \tag{22}$$

Of course, this simple model has important limitations. Undoubtedly, the lack of elastic behaviour is one of the most important. Thus, spring-back cannot be pre-

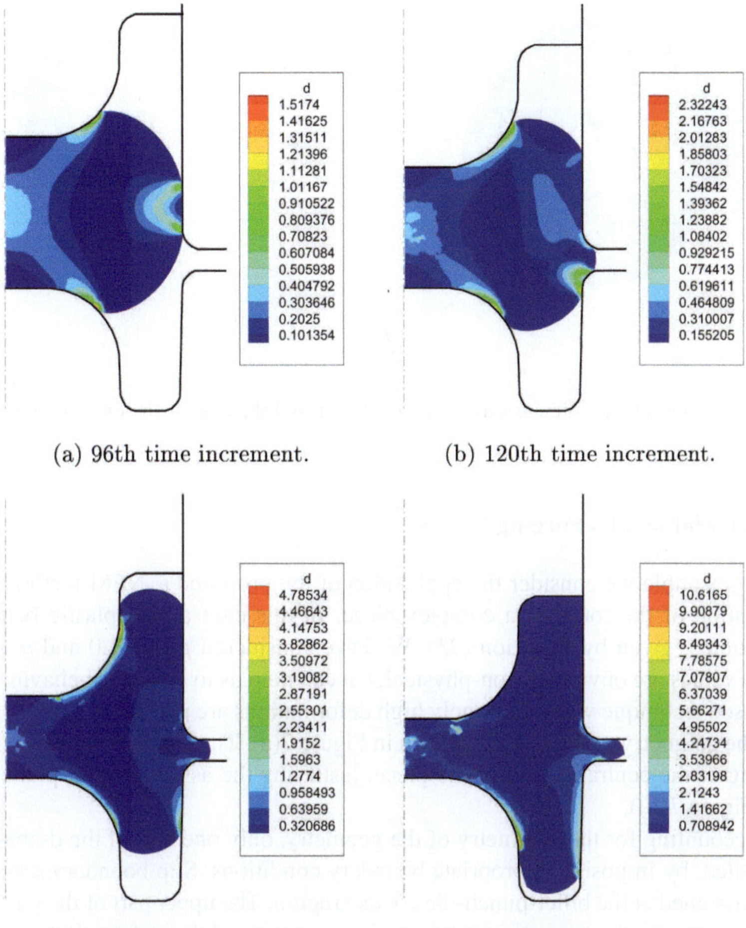

(a) 96th time increment. (b) 120th time increment.

(c) 150th time increment. (d) 168th time increment.

Fig. 9. Equivalent strain rate distribution (s^{-1}) throughout the forging process.

dicted. However, as mentioned before, it has rendered good results and seems to be widely accepted in the forming processes community [12, 23, 24].

This same model can be employed to model a wide variety of polymers, for instance, governed by a power law (also known as Norton–Hoff model):

$$\mu = \mu_0 \overline{d}^{\,n-1} \tag{23}$$

where n is the so-called sensitivity index. The interested reader can consult [17] for some applications on this class of problems.

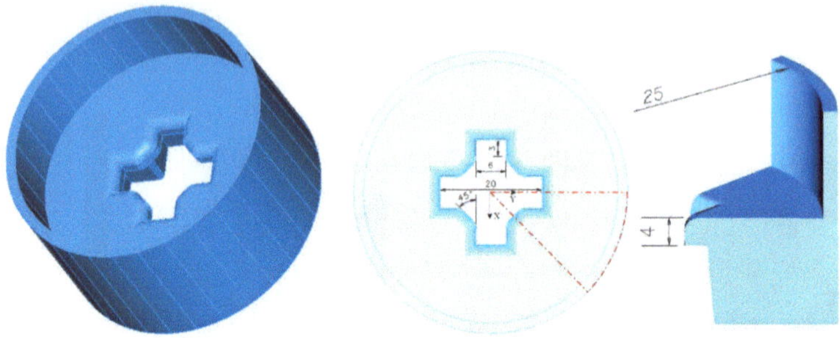

(a) (b)

Fig. 10. Geometry and dimensions of the container and the die for the cross-shaped profile problem.

4.2 Simulation of a Forging Process

In this example we consider the application of the proposed α-NEM method to the simulation of the forge of a complex piece. In this case a viscoplastic behaviour is assumed, given by Equation (23). We have considered $\mu_0 = 1.0$ and $n = 0.3$. These values are obviously non-physical, but can help us to show the behavior of the proposed technique when extremely high deformations are present.

The geometry of the piece is shown in Figure 7(a). The simulation deals with the forging of the central region of the piece, justifying the assumption of plane strain (see Figure 7(b)).

Accounting for the symmetry of the geometry, only one half of the domain was simulated, by imposing appropriate boundary conditions. Slip boundary conditions were assumed at the billet-punch-die contact region. The upper part of the punch, assumed perfectly rigid, moves towards the lower part, fixed throughout the simulation. Geometry and dimensions of the simulated region are shown in Figure 8.

The equivalent plastic strain at time steps 96, 120, 150 and 168 are shown in Figure 9. Very accurate results were obtained in spite of the large strains and displacements involved in the simulation.

It can be seen how the proposed method is a valuable tool to accurately predict the formation of the flash.

4.3 Simulation of the Extrusion of a Cross-Shaped Profile

In [7] an analysis is made of the process of extrusion of a cross-shaped profile, with particular interest on the effect of the misalignment between the die and the workpiece. Geometry and dimensions of the die are shown in Figure 10.

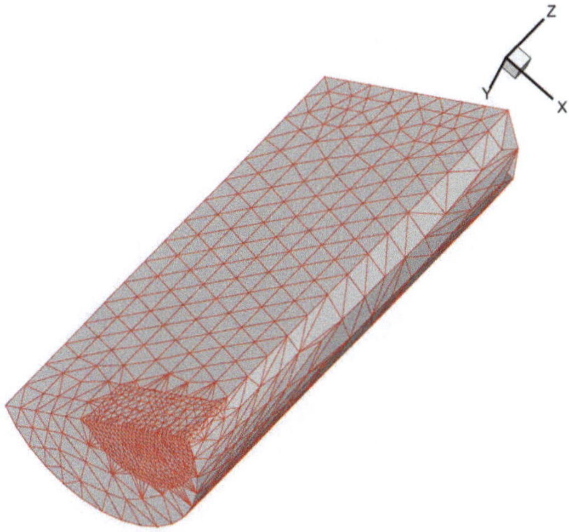

Fig. 11. Initial geometry of the mesh.

The hole of the die was misaligned 10 mm in order to study the influence of this defects in the resulting quality of the extruded profile. The initial mesh is shown in Figure 11. Only one half of the geometry was analyzed, by applying appropriate symmetry boundary conditions.

The material is assumed to be lead, which is able to flow at room temperature, and is therefore easily characterizable by simple experiments with an universal testing machine. In [7], the flow rule of lead at room temperature was adjusted to a Norton–Hoff law, giving

$$\sigma = 60\dot{\varepsilon}^{0.05} \text{ [MPa]} \tag{24}$$

in a uniaxial test.

Various snapshots of the mesh at different time steps are shown in Figure 12. The evolution of the equivalent strain rate is depicted in Figure 13.

Results are in good agreement with the experimental results performed by Filice and colleagues [7] and the resulting deviation of the profile is only slightly underestimated.

This approach notably simplifies those of the ALE methods, especially when large motion is produced in the direction perpendicular to the main flow of the extrudate, which is traditionally treated as purely Eulerian, as in [24]. This is the case when the flow is distorted by the misalignment of the die hole and the axis of the extruder.

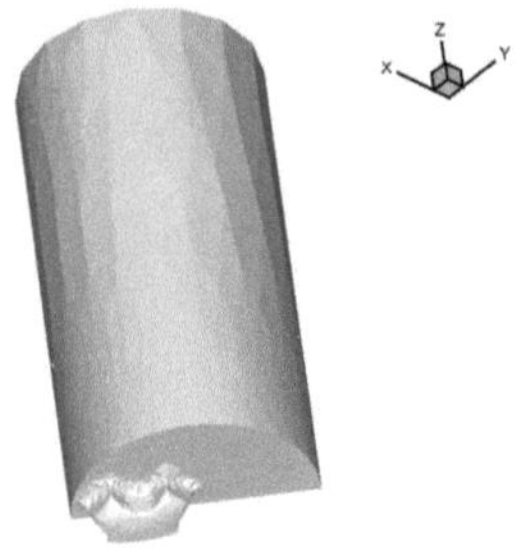

(a) 20th time increment.

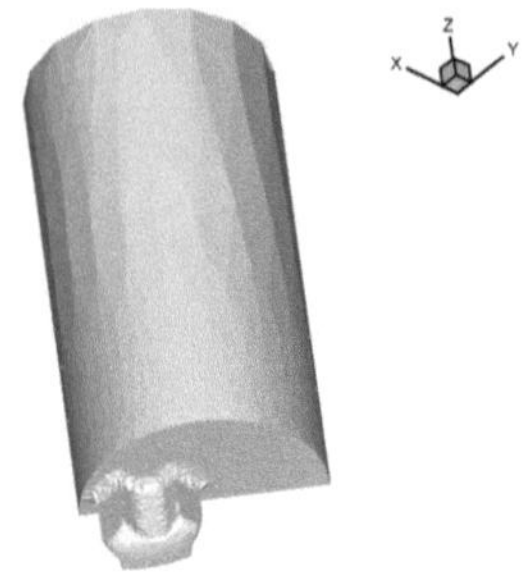

(b) 40th time increment.

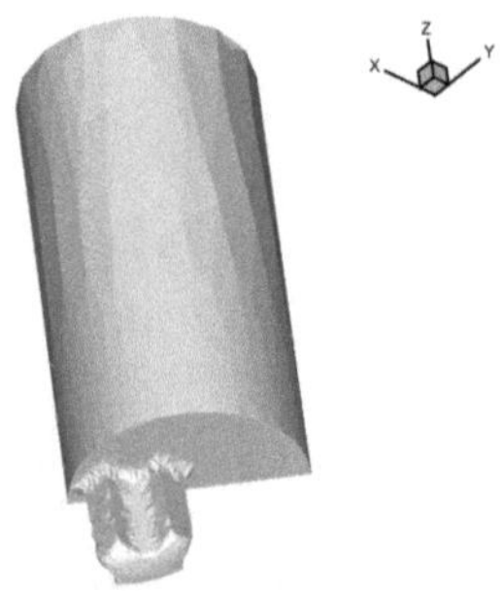

(c) 60th time increment.

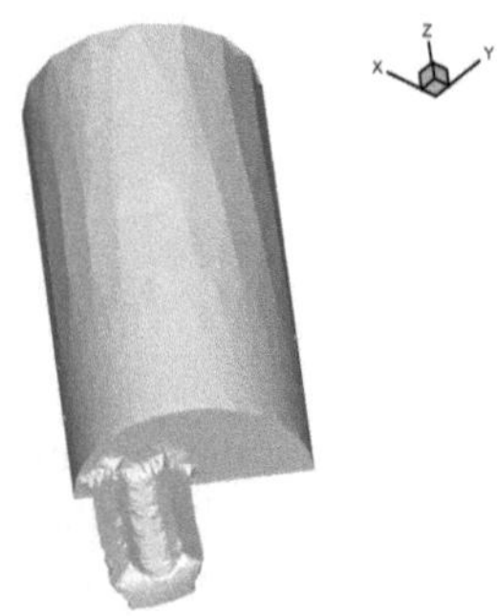

(d) 80th time increment.

Fig. 12. Evolution of the geometry of the extrudate throughout the process.

5 Conclusions

Forming processes usually involve large deformations and are therefore specially interesting as an objective for meshless methods. We have reviewed some of the more important characteristics of the α-shape-based Natural Element Method (α-NEM, see [3]) and studied the more interesting features of the method applied to such problems.

We believe that meshless methods and, in particular, the α-NEM are particularly well-suited to simulate that class of processes. As is well-known for all meshless methods, distortion of the cloud of points does not greatly affect the quality of the results, and therefore remeshing is avoided, in the sense that the cloud of nodes is – or can be – the same throughout the whole simulation. The connectivity is updated by the method in a process transparent to the user, thus greatly alleviating the user effort.

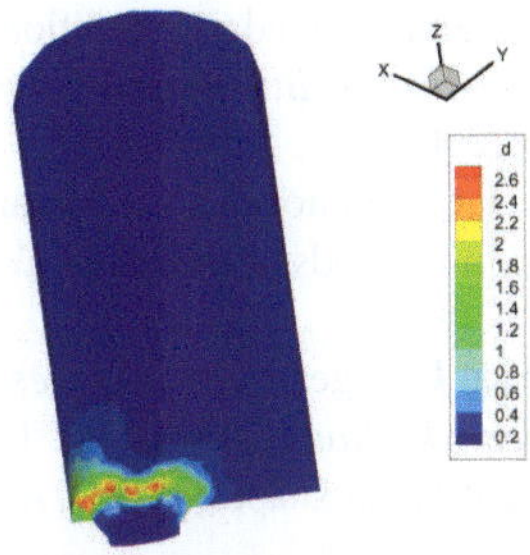

(a) 20th time increment.

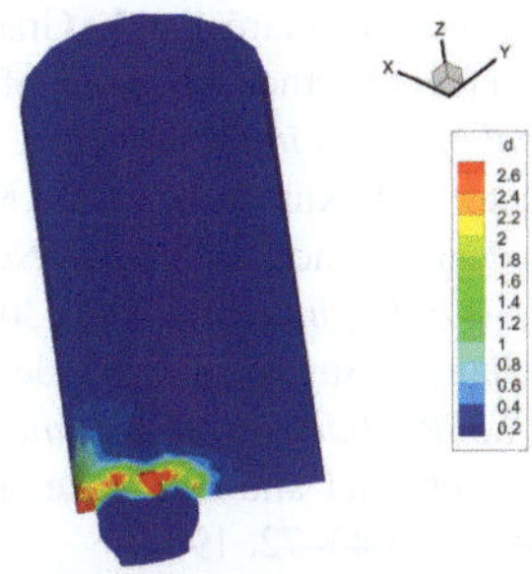

(b) 40th time increment.

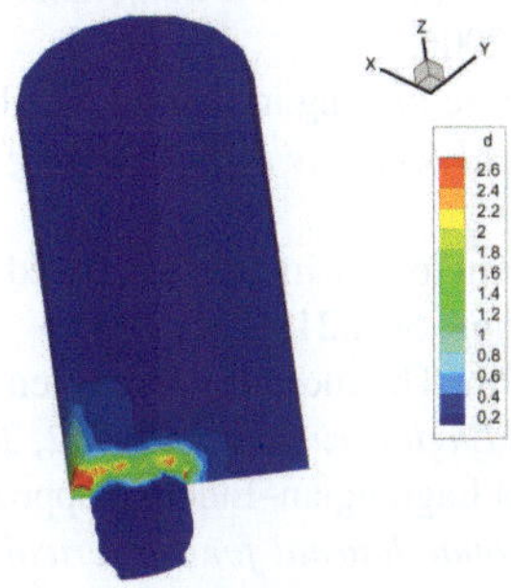

(c) 60th time increment.

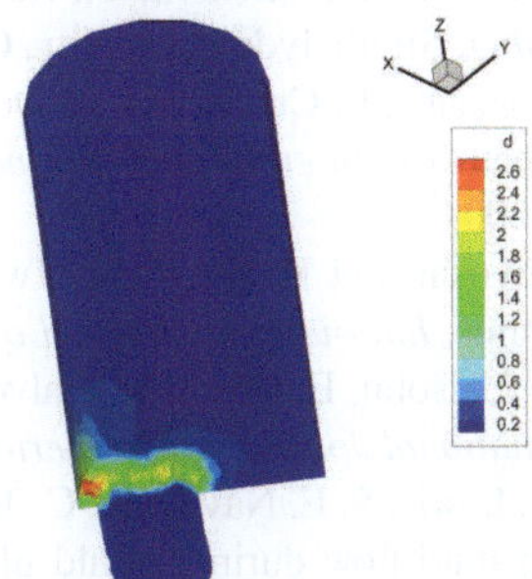

(d) 80th time increment.

Fig. 13. Evolution of the equivalent strain rate (s^{-1}).

Other aspects, such as an improvement of the speed of calculation are currently under investigation, since the computation of Sibson coordinates is known to be considerably harder than the computation of FE shape functions.

In sum, we think that meshless methods, and particularly the NEM, are a good choice for the simulation of problems involving large deformations in a Lagrangian framework.

References

1. E. Cueto, B. Calvo, and M. Doblaré. Modeling three-dimensional piece-wise homogeneous domains using the α-shape based Natural Element Method. *International Journal for Numerical Methods in Engineering*, 54:871–897, 2002.
2. E. Cueto, J. Cegoñino, B. Calvo, and M. Doblaré. On the imposition of essential boundary conditions in Natural Neighbour Galerkin methods. *Communications in Numerical Methods in Engineering*, 19(5):361–376, 2003.

3. E. Cueto, M. Doblaré, and L. Gracia. Imposing essential boundary conditions in the Natural Element Method by means of density-scaled α-shapes. *International Journal for Numerical Methods in Engineering*, 49-4:519–546, 2000.
4. E. Cueto, N. Sukumar, B. Calvo, M. A. Martínez, J. Cegoñino, and M. Doblaré. Overview and recent advances in Natural Neighbour Galerkin methods. *Archives of Computational Methods in Engineering*, 10(4):307–384, 2003.
5. B. Delaunay. Sur la Sphère Vide. A la memoire de Georges Voronoi. *Izvestia Akademii Nauk SSSR, Otdelenie Matematicheskii i Estestvennyka Nauk*, 7:793–800, 1934.
6. H. Edelsbrunner and E. Mücke. Three dimensional alpha shapes. *ACM Transactions on Graphics*, 13:43–72, 1994.
7. L. Filice, I. Alfaro, F. Gagliardi, E. Cueto, F. Micari, and F. Chinesta. Cross extrusion with asymmetric die: A first preliminary comparison between finite element and meshless formulation predictions. In A. Rosochowsky, editor, *ESAFORM Conference on Material Forming*, Strathclyde University, Glasgow, UK, 2006.
8. D. González, E. Cueto, and M. Doblaré. Volumetric locking in Natural Neighbour Galerkin methods. *International Journal for Numerical Methods in Engineering*, 61(4):611–632, 2004.
9. H. Hiyoshi and K. Sugihara. Two generalizations of an interpolant based on Voronoi diagrams. *International Journal of Shape Modeling*, 5(2):219–231, 1999.
10. S. R. Idelsohn, E. Oñate, N. Calvo, and F. Del Pin. The meshless finite element method. *International Journal for Numerical Methods in Engineering*, 58:893–912, 2003.
11. R. W. Lewis, S. E. Navti, and C. Taylor. A mixed Lagrangian–Eulerian approach to modelling fluid flow during mould filling. *International Journal for Numerical Methods in Engineering*, 25:931–952, 1997.
12. J. Lof and Y. Blokhuis. FEM simulations of the extrusion of complex thin-walled aluminium sections. *Journal of Materials Processing Technology*, 122:344–354, 2002.
13. P. Perzyna. Fundamental problems in visco-plasticity. In *Recent Advances in Applied Mechanics*. Academic Press, New York, 1966.
14. C. M. Sellars and W. J. M. Tegart. Hot workability. *International Metallurgical Review*, 17:1–24, 1972.
15. R. Sibson. A vector identity for the Dirichlet Tesselation. *Mathematical Proceedings of the Cambridge Philosophical Society*, 87:151–155, 1980.
16. R. Sibson. A brief description of natural neighbour interpolation. In *Interpreting Multivariate Data. V. Barnett (Ed.)*. John Wiley, 1981, pp. 21–36.
17. N. Sukumar, J. Dolbow, A. Devan, J. Yvonnet, F. Chinesta, D. Ryckelynck, P. Lorong, I. Alfaro, M. A. Martínez, E. Cueto, and M. Doblaré. Meshless methods and partition of unity finite elements. *Internatonal Journal of Forming Processes*, to appear, 2004.
18. N. Sukumar, B. Moran, and T. Belytschko. The Natural Element Method in solid mechanics. *International Journal for Numerical Methods in Engineering*, 43(5):839–887, 1998.
19. N. Sukumar, B. Moran, A. Yu Semenov, and V. V. Belikov. Natural Neighbor Galerkin methods. *International Journal for Numerical Methods in Engineering*, 50(1):1–27, 2001.
20. A. H. Thiessen. Precipitation averages for large areas. *Monthly Weather Report*, 39:1082–1084, 1911.
21. G. M. Voronoi. Nouvelles applications des paramètres continus à la théorie des formes quadratiques. Deuxième memoire: Recherches sur les parallélloèdres primitifs. *J. Reine Angew. Math.*, 134:198–287, 1908.

22. J. Yvonnet, D. Ryckelynck, P. Lorong, and F. Chinesta. A new extension of the Natural Element method for non-convex and discontinuous problems: The Constrained Natural Element method. *International Journal for Numerical Methods in Enginering*, 60(8):1452–1474, 2004.
23. O. C. Zienkiewicz and P. N. Godbolet. Flow of plastic and visco-plastic solids with special reference to extrusion and forming processes. *International Journal for Numerical Methods in Engineering*, 8:3–16, 1974.
24. O. C. Zienkiewicz, P. C. Pain, and E. Oñate. Flow of solids during forming and extrusion: Some aspects of numerical solutions. *International Journal of Solids and Structures*, 14:15–38, 1978.

New Advances in Meshless Methods: Coupling Natural Element and Moving Least Squares Techniques

F. Chinesta[1], J. Yvonnet[2], P. Villon[3], P. Breitkopf[3], P. Joyot[4], I. Alfaro[5] and E. Cueto[5]

[1]*LMSP UMR 8106 CNRS-ENSAM, 151 Boulevard de l'Hôpital, F-75013 Paris, France; E-mail: francisco.chinesta@paris.esnam.fr*
[2]*Université de Marne-la-Vallée, 5 Boulevard Descartes, F-77454 Marne-la-Vallée Cedex, France; E-mail: Julien.Yvonnet@univ-mlv.fr*
[3]*Université de Technlogies de Compiègne, BP 20529, F-60205 Compiègne Cedex, France; E-mail: pierre.villon@utc.fr*
[4]*LIPSI-ESTIA, Technopole Izarbel, F-64210 Bidart, France; E-mail: p.joyot@estia.fr*
[5]*Aragón Institute of Engineering Research, University of Zaragoza, Edificio Betancourt, María de Luna, 7, E-50018 Zaragoza, Spain; E-mail: ecueto@unizar.es*

Abstract. We explore in this work the connections between NN and MLS approximations, coming from the introduction of the NN approximation functions as the weights in the scope of MLS. Thus, it is easy to adjust the approximation consistency (with the possibility to enrich the approximation basis with some particular functions describing issues of the searched solution) in the framework of the MLS techniques, precribing exactly essential boundary conditions from the use of the NN approximation as MLS weight. This approach opens, as will be proved in the present paper, the way to a wide range of formulations: (i) NN collocation strategies; (ii) faster natural element discretizations; (iii) Hermite natural element formulations; (iv) hierarchical bubbles functions in the natural element method; and (v) and NN enriched approximations.

Key words: Meshless methods, Moving Least Squares, RKPM, Natural Neighbor approximations, Natural Element Method, Diffuse Finite Elements, Element Free Galerkin.

1 Introduction

The meshfree methods based on Moving Least Squares (MLS) approximation have been confronted to an active research during the last decade. These include Smooth Particle Hydrodynamics, Element Free Galerkin, Diffuse Elements, Reproducing Kernel Particle and other methods [4, 5, 12–14, 16]. However, one of the issues is the

V.M.A. Leitão et al. (eds), Advances in Meshfree Techniques, 97–121.

satisfaction of essential boundary conditions. This is due to the nature of the approximation itself. In fact, the MLS nodal domains of influence are the same as those of the corresponding weighting functions, who generally do not fit the boundary. The choice of neighboring nodes is also an issue. The approach based in considering the k closest nodes from any evaluation point results only in a C^0 continuity. Moreover, the geometrical complexity of the shape functions supports induces integration difficulties. Simpler integration and an arbitrary degree of continuity are obtained when nodes are associated with fixed, spherical or hexahedral domains of influence, whose optimal size constitutes the main difficulty of that approach. On the other hand, the Natural Neighbor (NN) approximation and associated family of computational methods [20] do not present these drawbacks. The boundary approximation is obtained naturally due to the fact that NN shape functions of internal nodes vanish at the boundary where only the boundary nodes contribute. The list of connected points – the natural neighbors – is also known in advance. However, the NN do not present all the advantages of the MLS. In particular, the shape function support is geometrically complex. Moreover, the NN shape functions have only C^0 continuity at the nodes and only linear consistency is guaranteed. The goal of the present paper is to connect the two approaches in order to get simultaneously the benefits of both NN and MLS approximations. The main idea lies in using NN functions as the weights in the scope of MLS. The two expected benefits are the imposition of the essential boundary conditions and a systematic framework for the search of neighbors nodes based on the Voronoi tessellation. In this way, the introduction of a visibility criterion in the NN framework, the so called constrained natural neighbour approximation (CNN) or the use of alpha-shapes – α-NEM – allow to remove the usual problems related to the application of the MLS in non-convex domains. Moreover, it allows to define nodal derivatives in the NN framework, which are required in thermomechanical simulations in order to update the internal variables when one proceeds using an updated Lagrangian formulation. This approach opens the way to a wide range of both variational and collocation formulations.

1.1 Meshless Techniques Based on the MLS Approximation: DFE and EFG Methods

Let the following approximation scheme:

$$\mathbf{u}^h(\mathbf{x}) = \mathbf{p}^T(\mathbf{x})\mathbf{a}(\mathbf{x}) \tag{1}$$

with $\mathbf{p}^T(\mathbf{x})$ a polynomial basis, i.e. $\mathbf{p}^T(\mathbf{x}) = [1, x, y, xy]$ and $\mathbf{p}^T(\mathbf{x}) = [1, x, y, xy, x^2, y^2]$ for a bilinear and quadratic basis, respectively, in 2D, and $\mathbf{a}(\mathbf{x})$ a vector of unknown coefficients. In order to determine $\mathbf{a}(\mathbf{x})$, we define the functional J that must be minimized with respect to $\mathbf{a}(\mathbf{x})$ [16]:

$$J = \frac{1}{2}\sum_{i=1}^{n} w_i(\mathbf{x})\left[\mathbf{p}^T(\mathbf{x}_i)\mathbf{a}(\mathbf{x}) - u_i\right]^2 \tag{2}$$

where u_i are the nodal unknowns associated with the neighbors nodes $\mathbf{x}_i$ of point $\mathbf{x}$ and $w_i(\mathbf{x})$ is a weighting function whose value decreases as the distance between $\mathbf{x}_i$ and $\mathbf{x}$ increases (see [5] for more details about properties of this function and the ones the most used). The minimization of J with respect to the unknown coefficient $a_j(\mathbf{x})$ leads to:

$$\frac{\partial J}{\partial a_j(\mathbf{x})} = \sum_{k=1}^{n} a_k \left[\sum_{i=1}^{n} w_i(\mathbf{x}) p_j(\mathbf{x}_i) p_k(\mathbf{x}_i) \right] - \sum_{i=1}^{n} w_i(\mathbf{x}) p_j(\mathbf{x}_i) u_i = 0 \tag{3}$$

which leads to the linear system:

$$\mathbf{A}(\mathbf{x})\mathbf{a}(\mathbf{x}) = \mathbf{B}(\mathbf{x})\mathbf{u} \tag{4}$$

where the matrix $\mathbf{A}(\mathbf{x})$ and $\mathbf{B}(\mathbf{x})$ are defined by:

$$A_{jk}(\mathbf{x}) = \sum_{i=1}^{n} w_i(\mathbf{x}) p_j(\mathbf{x}_i) p_k(\mathbf{x}_i) \tag{5}$$

$$B_{ij}(\mathbf{x}) = w_i(\mathbf{x}) p_j(\mathbf{x}_i) \tag{6}$$

Substituting $\mathbf{a}(\mathbf{x})$ in Equation (1), results in:

$$u^h(\mathbf{x}) = \mathbf{p}^T(\mathbf{x})\mathbf{A}^{-1}(\mathbf{x})\mathbf{B}(\mathbf{x})\mathbf{u} \tag{7}$$

By identification, the new shape functions are given by:

$$\boldsymbol{\psi}^T(\mathbf{x}) = \mathbf{p}^T(\mathbf{x})\mathbf{A}^{-1}(\mathbf{x})\mathbf{B}(\mathbf{x}) \tag{8}$$

The difference between the diffuse finite element and the element free Galerkin schemes comes from the evaluation of the shape function derivatives. In the first scheme only the term $\mathbf{p}^T(\mathbf{x})$ in Equation (8) is derived, whereas all terms depending on $\mathbf{x}$ are derived in the element free Galerkin approach.

1.2 Meshless Techniques Based on the Smooth Particles Approximation: The RKPM and the Enriched RKPM Methods

Let Ω be a 1D domain where the problem is defined (all the results have a direct 2D or 3D counterpart). The points within this domain will be noted by x or s.

1.2.1 Reproduction Conditions

The approximation $u^h(x)$ of $u(x)$ is built from the convolution integral

$$u^h(x) = \int_\Omega w(x-s,h)u(s)d\Omega \tag{9}$$

where $w(x-s,h)$ is the kernel function and h a parameter defining the size of the approximation support.

The main idea in the enriched RKPM method is to enforce the reproduction of a general function that we can write in the form of a polynomial plus another function noted by $u^e(x)$:

$$u^h(x) = a_0 + a_1 x + \cdots + a_n x^n + a_{n+1} u^e(x) \tag{10}$$

In the following paragraphs we analyze the required properties of the kernel function $w(x-s,h)$ for reproducing a function expressed by (10).

From Equation (9), the reproduction of a constant function a_0 is given by

$$\int_\Omega w(x-s,h) a_0 d\Omega = a_0 \tag{11}$$

which implies

$$\int_\Omega w(x-s,h) d\Omega = 1 \tag{12}$$

which constitutes the partition of unity.

Now, the required condition to reproduce a linear function $u^a(x) = a_0 + a_1 x$ is

$$\int_\Omega w(x-s,h)(a_0 + a_1 s) d\Omega = a_0 + a_1 x \tag{13}$$

By using the partition of unity (12), Equation (13) can be rewritten as

$$\begin{cases} \int_\Omega w(x-s,h) d\Omega = 1 \\ \int_\Omega w(x-s,h) s d\Omega = x \end{cases} \tag{14}$$

which implies the linear consistency of the approximation. Repeating this reasoning, we can write the n-order consistency as

$$\begin{cases} \int_\Omega w(x-s,h) d\Omega = 1 \\ \int_\Omega w(x-s,h) s d\Omega = x \\ \quad \vdots \\ \int_\Omega w(x-s,h) s^n d\Omega = x^n \end{cases} \tag{15}$$

and consequently, the reproduction of the function given by (10) implies

$$\int_\Omega w(x-s,h)(a_0 + a_1 s + \ldots + a_n s^n + a_{n+1} u^e(s)) d\Omega$$
$$= a_0 + a_1 x + \cdots + a_n x^n + a_{n+1} u^e(x) \tag{16}$$

from which it results

$$\begin{cases} \int_\Omega w(x-s,h)d\Omega = 1 \\ \int_\Omega w(x-s,h)sd\Omega = x \\ \vdots \\ \int_\Omega w(x-s,h)s^n d\Omega = x^n \\ \int_\Omega w(x-s,h)u^e(s)d\Omega = u^e(x) \end{cases} \tag{17}$$

In the original procedure proposed by Liu et al. [12] only n-order consistency was imposed, but it can not be directly used to enforce the reproduction condition associated with $u^e(x)$.

1.2.2 The Moment Matrix

We will note by $u^r(x)$ the approximation function verifying the conditions (17). Usually a cubic spline is considered as kernel function, and consequently the conditions given by Equation (17) are not satisfied. Liu et al. [12] propose the introduction of a correction function $C(x, x-s)$ for satisfying the reproduction conditions. In our case we consider the more general form $C(x, s, x-s)$ whose pertinence will be discussed later. Thus $u^r(x)$ will be expressed by

$$u^r(x) = \int_\Omega C(x,s,x-s)w(x-s,h)u(s)d\Omega \tag{18}$$

where $C(x, s, x-s)$ is assumed to have the following form

$$C(x,s,x-s) = \mathbf{H}^T(x,s,x-s)\mathbf{b}(x) \tag{19}$$

where $\mathbf{H}^T(x, s, x-s)$ represents the vector containing the functions considered in the approximation basis, and $\mathbf{b}(x)$ is a vector containing unknown functions that will be determined for satisfying the reproduction conditions. Thus, Equation (17) can be rewritten as

$$\begin{cases} \int_\Omega \mathbf{H}^T(x,s,x-s)\mathbf{b}(x)w(x-s,h)d\Omega = 1 \\ \int_\Omega \mathbf{H}^T(x,s,x-s)\mathbf{b}(x)w(x-s,h)sd\Omega = x \\ \vdots \\ \int_\Omega \mathbf{H}^T(x,s,x-s)\mathbf{b}(x)w(x-s,h)s^n d\Omega = x^n \\ \int_\Omega \mathbf{H}^T(x,s,x-s)\mathbf{b}(x)w(x-s,h)u^e(s)d\Omega = u^e(x) \end{cases} \tag{20}$$

In fact, the reproduction conditions must be enforced in a discrete form. For this purpose we consider N points (also refereed as nodes) which allow to compute the discrete form of Equation (20), i.e.

$$\begin{cases} \sum_{i=1}^{N} \mathbf{H}^T(x, x_i, x - x_i)\mathbf{b}(x)w(x - x_i, h)\Delta x_i = 1 \\ \sum_{i=1}^{N} \mathbf{H}^T(x, x_i, x - x_i)\mathbf{b}(x)w(x - x_i, h)x_i\Delta x_i = x \\ \vdots \\ \sum_{i=1}^{N} \mathbf{H}^T(x, x_i, x - x_i)\mathbf{b}(x)w(x - x_i, h)x_i^n\Delta x_i = x^n \\ \sum_{i=1}^{N} \mathbf{H}^T(x, x_i, x - x_i)\mathbf{b}(x)w(x - x_i, h)u^e(x_i)\Delta x_i = u^e(x) \end{cases} \tag{21}$$

that in a matrix form results

$$\left[\sum_{i=1}^{N} \mathbf{R}(x_i)\mathbf{H}^T(x, x_i, x - x_i)w(x - x_i, h)\Delta x_i\right]\mathbf{b}(x) = \mathbf{R}(x) \tag{22}$$

where $\mathbf{R}(x)$ is the reproduction vector

$$\mathbf{R}^T(x) = \left[1, x, \ldots, x^n, u^e(x)\right] \tag{23}$$

Equation (22) allows the computation of vector $\mathbf{b}(x)$,

$$\mathbf{b}(x) = \mathbf{M}(x)^{-1}\mathbf{R}(x) \tag{24}$$

where the moment matrix $\mathbf{M}(x)$ is defined by

$$\mathbf{M}(x) = \sum_{i=1}^{N} \mathbf{R}(x_i)\mathbf{H}^T(x, x_i, x - x_i)w(x - x_i, h)\Delta x_i \tag{25}$$

This moment matrix differs from the usual moment matrix proposed in [12], and in fact it becomes non symmetric.

1.2.3 Discrete Form of the Approximation Function

The discrete form $u^r(x)$ of $u^h(x)$ derives from Equations (18), (19) and (24)

$$\begin{aligned} u^r(x) &\cong \sum_{i=1}^{N} \mathbf{H}^T(x, x_i, x - x_i)\mathbf{M}(x)^{-1}\mathbf{R}(x)w(x - x_i, h)u(x_i)\Delta x_i \\ &= \sum_{i=1}^{N} \psi_i(x)u_i \end{aligned} \tag{26}$$

where ψ_i is the enriched RKP approximation shape function

$$\psi_i(x) = \mathbf{H}^T(x, x_i, x - x_i)\mathbf{M}(x)^{-1}\mathbf{R}(x)w(x - x_i, h)\Delta x_i \tag{27}$$

As in the classical RKPM we take $\Delta x_i = 1$. Different quadrature rules exist and they have been tested without a significant incidence on the reproducing condition accuracy.

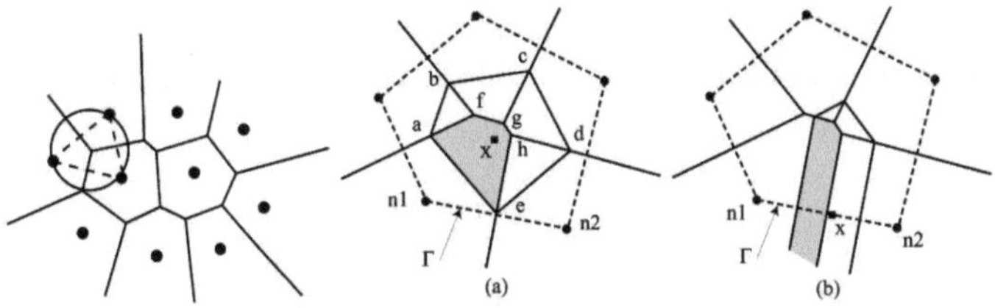

Fig. 1. (Left) Voronoi diagram, Delaunay triangle and Delaunay circle. (Right) Construction of the Sibson shape functions.

1.3 Meshless Techniques Based on the Natural Neighbor Approximation: The NEM Method

We briefly touch upon the foundation of Sibson's natural neighbor (NN) coordinates (shape functions) that are used in the natural element method. For a more in-depth discussion on the Sibson interpolant and its application for solving second-order partial differential equations, the interested reader can refer to Sambridge and Braun [19] and Sukumar et al. [20]. The NEM interpolant is constructed on the basis of the Voronoi diagram (see Figure 1). The Delaunay tessellation is the topological dual of the Voronoi diagram.

Consider a set of nodes $S = \{n_1, n_2, \ldots, n_N\}$ in $\Re^{dim}$. The Voronoi diagram is the subdivision of $\Re^{dim}$ into regions T_i (Voronoi cells) defined by:

$$T_i = \{\mathbf{x} \in \Re^{dim} : d(\mathbf{x}, \mathbf{x}_i) < d(\mathbf{x}, \mathbf{x}_j), \forall j \neq i\}, \quad \forall\, i \tag{28}$$

The Sibson coordinates of $\mathbf{x}$ with respect to a natural neighbor n_i (see Figure 1 (right)) is defined as the ratio of the overlap area (volume in 3D) of their Voronoi cells to the total area (volume in 3D) of the Voronoi cell related to point $\mathbf{x}$. If we consider the 2D example of Figure 1(a), we have:

$$\phi_1(\mathbf{x}) = \frac{Area(afghe)}{Area(abcde)} \tag{29}$$

Remark. From now on, we denote by $\phi_i(\mathbf{x})$ the shape functions related to the NEM, whereas $\psi_i(\mathbf{x})$ is used to denote the ones associated with the MLS or RKPM techniques.

If the point $\mathbf{x}$ coincides with the node n_i, i.e. $(\mathbf{x} = \mathbf{x}_i)$, $\phi_i(\mathbf{x}_i) = 1$, and all other shape functions are zero, i.e. $\phi_j(\mathbf{x}_i) = \delta_{ij}$ (δ_{ij} being the Kronecker delta). The properties of positivity, interpolation, and partition of unity are then verified [20]:

$$\begin{cases} 0 \leq \phi_i(\mathbf{x}) \leq 1 \\ \phi_i(\mathbf{x}_j) = \delta_{ij} \\ \sum_{i=1}^{n} \phi_i(\mathbf{x}) = 1 \end{cases} \tag{30}$$

where n is the number of neighbor nodes related to point $\mathbf{x}$.

The natural neighbor shape functions also satisfy the local coordinate property [22], namely:

$$\mathbf{x} = \sum_{i=1}^{n} \phi_i(\mathbf{x})\mathbf{x}_i \tag{31}$$

which combined with Equation (30), implies that the natural neighbor interpolant spans the space of linear polynomials (linear completeness).

Sibson natural neighbor shape functions are C^1 at any point except at the nodes, where they are only C^0. The C^1 continuity away from the nodes can be improved by using special classes of natural neighbor shape functions [11].

The support (domain of influence) of a shape function ϕ_i is the union of the Delaunay spheres (circumscribing the Delaunay tetrahedrons) containing the node n_i. This support is thus not radial and automatically adapts to the relative position of n_i and its neighbors, whether is the density or the regularity of the local nodal distribution.

Another important property of this interpolant is its strict linearity over the boundary of convex domains. The proof can be found in Sukumar et al. [20]. An illustration is depicted in Figure 1: as the areas associated to points on the boundary become infinite, the contribution of internal points vanish in the limit when the point approaches the convex boundary, and the shape functions associated with nodes n_1 and n_2 become linear on the segment $(n_1 - n_2)$. This is not true in the case of non convex boundaries, and an appropriate treatment must be introduced to maintain this property over non-convex boundaries [9, 24]. In tandem with the delta Kronecker property, essential boundary conditions can thus be enforced directly, as in the finite element method. This property also guarantees strict continuity of the approximation across material interfaces [23], which is an issue in most meshfree methods.

Consider an interpolation scheme for a vector-valued function $\mathbf{u}(\mathbf{x}) : \Omega \subset \Re^2 \rightarrow \Re$, in the form:

$$\mathbf{u}^h(\mathbf{x}) = \sum_{i=1}^{n} \phi_i(\mathbf{x})\, \mathbf{u}_i \tag{32}$$

where $\mathbf{u}_i$ are the nodal values of the field at the n natural neighbor nodes of point $\mathbf{x}$, and $\phi_i(\mathbf{x})$ are the shape functions associated with each neighbor node. It is noted that Equation (32) defines a local interpolation scheme. Thus, the trial and test functions used in the discretization of the variational formulation describing the problems treated in this paper take the form of Equation (32).

2 Coupling NN and MLS Approximations

In this section, that constitutes the main contribution of this paper, different hybrid schemes combining the natural neighbor and the moving least squares interpolations will be proposed and analyzed.

2.1 Defining Natural-Neighbor Collocation Schemes

The proposed approximation scheme consists in using the NN shape functions as weights in the MLS approximation. In this way we want to increase the consistency of the NN approximation in an adjustable manner and get simpler formulas for computing the derivatives. We will show firstly, that this approach does not change the shape functions when choosing linear polynomial basis $\mathbf{P}^T(\mathbf{x}) = \{1, x, y\}$. When we minimize the usual moving least square criterion

$$J = \frac{1}{2}\sum_{i=1}^{n} w_i(\mathbf{x})\left[\mathbf{p}^T(\mathbf{x}_i)\mathbf{a}(\mathbf{x}) - u_i\right]^2 \tag{33}$$

with $w_i(\mathbf{x})$ the NN shape function, i.e. $w_i(\mathbf{x}) = \phi_i(\mathbf{x})$, it results (see [7] for details) $\Psi_i(\mathbf{x}) = \phi_i(\mathbf{x})$ and that diffuse derivatives are discontinuous at the nodes. The same results are obtained by considering quadratic reproduction conditions, i.e. $P^T(\mathbf{x}) = \{1, x, y, x^2, xy, y^2\}$.

Thus, the use of collocation techniques becomes delicate. To circumvent this difficulty, we consider another cloud of auxiliary points $\mathbf{x}_j^*$, as in the double grid technique [6]. Now, the approximation field derivatives can be defined at those points, and then the nodal approximation derivatives defined at nodes using the moving least squares technique.

If we denote by u^h the natural neighbor approximation, its derivative can by calculated at points $\mathbf{x}_j^*$. Now, the diffuse derivatives can be computed at any point $\mathbf{x}$ using the standard MLS technique, from the functionals:

$$J = \frac{1}{2}\sum_{j=1}^{m} w_j(\mathbf{x})\left[\mathbf{p}^T(\mathbf{x}_j^*)\mathbf{a}(\mathbf{x}) - \frac{\partial u^h}{\partial x}(\mathbf{x}_j^*)\right]^2 \tag{34}$$

and

$$J = \frac{1}{2}\sum_{j=1}^{m} w_j(\mathbf{x})\left[\mathbf{p}^T(\mathbf{x}_j^*)\mathbf{a}(\mathbf{x}) - \frac{\partial u^h}{\partial y}(\mathbf{x}_j^*)\right]^2 \tag{35}$$

where m is the number of auxiliary points. Thus, after minimization, the coefficients $\mathbf{a}(\mathbf{x})$ of both approximations are obtained, and then the shape function diffuse derivatives, allowing to define the diffuse derivatives at any point $\mathbf{x}$:

$$\frac{\delta^2 u}{\delta x^2}(\mathbf{x}) = \sum_{j=1}^{m} \Psi_{x,j}(\mathbf{x}) \frac{\partial u^h}{\partial x}(\mathbf{x}_j^*) \tag{36}$$

and

$$\frac{\delta^2 u}{\delta y^2}(\mathbf{x}) = \sum_{j=1}^{m} \Psi_{y,j}(\mathbf{x}) \frac{\partial u^h}{\partial y}(\mathbf{x}_j^*) \tag{37}$$

where $\Psi_{x,j}(\mathbf{x})$ and $\Psi_{y,j}(\mathbf{x})$ denote the diffuse shape function derivatives related to point $\mathbf{x}_j^*$ with respect to the x and y coordinates, evaluated at point $\mathbf{x}$.

Now, considering that:

$$\frac{\partial u^h}{\partial x}(\mathbf{x}_j^*) = \sum_{i=1}^{n} \phi_{x,i}(\mathbf{x}_j^*) u_i \tag{38}$$

and

$$\frac{\partial u^h}{\partial y}(\mathbf{x}_j^*) = \sum_{i=1}^{n} \phi_{y,i}(\mathbf{x}_j^*) u_i \tag{39}$$

where $\phi_{x,i}(\mathbf{x}_j^*)$ and $\phi_{y,i}(\mathbf{x}_j^*)$ denote the Natural Neighbor shape function derivatives related to node $\mathbf{x}_i$ with respect to the x and y coordinates evaluated at point $\mathbf{x}_j^*$, Equations (36) and (37) could be used to define collocation schemes.

Remark. When the point $\mathbf{x}$ approaches to $\mathbf{x}_i$, Equations (36) and (37) give the nodal diffuse derivatives, that can be used in the postprocessing or in the context of a collocation technique.

Linear convergence of the second derivative may be demonstrated for an appropriate election of the auxiliary points. For this purpose we locate the auxiliary points (for quadratic approximation consistency) verifying $\forall \mathbf{x}_j^*$ the following conditions:

$$\sum_{i=1}^{n} \left\{ \phi_{x,i}(\mathbf{x}_j^*) u_i - \phi_i(\mathbf{x}^*) u_{x,i} \right\} = 0 \tag{40}$$

and

$$\sum_{i=1}^{n} \left\{ \phi_{y,i}(\mathbf{x}_j^*) u_i - \phi_i(\mathbf{x}^*) u_{y,i} \right\} = 0 \tag{41}$$

where n is the number of neighbor nodes of point $\mathbf{x}_j^*$, and

$$\begin{cases} u_i = u(\mathbf{x}_i) = a + bx_i + cy_i + dx_i^2 + ex_i y_i + fy_i^2 \\ u_{x,i} = \frac{\partial u}{\partial x}(\mathbf{x}_i) = b + 2dx_i + ey_i \\ u_{y,i} = \frac{\partial u}{\partial y}(\mathbf{x}_i) = c + 2fy_i + ex_i \end{cases} \tag{42}$$

This strategy was successfully applied for solving second order partial differential equations using a collocation discretization in [7].

2.2 Faster Natural-Neighbor Interpolation Formulas

The natural neighbor meshfree method provides equivalent quality compared to quadrilateral/hexaedral finite elements, but only uses the Delauny triangulation (which is automatic and unique for a given cloud of nodes), to construct the shape functions, which avoids the burden of mesh generation with these elements. Furthermore, as the shape functions satisfy the Kroenecker delta property, the imposition of essential boundary conditions is direct, unlike in the vast majority of meshless methods. Nevertheless, the computation of 3D natural neighbor shape functions is complex and costly, involving geometric constructions in the Voronoi diagram.

The computation of the natural neighbor shape functions is not direct and requires some geometric operations (intersection, volume and area computations) at each integration point. A classical algorithm for the computation of the shape functions at a point $\mathbf{x}$ involves the following steps: (a) Find the natural neighbor of the point $\mathbf{x}$; (b) Construct the new Voronoi cell associated with point $\mathbf{x}$; (c) Compute the volumes or areas associated with Voronoi cells entities used in the shape function computations; and (d) Compute the shape functions. Step (a) can be performed in constant time by performing local search in the Voronoi diagram. In our experience, steps (b) and (c) are the most expensive from a CPU point of view. In the next section, we propose new natural neighbor shape functions which avoid the geometric operations involved in the steps (b) and (c).

For this purpose, a particular weight function $w_i(\mathbf{x})$ based on the Delaunay spheres is used, which posses the main features of natural neighbor shape functions support [25]. The introduction of this particular weight in the EFG methology leads to shape functions which posses the same properties of the natural neighbor shape functions (i.e. interpolation and connectivity based on the natural neighbors, and linear consistency), but without any geometric construction based on the Voronoi diagram, which simplifies the extension of the method to the 3D case and reduces the computational costs.

2.2.1 Pseudo Natural Neighbor Weight Functions

When we consider the MLS technique, summarized in the first section of this paper, the approximation of a field $u(\mathbf{x})$, $u^h(\mathbf{x})$, can be written as:

$$u^h(\mathbf{x}) = \mathbf{p}^T(\mathbf{x})\mathbf{A}^{-1}\mathbf{B}\mathbf{u} = \boldsymbol{\psi}^T(\mathbf{x})\,\mathbf{u} \tag{43}$$

where $\boldsymbol{\psi}(\mathbf{x})$ is the vector containing the shape functions associated with neighbors of point x. In the following, we are interested in defining an appropriate weight function $w_i(\mathbf{x})$ such as the resulting shape functions satisfy: (a) the Kroenecker delta property

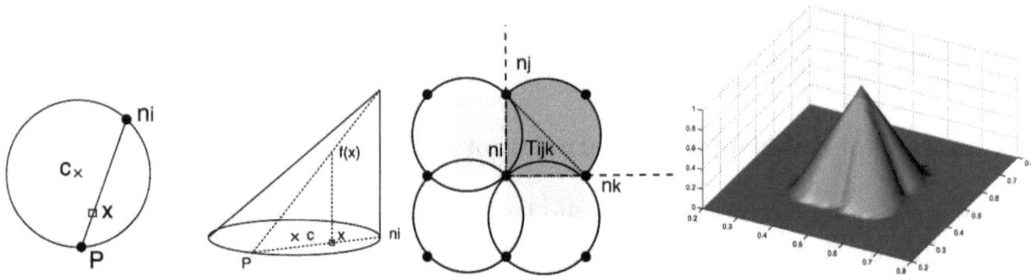

Fig. 2. (Left) Eccentric conical function. (Center) Support of a cone function portion. (Right) Weight function.

($\psi_i(\mathbf{x}_j) = \delta_{ij}$); (b) the linear consistency (which is automatically satisfied in the MLS framework); (c) the linearity of the shape functions on the domain boundary; and (d) $\psi_i(\mathbf{x}_j)$ vanishes on the Delaunay spheres containing the node n_i.

The following weight function is proposed to satisfy the former conditions. The definition is given here in 2D, but is straightforward in 3D. Let a cone function which basis matches one of the Delaunay circle containing the point x, and where the projection of the tip matches the node n_i (see Figure 2).

The value of the conic function computed at point x is given by:

$$f(\mathbf{x}) = \frac{\|\mathbf{n_i P}\| - \|\mathbf{n_i x}\|}{\|\mathbf{n_i P}\|} \tag{44}$$

with:

$$\mathbf{n_i P} = -2\left(\frac{\mathbf{cn_i} \cdot \mathbf{n_i x}}{\mathbf{n_i x} \cdot \mathbf{n_i x}}\right)\mathbf{n_i x} \tag{45}$$

In order to avoid the overlapping of cone functions whereas conserving the continuity of the weight function, a cone portion is associated with each of the Delaunay triangles connected to node n_i. The cone function is thus non-zero if a point x belong to the intersection between the Delaunay circumcircle and the portion of the plane such as any point in the basis formed by the origin node n_i and the vectors $n_i - n_j$ and $n_i - n_k$ has positive coordinates in this basis. n_j et n_k are the other two vertices of the triangle (see Figure 2). Due to the particular shape of its support, this weight function guarantees interpolation conditions ($w_i(\mathbf{x}_j) = \delta_{ij}$), as Delaunay circles passes through the nodes. Furthermore, the properties of positiveness and monotonically decreasing are verifed. As the cone functions are linear between two nodes, the continuity of the weight function is guaranteed.

In order to guarantee strict linearity of the shape functions over the boundaries of the domain, the shape functions associated with interior node must vanish on the external boundaries. For this purpose, we multiply the weight function in Equation (44) by a function $\Xi(\mathbf{x})$ which vanish over the external boundaries. A simple solution is to define $\Xi(\mathbf{x})$ by:

$$\Xi(\mathbf{x}) = \mathbf{N}^T(\mathbf{x})\boldsymbol{\delta} \tag{46}$$

where $\mathbf{N}(\mathbf{x})$ is the vector containing the linear finite element shape functions associated with the Delaunay triangles, and $\boldsymbol{\delta}$ the nodal values of a field taking a unit value inside the domain, i.e. $\delta_i = 1$ if $\mathbf{x}_i \in \Omega$, vanishing on the domain boundary $\Gamma \equiv \partial\Omega$, i.e. $\delta_i = 0$ if $\mathbf{x}_i \in \Gamma$.

2.2.2 Numerical Example

The following Poisson's problem is considered in a 3D unit cube:

$$\begin{cases} \Delta u = 0 \; in \; \Omega =]0, 1[^3 \\ u(\mathbf{x}) = u_g(\mathbf{x}) = 2x^2 - y^2 - z^2 \;\; \text{on} \;\; \Gamma = \partial\Omega \end{cases} \tag{47}$$

whose exact solution results:

$$u^{ex}(\mathbf{x}) = 2x^2 - y^2 - z^2 \;\; \text{in} \;\; \Omega \tag{48}$$

The weak form associated with the problem defined in Equation (47) is expressed by:

Find $u \in H^1(\Omega)$ $(u = u_g$ *on* $\Gamma)$ *such that:*

$$\int_\Omega \nabla u^* \cdot \nabla u \; d\Omega = 0, \;\; \forall u^* \in H_0^1(\Omega) \tag{49}$$

where $H^1(\Omega)$ and $H_0^1(\Omega)$ are usual Sobolev functional spaces.

The problem has been solved by using several refined meshes: $3 \times 3 \times 3$, $5 \times 5 \times 5$, $7 \times 7 \times 7$ and $10 \times 10 \times 10$ nodes. The energy norm has been computed to determine the convergence of the solution. Results are depicted in Figure 3. A comparison between the computational times associated with standard Sibson and pseudo-NEM shape functions is depicted in Figure 3 (right).

2.2.3 CPU Time Comparison between Different NN Approximations

As mentioned before, CPU time is one of the major drawbacks of the NEM, when compared to that of the FEM. In order to compare how important this cost could be, we analyze here a problem with know analytical solution, solved through the four different techniques, namely, NEM-Sibson, NEM-Laplace, Pseudo-NEM and, of course, FEM.

The problem here considered is the compression of a cubic block of a linear elastic material (or, equivalently, a stress *patch test*). A displacement of 0.01 on z direction is prescribed on the top face and the nodes of the bottom face are constrained in the z direction, see Figure 4.

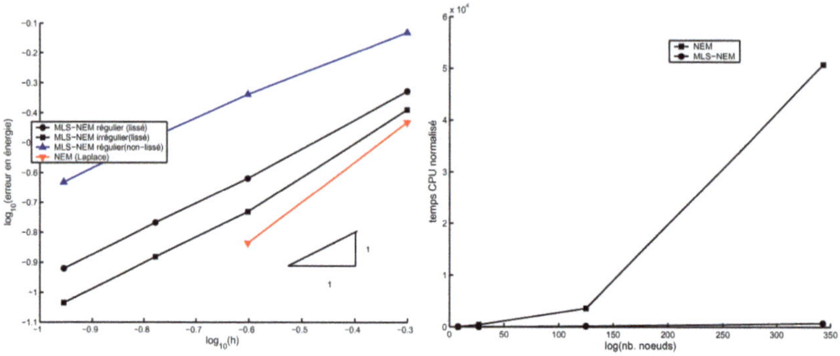

Fig. 3. (Left) Convergence analysis of the Poisson's problem. (Right) Comparison of CPU times of Sibson-NEM and pseudo-NEM discretizations.

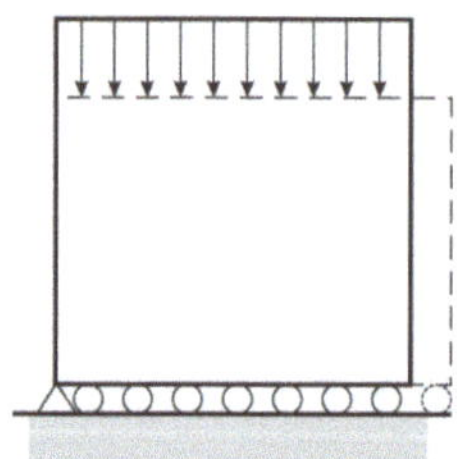

Fig. 4. Geometry of the compression test.

Different clouds of nodes have been employed, both in regular and irregular distributions, in an attempt to check wether the relative location of the nodes has an influence on the neighbour search time.

In the FEM context, Delaunay triangulation has been used, and three integration points are used in each tetrahedron. All meshless simulations use the same integration points.

Figure 5 shows the time needed to solve different problems. It is obvious that solving the problem using Sibson's shape functions (denoted by NEM-S) employs the highest amount of time, increasing very fastly with the number of nodes. Furthermore, the time needed for irregular distributions is higher than regular ones, since regular ones have less Delaunay tetrahedra and thus less integration points.

It is also interesting to evaluate the ratio of total time employed in calculating the shape functions and their derivatives, compared with the CPU time of the whole simulation. This is shown in Figure 6, where the pseudo NEM (P-NEM) approach needs less than 10% of total time and decreases comparatively when the size of the problem increases. Laplace shape functions (NEM-L) also decreases, but takes about 30% of the total time. Again, Sibson approach is very time consuming.

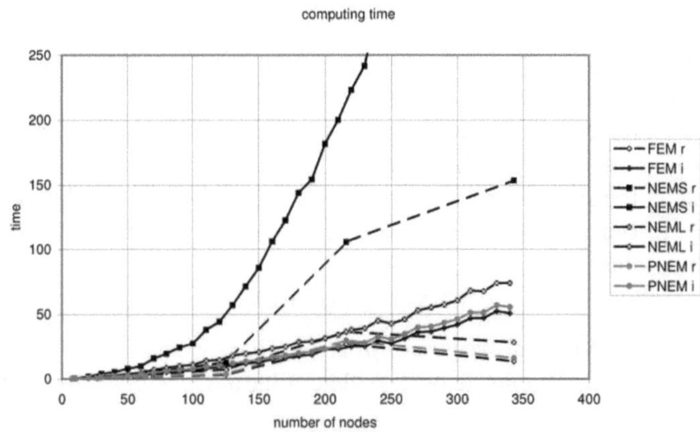

Fig. 5. Total computing time comparison.

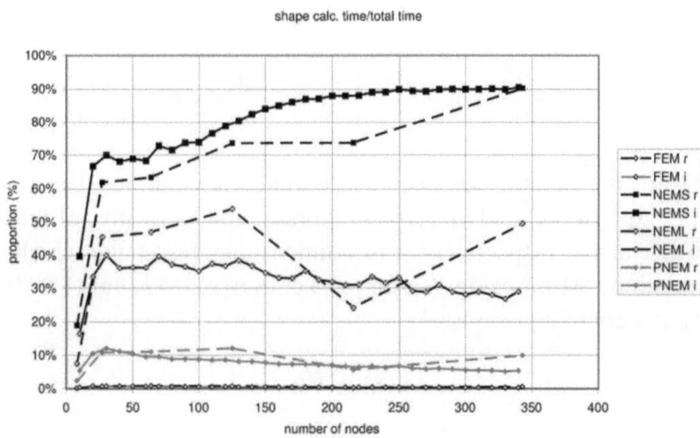

Fig. 6. Time employed in shape function calculation vs. total simulation time.

2.3 Hermite Natural Element Formulation

In this section, quadratic approximation consistency is achieved through a diffuse Hermite interpolation [18], by using natural neighbor weights in the moving least square approximation. Compared to standard moving least square method, the minimization is performed both with respect to the primary variable, and the diffuse spatial derivatives. For this purpose, we consider an interpolation scheme in the form:

$$u^h(\mathbf{x}) = \sum_{i=1}^{n} \psi_i(\mathbf{x})u_i + \sum_{i=1}^{n} \psi_i^x(\mathbf{x})\frac{\partial u_i}{\partial x} + \sum_{i=1}^{N} \psi_i^y(\mathbf{x})\frac{\partial u_i}{\partial y} \tag{50}$$

where $\psi_i(\mathbf{x})$ are the shape function associated with the unknown variable u_i, $\psi_i^x(\mathbf{x})$ and $\psi_i^y(\mathbf{x})$ are the shape function associated with the space derivative of u_i with respect to x and y, respectively. For the sake of simplicity, we consider the sum in Equation (50) extended to all the nodes N instead to the n natural neighbors, but both expressions are equivalent because the shape functions related to non-neighbor nodes vanish. In the above framework, u_i, $\partial u_i/\partial x$ and $\partial u_i/\partial y$ are unknown (degrees of freedom). In order to construct the shape functions, we consider the following approximation scheme:

$$u^h(\mathbf{x}) = \mathbf{p}(\mathbf{x})^T\mathbf{a}(\mathbf{x}) \tag{51}$$

where $\mathbf{p}(\mathbf{x})$ is a polynomial basis, i.e. $\mathbf{p}(\mathbf{x}) = \{1, x, y, xy, x^2, y^2\}$ and $\mathbf{a}(\mathbf{x})$ is a vector of unknown coefficients. In order to determine $\mathbf{a}(\mathbf{x})$, we consider the following functional:

$$J = \frac{1}{2}\sum_{i=1}^{N} w_i(\mathbf{x})\left\{\left[\mathbf{p}^T(\mathbf{x})\mathbf{a} - u_i\right]^2 + \right.$$
$$\left. + \alpha\left[\frac{\partial \mathbf{p}^T}{\partial x}(\mathbf{x})\mathbf{a} - \frac{\partial u_i}{\partial x}\right]^2 + \alpha\left[\frac{\partial \mathbf{p}^T}{\partial y}(\mathbf{x})\mathbf{a} - \frac{\partial u_i}{\partial y}\right]^2\right\} \tag{52}$$

where $w_i(\mathbf{x})$ are the natural neighbor shape functions computed at point x, i.e. $w_i(\mathbf{x}) = \phi_i(\mathbf{x})$, $\partial\mathbf{p}^T/\partial x(\mathbf{x})$ and $\partial\mathbf{p}^T/\partial y(\mathbf{x})$ represent the derivative of the basis $\mathbf{p}(\mathbf{x})$ with respect to x and y, respectively. α is a dimensional parameter which is fixed to 1 in our simulations. Minimizing J with respect to $\mathbf{a}(\mathbf{x})$, $(\partial J/\partial \mathbf{a}(\mathbf{x}) = 0)$, leads to the following system of equations:

$$\mathbf{A}\mathbf{a}(\mathbf{x}) = \mathbf{B}\mathbf{q} \tag{53}$$

with

$$\mathbf{q} = \left\{u_1, \frac{\partial u_1}{\partial x}, \frac{\partial u_1}{\partial y}, u_2, \frac{\partial u_2}{\partial x}, \frac{\partial u_2}{\partial y}, \ldots, u_N, \frac{\partial u_N}{\partial x}, \frac{\partial u_N}{\partial y}\right\}.$$

Derivatives of the shape functions are obtained through standard procedure [4], involving the derivative of the weight functions $w_i(\mathbf{x})$. The derivatives involved in $\mathbf{q}$ are in fact pseudo-derivatives (diffuse derivatives) and they cannot be used for discretizing variational formulations, however they can be used in a collocation framework. Closed form of Sibson shape functions derivatives can be found in [17]. The obtained shape functions are depicted in Figure 7.

According to Equation (52), the new degrees of freedom associated with the derivatives can be interpreted like pseudo-derivatives which do not coincide with the real derivatives. Thus, imposition of essential boundary conditions becomes delicate. Nevertheless, in order to investigate the accuracy of the technique without being polluted by this issue, we consider in next section a Poisson's whose solution and its derivatives on the boundary vanish.

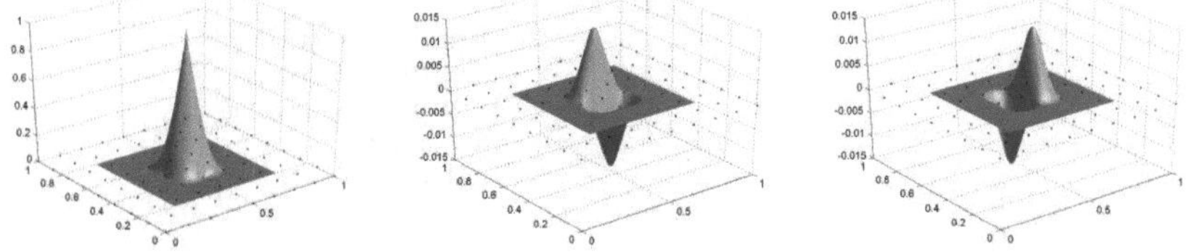

Fig. 7. Hermite natural neighbor shape functions.

2.3.1 Numerical Example

The boundary value problem is defined by:

$$\begin{cases} -\Delta u = f \text{ in } \Omega =]0,1[\times]0,1[\\ u = u_g \quad \text{ on } \Gamma \equiv \partial\Omega \end{cases} \tag{54}$$

we consider from now on:

$$\begin{cases} u_g = 0, \\ f = 4\pi^2 \{2\cos(2\pi x)\cos(2\pi y) - \cos(2\pi x) - \cos(2\pi y)\} \end{cases} \tag{55}$$

whose exact solution results in:

$$u^{ex}(\mathbf{x}) = \{1 - \cos(2\pi x)\}\{1 - \cos(2\pi y)\} \tag{56}$$

The weak form associated with Equation (54) is given by:

Find $u \in H_0^1(\Omega)$ such that:

$$\int_\Omega \nabla u \cdot \nabla \delta u \, d\Omega = \int_\Omega f \delta u \, d\Omega, \quad \forall \, \delta u \in H_0^1(\Omega) \tag{57}$$

where $H_0^1(\Omega)$ is the usual Sobolev functional space. The Hermite-NEM interpolation just described is used for approximating the trial and test functions u and δu, respectively, which are built with the only contribution of internal nodes.

The error using the energy norm is computed according to:

$$\left\| \mathbf{u} - \mathbf{u}^h \right\|_{E(\Omega)} = \left(\frac{1}{2} \int_\Omega (\nabla u^{ex} - \nabla u^h)^T (\nabla u^{ex} - \nabla u^h) \right)^{1/2} \tag{58}$$

For the evaluation of both Equation (57) and (58), the Voronoi cells associated with each node are triangulated and a Gauss quadrature scheme is applied in each subtriangle, using 3, 6 and 12 integration points. Figure 8 compares the accuracy of the Hermite-NEM (H-NEM) approximation with the standard Sibson-NEM. If only three Gauss points quadrature is used, the accuracy of the H-NEM exceeds the accuracy of the NEM, but the difference in the convergence rate is not significant. If a fine enough quadrature scheme is applied (6 points or more), the H-NEM reaches, as expected, a second-order convergence rate.

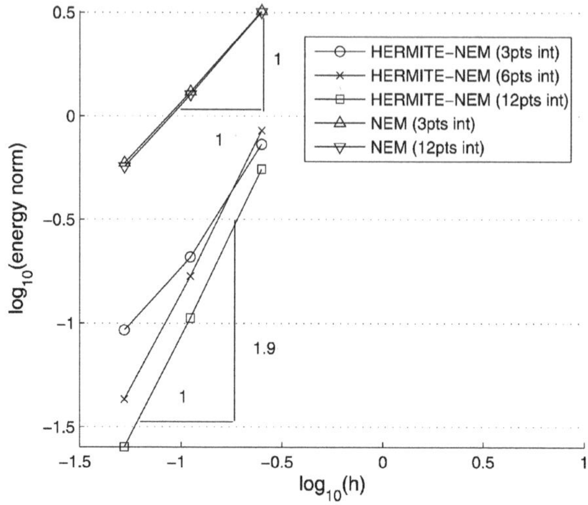

Fig. 8. Convergence analysis using the energy norm for the 2D Poisson's problem.

2.4 Hierarchical Bubbles Functions in the Natural Element Method

The interest of this approach lies in the construction of mixed approximations to be applied in the stable discretization of mixed variational formulations as encountered in mechanics of incompressible media where the mixed approximation must verify the well known LBB condition.

Consider an open bounded domain $\Omega \in \Re^{dim}$ with boundary Γ, dim being the space dimension. Assume that Ω is discretized by a set of nodes S. Let $D(S)$ the simplicial complex associated with the Delaunay tesselation of S. A simplicial complex K in $\Re^{dim}$ is a collection of simplices (hypertetrahedra) in $\Re^{dim}$ such that:

(i) Every face of a simplex K is in K;
(ii) The intersection of any two simplices of K is a face of each of them [15].

If we denote F_k the set of $k - simplices$ $(0 \leq k \leq 3)$, in R^3 the Delaunay tessellation $D(S)$ will be defined as the simplicial complex defined by the tetrahedra in F_3, the triangles in F_2, the edges in F_1, and the vertices in F_0. We denote these collections $T(S)$, $F(S)$, $E(S)$ and $V(S)$, respectively.

In order to construct richer approximations, new shape functions can be associated with the different *k-simplices*. The case $1 < k < 3$ is related to the concept of hierarchical methods [27]. The concept of hierarchical bubble shape functions is a very simple way to construct richer approximations. The extension to meshfree methods is not an easy matter in general, in the absence of topology related to some elements. In the natural element, the underlying Delaunay triangulation allows the use of such approach.

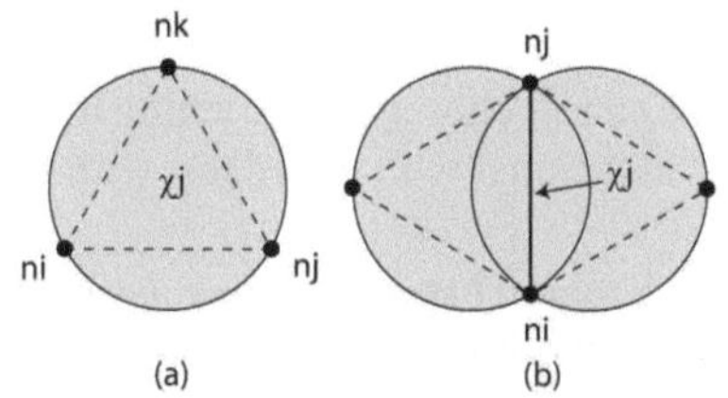

Fig. 9. Supports of the bubble shape functions associated with the Delaunay k-simplex; (a) support of a Delaunay triangle $\chi_j\ n_i - n_j - n_k$; (b) support of a Delaunay edge $\chi_j\ n_i - n_j$.

The key idea is to associate new shape functions to the *k-simplices* of the Delaunay tessellation, i.e. tetrahedra $T' \in T(S)$, triangular facets $F' \in F(S)$ and edges connecting two nodes in the Delaunay triangulation $E' \in E(S)$ [26].

2.4.1 b-NEM Approximation

A k-simplex (K-S) (vertex, edge, triangular facet or tetrahedron) is generated by $K = k + 1$ vertices ($k = 0$, 1, 2 and 3, respectively). The bubble shape function of an entity χ_j generated by K vertices is computed like:

$$\phi_j^*(\mathbf{x}) = \prod_{p=1}^{K} \phi_p(\mathbf{x}) \tag{59}$$

where $\phi_p(\mathbf{x})$ is the NEM shape function associated with node n_p computed at point $\mathbf{x}$.

The support of a K-S generated by K vertices (nodes) in S is the union of the Delaunay spheres containing the K nodes. It results, in 2D:

(i) if χ_j is a Delaunay triangle ($\chi \in F(S)$) ($k = 2$), the support of χ_j is composed with one circle containing the 3 generating nodes of the triangle (see fig. 9 (a));
(ii) if χ_j is an edge of a Delaunay triangle ($\chi \in E(S)$) ($k = 1$), the support of χ_j is composed with the union of two circles (if $\chi_j \notin \Gamma$), or one circle if $\chi_j \in \Gamma$ (see Figure 9(b)), containing the 2 generating nodes of χ_j.

We now consider the following approximation scheme:

$$\mathbf{u}^h(\mathbf{x}) = \sum_{i=1}^{n} \phi_i(\mathbf{x})\, \mathbf{u}_i + \sum_{j=1}^{m} \phi_j^*(\mathbf{x})\, \gamma_j \tag{60}$$

where n is the number of natural neighbors of point $\mathbf{x}$, $\phi_i(\mathbf{x})$ is the NEM shape function related to node $n_i \in S$ computed at point $\mathbf{x}$, $\phi_j^*(\mathbf{x})$ is the bubble shape function defined in Equation (59) associated with the m influent K-S, and γ_j is an additional degree of freedom.

Remarks.

(i) Different combinations can be chosen for enriching the approximation, i.e. using only bubble functions associated with the edges, with the Delaunay triangles, or both.
(ii) The evaluation of the bubble shape functions associated with the K-S is not costly as it only requires the product of available NEM shape functions computed at point $\mathbf{x}$.
(iii) Despite that the approximation scheme defined in Equation (60) is richer than standard NEM approximation, it does not satisfy any polynomial reproducing property other than the linear consistency.
(iv) In this paper, two approximations schemes are investigated and compared: (i) one using bubble functions associated with the Delaunay triangles (that we call b1-NEM); and (ii) one using bubble functions associated with the Delaunay edges (called b2-NEM).

2.4.2 b-NEM with Reproducing Properties

In this section we proceed to correct the shape functions previously constructed defining the approximation scheme (60) within a standard moving least squares framework, in order to evaluate the benefits provided by the higher approximation consistency. The MLS procedure has been summarized in the first section. Let $w_i(\mathbf{x})$ some weight function either associated with a standard or a bubble-NEM shape function, i.e. $w_i(\mathbf{x}) = \phi_i(\mathbf{x})$ or $w_i(\mathbf{x}) = \phi_i^*(\mathbf{x})$, computed at point $\mathbf{x}$.

The MLS procedure leads to:

$$u^h(\mathbf{x}) = \mathbf{p}^T(\mathbf{x})\mathbf{A}^{-1}(\mathbf{x})\mathbf{B}(\mathbf{x})\mathbf{u} \tag{61}$$

where we can identify the vector containing the approximation shape functions:

$$\boldsymbol{\psi}^T(\mathbf{x}) = \mathbf{p}^T(\mathbf{x})\mathbf{A}^{-1}(\mathbf{x})\mathbf{B}(\mathbf{x}) \tag{62}$$

As just commented, the reproducing b-NEM shape functions are computed by setting $w_i(\mathbf{x}) = \left\{\phi_i(\mathbf{x});\ \phi_j^*(\mathbf{x})\right\}$, $\phi_i(\mathbf{x})$ and $\phi_j^*(\mathbf{x})$ being the shape functions defined in (29) and (59).

Remark. The main difference between the reproducing-b-NEM and the b-NEM without additional reproducing properties is that physical coordinates must be associated with each K-S shape function, in order to evaluate the terms $p_j(\mathbf{x}_i)$ and $p_k(\mathbf{x}_i)$ in Equations (5) and (6). A simple solution is to consider the K-S centroid coordinates.

In the following, the b1-NEM and b2-NEM schemes just described are corrected using the MLS procedure just described. In the most unfavourable case a point $\mathbf{x}$ is influenced by four shape functions in the b1-NEM (3 NEM shape functions, and 1 bubble shape function associated with the Delaunay triangle), and being these weight functions independent, the method is stable if the basis $\mathbf{p}^T(\mathbf{x})$ contains 4 monomials. We call b1-NEM$^+$ the enrichment of the b1-NEM from $\mathbf{p}^T(\mathbf{x}) = \{1, x, y, xy\}$. Following similar assumptions, b2-NEM$^+$ results from the enrichment of the b2-NEM using $\mathbf{p}^T(\mathbf{x}) = \{1, x, y, xy, x^2, y^2\}$.

We have shown in [26] that essential boundary conditions can be enforced directly in all the proposed approximation schemes, as the bubble-NEM shape functions vanish over all external boundaries. For further details, see the proofs for the different schemes in that paper.

2.4.3 Natural Element Discretization

We consider the usual mixed variational formulation of the incompressible linear elastostatics problem where displacement trial and test functions are interpolated using the same shape functions, as the same for the pressure trial and test functions. In the following, the pressure is interpolated using the standard (Sibson) NEM shape functions, while the displacements are interpolated using the b-NEM or the b-NEM$^+$ shape functions previously defined (see [26] for more details).

In order to perform the inf-sup test a sequence of successive refined meshes is considered (uniform distributions) according to the procedure proposed in [2, 8]. The objective is to monitor the inf-sup values, $\lambda_{\min}$, when h decreases. If $\log \lambda_{\min}$ decreases with $logh$ the approximation scheme does not pass the LBB numerical test, which requires that $\log \lambda_{\min}$ remains bounded by a positive constant when $\log h$ decreases.

Figure 10 shows numerical test comparing some mixed NEM approximation schemes, i.e. b-NEM/NEM, NEM/Thiessen [21] (NEM approximation for the displacements and constant pressure within each Voronoi cell), and the P1/P0 and P2/P1 mixed FEM approximation schemes. The FEM computations are carried out using directly the Delaunay triangles. As claimed in other previous works [10], the mixed NEM/Thiessen approximation scheme does not pass the numerical inf/sup test. The mixed FEM P1/P0 also violates the LBB condition [8]. All the bubble-NEM schemes are clearly LBB compliant, being the results similar to the ones computed by using the P2/P1 FEM, which satisfy the LBB condition.

As noticed previously, the NEM shape functions only possesses linear completeness [22]. The enrichment of bubble in the context of MLS does not seem to increase the convergence rate with standard integration despite the proved increase in the approximation consistency. The reasons of this strange behavior constitutes a work in progress.

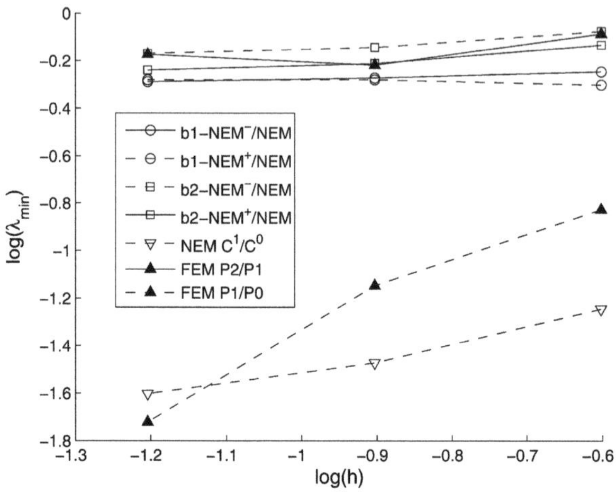

Fig. 10. inf-sup numerical test.

2.5 Natural Neighbor Interpolation with Discontinuous Derivatives

To define NN-approximations with discontinuous derivatives we could proceed in the context of the partition of unity (as in the extended finite element technique) [1]. However, in this work we propose an enrichment that does not involve additional degrees of freedom. For this purpose we start introducing the enriched reproducing kernel particle method, that by introducing the NN-interpolation as kernel function leads to NN-interpolation functions with discontinuous derivatives.

2.5.1 Introducing NN Approximations into E-RKPM: The Enriched NEM (E-NEM)

We consider a level set description $\Theta(\mathbf{x})$ of an interface where the field normal derivatives (with respect to the interface) are discontinuous. Now, we can introduce as enrichment function $u^e(x)$ the following function:

$$u^e(\mathbf{x}) = H_0(\Theta(\mathbf{x}))\Theta(\mathbf{x}) \tag{63}$$

where

$$\Theta(\mathbf{x}) = \begin{cases} \Theta(\mathbf{x}) < 0 \text{ if } \mathbf{x} \in \Omega_1 \\ \Theta(\mathbf{x}) > 0 \text{ if } \mathbf{x} \in \Omega_2 \\ \Theta(\mathbf{x}) = 0 \text{ if } \mathbf{x} \in \Gamma_d \end{cases} \tag{64}$$

and

$$\begin{cases} H_0(\Theta(\mathbf{x})) = 1 \; if \; \Theta(\mathbf{x}) \geq 0 \\ H_0(\Theta(\mathbf{x})) = 0 \; if \; \Theta(\mathbf{x}) < 0 \end{cases} \tag{65}$$

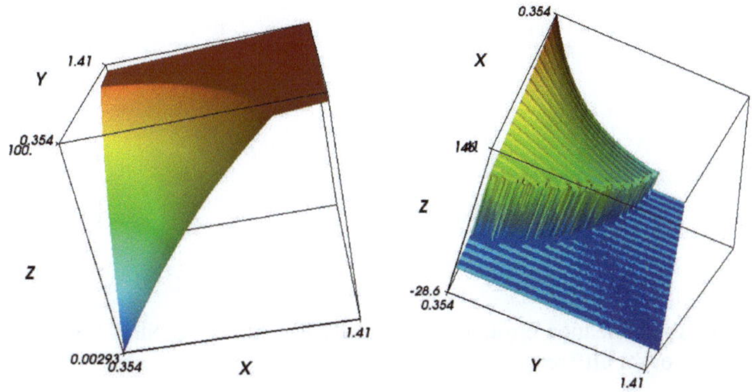

Fig. 11. (Left) Enriched Natural Neighbor approximation with discontinuous normal derivatives across a circular interface. (Right) x-derivative of the temperature field.

Now, we consider a linear consistency enriched with the function given by Equation (63) and the kernel function $w(\mathbf{x} - \mathbf{x}_i, h) = \phi_i(\mathbf{x})$ (the natural neighbor shame functions). The resulting approximation shape functions have the linear consistency but allows also to reproduce discontinuous normal derivatives across the interface Γ_d.

To illustrate the capabilities of the proposed technique we consider the exact solution of the Laplace's problem (modelling the temperature distribution in a steady heat transfer problem) defined in a bi-material consisting of two cylinders with different thermal conductivities. The reproduction tests have been carried out using the E-RKPM as well as the E-NEM, where the circular interface was modelled from the distance to that interface that multiplies the Heaviside's function related to that distance. Figure 11 illustrate a detail of the reconstructed temperature field where we can notice the accurate interface description. The discontinuity in the field derivatives is accurately accounted, as suggested by the representation of the x-derivative depicted in Figure 11.

Finally, in order to quantify the results accuracy we compare in Figure 12 the error (using the two usual norms) using the E-RKPM and the E-NEM techniques. In Figure 12 (right) we can notice that the E-NEM error is not affected by the slope change across the interface, that increases with the difference of thermal conductivities (for $K1 = 10$ the conductivities ratio is 10 whereas it is of 100 for $k1 = 100$.

3 Conclusions

We have explored the connections between NN and MLS approximations, coming from the introduction of the NN approximation functions as the weights in the scope of MLS. Thus, we can adjust the approximation consistency (with the possibility

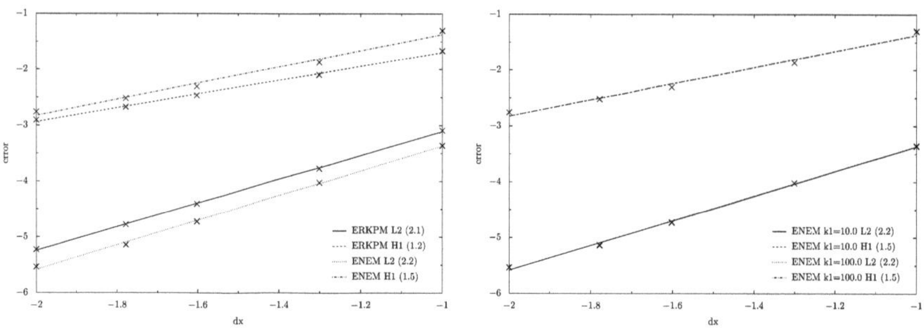

Fig. 12. (Left) Approximation errors using the E-RKPM and the E-NEM. (Right) E-NEM approximation error for different conductivities ratios.

to enrich the approximation basis with some particular functions describing issues of the searched solution) in the framework of the MLS techniques, imposing exactly essential boundary conditions from the use of the NN approximation as MLS weight.

This approach opens, as proved in the present paper, the way to a wide range of formulations: (i) NN collocation strategies; (ii) faster natural element discretizations; (iii) Hermite natural element formulations; (iv) hierarchical bubbles functions in the natural element method; and (v) and NN enriched approximations.

References

1. Babuška I. and Melenk J.M. *Computer Methods in Applied Mechanics and Engineering*, 4:289–314, 1996.
2. Bathe K.J. *Finite Element Procedures*. Prentice Hall, 1986.
3. Belikov V.V., Ivanov V.D., Kontorovich V.K., Korytnik S.A. and Semenov A.Y. *Computational Mathematics and Mathematical*, 37:9–15, 1997.
4. Belytschko T., Lu Y.Y. and Gu L. *International Journal for Numerical Methods in Engineering*, 37:229–256, 1994.
5. Belytschko T., Krongauz Y., Organ D., Fleming M. and Krysl P. *Computer Methods in Applied Mechanics and Engineering*, 139:3–47, 1996.
6. Breitkopf P., Touzot G. and Villon P. *Computational Mechanics*, 25:199–206, 2000.
7. Breitkopf P., Chinesta F., Villon P., Rassineux A. and Yvonnet J. A mixed natural neighbor and diffuse element framework for meshfree methods development. In *Computational Mechanics WCCM VI*, in conjunction with APCOM'04, Beijing, China, Tsinghua University Press & Springer-Verlag, 2004.
8. Chapelle D. and Bathe K.J. *Computers & Structures*, 48:745–760, 1993.
9. Cueto E., Doblaré M. and Gracia L. *International Journal for Numerical Methods in Engineering*, 49:519–546, 2000.
10. Gonzalez D., Cueto E. and Doblare M. *International Journal for Numerical Methods in Engineering*, 61:611–632, 2004.
11. Hiyoshi H. and Sugihara K. *Computational Geometry*, 22:167–183, 2002.

12. Liu W.K., Jun S. and Zhang Y.F. *International Journal for Numerical Methods in Fluids*, 21:1081–1106, 1995.
13. Liu W.K., Chen Y., Jun S., Chen J.S., Belytschko T., Pan C., Uras R.A. and Chang C.T. *Archives of Computational Methods in Engineering: State of the Art Reviews*, 3:3–80, 1996.
14. Lucy L.B. *The Astronomic Journal* 88:1013–1024, 1977.
15. Munkres J.R. *Elements of Algebraic Topology*. Perseus Press, 1993.
16. Nayroles B., Touzot G. and Villon P. *Computational Mechanics*, 10:307–318, 1992.
17. Piper B. *Computing Suppl.* 8:227–239, 1993.
18. Rassineux A., Villon P., Savignat J.M. and Stab O. *International Journal for Numerical Methods in Engineering*, 49:10–20, 2000.
19. Sambridge M., Braun J. and McQueen M. *Geophys. J. Int.*, 122:837–857, 1995.
20. Sukumar N., Moran B. and Belytschko T. *International Journal for Numerical Methods in Engineering*, 43:839–887, 1998.
21. Sukumar N. The natural element method in solid mechanics. Ph.D. Thesis, Northwestern University, Evanston, IL, 1998.
22. Sibson R. *Math. Proc. Camb. Phil. Soc.*, 87:151–155, 1980.
23. Yvonnet J., Chinesta F., Lorong P. and Rynckelynck D. *International Journal of Thermal Sciences*, 44:559–569, 2005.
24. Yvonnet J., Ryckelynck D., Lorong P. and Chinesta F. *International Journal for Numerical Methods in Engineering*, 60:1451–1474, 2004.
25. Yvonnet J. and Chinesta F. An hybrid element free Galerkin and natural element Meshfree method for direct imposition of boundary conditions and faster three-dimensional computations. In *Third MIT Conference on Computational Fluid and Solid Mechanics*, MIT, Cambridge, MA, 2005.
26. Yvonnet J., Villon P. and Chinesta F. *International Journal for Numerical Methods in Engineering*, in press, 2006.
27. Zienkiewicz O.C., De J.P., Gago S.R. and Kelly D.W. *Computers & Structures*, 16:53–65, 1983.

Eliminating Shear-Locking in Meshless Methods: A Critical Overview and a New Framework for Structural Theories

Carlos Tiago and Vitor M.A. Leitão

Instituto Superior Técnico, Universidade Técnica de Lisboa, Av. Rovisco Pais, 1049-001 Lisboa, Portugal; E-mail: {carlos.tiago, vitor}@civil.ist.utl.pt

Abstract. Meshfree methods are recent additions to the family of numerical techniques for the solution of partial differential equations. Issues like shear locking are yet to be clarified. Earlier expectations that these methods would be free from locking problems, were not confirmed. In particular, it is shown that, for the case of moderately thick structural beam and plate theories, if the same approximation is used for both displacement and rotation(s), then shear locking occurs.

Considering that reduced integration is not an option in the present context, a review of the available methodologies is presented. It is proven that one of these methodologies, that of consistent fields, i.e., where the rotation(s) approximation(s) are obtained from the displacement approximation, always leads to linearly dependent approximations.

As an alternative, a shear deformable framework for structural theories is presented where, instead of the rotations, the shear strain(s) are approximated in addition to that of the displacement field. The particularization for beams and plates is presented and a comparison with the traditional thin (irreducible) and moderately thick (shear deformable) theories is made. One aspect to emphasize is the presence of second order differentials in the dual pair of equilibrium and compatibility operators.

Key words: Meshless approximations, shear locking, structural models.

1 Introduction

1.1 A State-of-the-Art Review

The presence of locking (whether is shear, membrane or volumetric) in the numerical solutions can lead to totally erroneous solutions.

In particular, shear-locking is a well studied phenomenon in the conventional displacement approach of the Finite Element Method (FEM). In this case the problem is caused by the use, and eventually abuse, of the *same* interpolation functions for

V.M.A. Leitão et al. (eds), *Advances in Meshfree Techniques*, 123–145.

all the generalized displacement fields. Although this approximation criteria is very simple to implement and computationally inexpensive, it can not reproduce the real behaviour of slender structures.

Although not in a very elegant manner (and lacking some mathematical background), the reduced integration technique proposed by Zienkiewicz et al. [27] in the early 70s has been, in general, the preferred option to alleviate this problem. The mathematical foundations of this procedure were latter given by Malkus and Hughes [18] and is based on the *equivalence theorem* between the reduced integration displacement based approach and certain mixed models derived, e.g., from the Hellinger-Reissner functional. Thus, the reduced/selective integration procedure can be interpreted just as an efficient way to implement mixed methods.

The use of the *equivalence theorem* requires the construction of an auxiliary field with nodal points, in each element, located on the sample points of the exact integration rule. Hence, this procedure is, in general, impracticable for meshless methods due to the inexistence of such integration rule for the usual non-scattered data approximations. Examples of these approximations are provided by the Moving Least Squares (MLS) [14], *hp*-clouds families $\mathcal{F}_N^{k,p}$ [9], the natural neighbour coordinates (or Sibson coordinates) [22] or the reproducing kernel particle interpolation functions [15], which are associated with the EFG method [3], the *hp*-cloud method [9], the Natural Element Method (NEM) [22] and the Reproducing Kernel Particle Method (RKPM) [15], respectively.

Other common alternatives used in the FEM, like the Assumed Natural Strain (ANS) method [12] and the Enhanced Assumed Strain (EAS) method [21] are also unsuitable for meshless methods.

Thus, alternative procedures were proposed to circumvent the shear-locking problem. In the following, a brief review is presented:

- Increase of the degree of basis functions. It is well known that the increase of the degree of the interpolation functions can alleviate the locking effects in the traditional FEM. However, the convergence rate of the approximation is not optimal and there will exist spurious oscillations in the shear forces [20]. The equivalence in some meshless methods is the increase of the number of terms in the enrichment basis functions, i.e., *p*-refinement. This approach, only available for meshless methods with *p* capabilities, was investigated by Mendonça et al. [6] to solve shear deformable beams and by Garcia et al. [10] to analise plates by the *h-p* cloud method.
- Nodal integration. Several nodal schemes were devised for the underintegration of the weak form. Usually this procedure suffers from spurious singular modes, as noted by Beissel and Belytschko [2] and it requires some sort of stabilization. Wang and Chen [26] used a *curvature smoothing* to solve shear deformable beams and plates.
- Mixed formulation. With a mixed formulation, based on independent approximations of some interior fields, the volumetric locking can be eliminated, as shown

by Dolbow and Belytschko [7] in the EFG context and De and Bathe [5] in the Method of Finite Spheres (MFS)[1] framework. This idea could also be extended to the shear-locking problem.

- Consistency paradigm. This idea is based on the *field-consistency paradigm* suggested by Prathap [20] (sometimes referred as *unequal order of interpolation*) and was developed in the meshless framework by Donning and Liu [8] resorting to cardinal splines for approximation. Of course, the original expression *unequal order of interpolation* has to be restated in the context of the meshless methods, as most of the meshless functions do not posses interpolatory character. This approach was later revisited by Kanok-Nukulchai et al. [13] and Tiago and Leitão [23] in the EFG context.
- Change of variables. By a simple change of the independent variables, it is possible to construct a locking-free formulation, as the model proposed by Cho and Atluri [4] for the analysis of Timoshenko beams by the Meshless Local Petrov–Galerkin (MLPG) [1] method. This change does *not* increase the total number of degrees of freedom.

As could be anticipated, the *increase of the degree of basis functions* does not eliminate completely the shear-locking, as can be seen in [10].

Although the *nodal integration* sounds very appealing to use in conjunction with a meshless method, this approach can lead to singularities in the system matrix if special procedures, like the addition of stabilization terms to the energy functional, are not employed. Sometimes these stabilization terms involve *second order* derivatives [2], which are specially expensive in MLS approximations. The weights of nodal integration rules are, sometimes, based on the Voronoy diagram, which is, in fact, a sort of cell structure. Another important fact is the complexity of the approximation functions, which leads to awkward integrals, when used in conjunction with a weak form. Thus, a poor integration of this form leads, in general, to unsatisfactory results.

The *mixed formulation* will work correctly as long as the Ladyzhenskaya–Babŭska–Brezzi (LBB) stability criterion is satisfied. In general, this is not a trivial task but efforts have been made in this direction in the aforementioned works [7, 5] to identify the right combinations of approximations of the displacement and the pressure fields that pass the LBB condition. Also, there is an increase in the dimension on the problem.

Tiago and Leitão [23] proved that the consistency approach has a side effect: gives rise to rank deficient systems of equations.

The change of variables approach seams very promising since it opens the possibility of employing a locking-free formulation but still using the *same* approxim-

[1] The Method of Finite Spheres uses the Galerkin method with a particular set of the *h-p* cloud approximation families, $\mathcal{F}_N^{k,p}$. More precisely, it employs $k = 0$, i.e. the partition of unity is the Shepard function. It can also be shown that in MFS the essential boundary conditions are implicitly imposed via the same modified variational principle as presented by Lu et al. [17].

ation functions for all the unknown fields. Surprisingly, in the original work of Cho and Atluri [4] this possibility was not explored, since *different*[2] approximations were used for the generalized displacements and strains.

1.2 The Present Contribution

In Section 2 a unified framework for the derivation of structural theories is exposed. The interest variables and operators of the structural theories are derived from the three-dimensional continuum using the Principle of Virtual Work (PVW). The advantages of proceeding in this way are (i) a clear understanding of the assumptions needed, (ii) a distinction is made between the generalized displacements in the domain and the generalized displacements needed to impose the boundary conditions, here called pseudo-generalized displacements and (iii) all the loading terms are naturally taken into account in a straightforward manner. As an example, the classical and first order models for beams and plates are rewritten in the aforementioned framework.

In Section 3 the consistency approach is thoroughly analyzed. The reproducing properties of the spaces spanned by the consistent approximations for the particular cases of MLS/RKPM nodal functions are characterized. A proof of the linear dependence of the consistent approximation is given and the *exact* number of linear dependencies appearing in the approximation is identified. An alternative interpretation, in which the compatibility operator is modified in order to allow the use of equal approximations for all the generalized displacements, is also presented.

In Section 4 a Displacement-Strain Primary Variables (DSPV) formulation is developed. The motivation to develop a formulation in which the unknowns are the transversal displacements and strains is the possibility of generating a locking-free model that can be discretized and implemented in order to obtain solutions of plates independently of their dimensions, i.e., they may should be regarded as thin or thick and, at the same time, avoid the use of any additional resource in the approximation. In fact, the objective is to employ the *same* approximation functions for all the unknown generalized displacements. This procedure provides an inherited elegance and simplicity of implementation. This formulation is a generalization of the change of variables procedure whose enhancements are the use of the same framework for both beams and plates and the unveil of the true meaning of all the variables at stake. Notice that the original formulation was developed only for beams and the actual problem being solved (governing equations and boundary conditions) was not presented.

[2] Although both approximations may share the designation GMLS they are, in fact, different. This is due to the different norms used in the approximations functionals for displacements (H^1) and strains (L_2).

As for the implementation aspects, the discretization should be exactly the same for all the generalized displacements thus (i) simplifying the implementation and (ii) increasing the performance.

By approximating directly the displacements and strains the model can reproduce exactly the classical limit models[3] by a simple direct elimination of the strains, which is not possible in a model whose primary variables are the displacement and the rotations (the first order theories).

A crucial aspect is the one order increase, relatively to the first order C^0 formulations, of the required continuity of the approximation for the displacement, which is now C^1. If in FEM this would be (almost) prohibitive (for problems with dimensions greater than one), with meshless approximations this is trivial to achieve. Hence, this approach is especially suited for meshless approximations.

The complete set of associated equations will be derived from the general framework, clearly showing the meaning of all relations. The DSPV formulation includes the classical theories as particular cases, i.e., the compatibility and equilibrium differential operators of second order that appear in the classical theories are also present in this theory.

The shear-locking phenomenon is, usually, associated with high slenderness ratios, typical of structures requiring geometrically non-linear analysis. Although only *linear* structural theories will be addressed here, the use of some of the concepts could be extended to geometrically non-linear analysis of structures, as in [24] for beams and [25] for shells.

A comparison of the formulations (classical, first-order and DSPV) for beams and plates in terms of the variables and operators is also made.

1.3 Notation

Summation convention over repeated indices is adopted in the entire text, with italic Greek indices ranging from 1 to 2 and italic Latin subscripts from 1 to 3.

The notation

$$\partial_i = \frac{\partial}{\partial x_i} \quad \text{and} \quad \partial_{ij} = \frac{\partial^2}{\partial x_i \partial x_j}$$

for derivatives is sometimes used along the text. Comma derivative notation is also implied. The outward normal vector components are denoted by n_i.

$\left[f \right](\mathbf{x}) = f(\mathbf{x}^+) - f(\mathbf{x}^-)$ denotes the jump in the function f at point $\mathbf{x}$.

A multi-index notation is used in Section 3. Let $\boldsymbol{\alpha} = (\alpha_1, \alpha_2, \ldots, \alpha_d)$ and $\boldsymbol{\beta} = (\beta_1, \beta_2, \ldots, \beta_d)$ be multi-indexes such that $\alpha_i, \beta_i \in \mathbb{Z}_0^+$ and $\Omega \subset \mathbb{R}^d$ be an open set, where d is the dimension of the problem. The multi-index length of $\boldsymbol{\alpha}$ is

---

[3] Along the text the terminology *classical limit* is used to designate the classical Euler–Bernoulli beam and the Kirchhoff–Love plate theories by opposition to the *first order* nomenclature, used to coin the Timoshenko beam and the Reissner–Mindlin plate models.

$|\boldsymbol{\alpha}| = \sum_{i=1}^{d} \alpha_i$, its factorial is $\boldsymbol{\alpha}! = \alpha_1!\alpha_2!\ldots\alpha_d!$, the delta Kronecker symbol of $\boldsymbol{\alpha\beta}$ is $\delta_{\boldsymbol{\alpha\beta}} = \delta_{\alpha_1\beta_1}\delta_{\alpha_2\beta_2}\ldots\delta_{\alpha_d\beta_d}$, $\boldsymbol{x}^{\boldsymbol{\alpha}}$ stands for

$$\mathbf{x}^{\boldsymbol{\alpha}} = x_1^{\alpha_1} x_2^{\alpha_2} \ldots x_d^{\alpha_d} \tag{1}$$

and $D^{\boldsymbol{\alpha}} u(\mathbf{x})$ is the Fréchet derivative of the function u at point $\boldsymbol{x}$, given by the tensor

$$D^{\boldsymbol{\alpha}} u(\mathbf{x}) = \frac{\partial^{|\boldsymbol{\alpha}|} u(\mathbf{x})}{\partial^{\alpha_1} x_1 \partial^{\alpha_2} x_2 \ldots \partial^{\alpha_d} x_d} \tag{2}$$

2 The Model Problem

2.1 The General Pattern

Assuming physical and geometrical linearity, the kinematic requirements, the equilibrium conditions and the material law for the three-dimensional continuum are

$$\boldsymbol{\varepsilon} = \mathbf{D}\mathbf{u} \tag{3a}$$

$$\mathbf{D}^* \boldsymbol{\sigma} + \bar{\mathbf{f}} = \mathbf{0} \qquad \text{in } V \tag{3b}$$

$$\boldsymbol{\sigma} = \mathbf{C}\boldsymbol{\varepsilon} \tag{3c}$$

complemented by the boundary conditions

$$\mathbf{u} = \bar{\mathbf{u}} \qquad \text{on } \partial V_u \tag{4a}$$

$$\mathbf{N}^T \boldsymbol{\sigma} = \bar{\mathbf{t}} \qquad \text{on } \partial V_t \tag{4b}$$

and $\partial V = \overline{\partial V_u \cup \partial V_t}$, $\overline{V} = V \cup \partial V$, $\partial V_u \cap \partial V_t = \emptyset$, where V denotes the domain, ∂V_u the Dirichlet boundary and ∂V_t the Neumann boundary. Moreover, $\boldsymbol{\varepsilon}$ is the (infinitesimal) strain vector, $\boldsymbol{\sigma}$ is the (Cauchy) stress vector, $\mathbf{u}$ is the displacement vector, $\bar{\mathbf{u}}$ is the prescribed displacement vector, $\bar{\mathbf{f}}$ is the prescribed body force vector, $\bar{\mathbf{t}}$ is the prescribed traction vector, $\mathbf{C}$ is a matrix of elastic coefficients, $\mathbf{D}$ is a matrix of differential operators, $\mathbf{D}^*$ is the adjoint of $\mathbf{D}$ and $\mathbf{N}$ is a matrix of the outward normal components associated to $\mathbf{D}$. For details on the sign convention, the collecting of the variables and operators on the sets of equations (3) and (4), see [19].

Let $\mathcal{T}$ and $\mathcal{W}$ be appropriate trial and weighting spaces according to the context where they are mentioned that satisfy the Dirichlet boundary conditions and the homogeneous Dirichlet boundary conditions, respectively. For definition of these spaces see, e.g., Hughes [11].

The PVW reads

$$\int_V \delta\boldsymbol{\varepsilon}^T \boldsymbol{\sigma} \mathrm{d}V - \int_V \delta\mathbf{u}^T \bar{\mathbf{f}} \mathrm{d}V - \int_{\partial V_t} \delta\mathbf{u}^T \bar{\mathbf{t}} \mathrm{d}\partial V_t = 0 \qquad \forall\, \delta\mathbf{u} \in \mathcal{W} \tag{5}$$

In general, a structural model is governed by the following set of compatibility equations, equilibrium conditions and constitutive relations:

$$\mathbf{e} = \boldsymbol{\partial}\mathbf{d} \tag{6a}$$

$$\boldsymbol{\partial}^*\mathbf{s} + \overline{\mathbf{p}} = \mathbf{0} \qquad \text{in } \Omega \tag{6b}$$

$$\mathbf{s} = \mathbf{E}\mathbf{e} \tag{6c}$$

and boundary conditions

$$\mathbf{d} = \overline{\mathbf{d}} \qquad \text{on } \partial\Omega_u, \tag{7a}$$

$$\mathbf{M}^T\mathbf{s} = \overline{\mathbf{b}} \qquad \text{on } \partial\Omega_t. \tag{7b}$$

and $\overline{\Omega} = \Omega \cup \partial\Omega$, $\partial\Omega = \overline{\partial\Omega_u \cup \partial\Omega_t}$, $\partial\Omega_u \cap \partial\Omega_t = \emptyset$, where $\partial\Omega_u$ denotes the Dirichlet boundary and $\partial\Omega_t$ the Neumann boundary. The precise definition of each identity in the previous two sets of equations will be given in Section 2.2 for the classical and first order models and in Section 4 for the DSPV model. These definitions are based on the relations between the three-dimensional continuum and the structural theories. For now, the sets of equations (6) and (7) are only the generalized counterpart of (3) and (4), respectively.

The kinematics may be expressed as

$$\mathbf{u} = \mathbf{A}\mathbf{r} \tag{8a}$$

$$\mathbf{r} = \mathbf{L}\mathbf{d} \tag{8b}$$

$$\boldsymbol{\varepsilon} = \mathbf{B}\mathbf{e} \tag{8c}$$

where $\mathbf{d}$ is the generalized displacement vector, $\mathbf{r}$ is a generalized pseudo-displacement vector, $\mathbf{e}$ is the generalized strain vector, $\mathbf{A}$ and $\mathbf{B}$ are matricial operators and $\mathbf{L}$ is a differential operator. The introduction of the generalized pseudo-displacement vector becomes necessary as the kinematic boundary conditions are *not always* expressed in terms of the generalized displacements.

Let us assume the domain V and the boundary ∂V have the particular forms:

$$V = \Omega \times S \tag{9a}$$

$$\partial V = \partial\Omega \times S \cup \Omega \times \partial S \tag{9b}$$

and

$$\partial V_t = \partial\Omega_t \times S \cup \Omega \times \partial S \tag{10a}$$

$$\partial V_u = \partial\Omega_u \times S \tag{10b}$$

Then, the internal virtual work (first term of the first member in (5)) reads[4]

[4] The letter δ is used to denote both the variational operator and the delta Kronecker symbol. The meaning in each type occurrence should be inferred from the context.

$$\begin{aligned}\int_V \delta\boldsymbol{\varepsilon}^T\boldsymbol{\sigma}\mathrm{d}V &= \int_V (\mathbf{D}\delta\mathbf{u})^T\boldsymbol{\sigma}\mathrm{d}V = \int_V (\mathbf{DA}\delta\mathbf{r})^T\boldsymbol{\sigma}\mathrm{d}V = \int_V (\mathbf{DAL}\delta\mathbf{d})^T\boldsymbol{\sigma}\mathrm{d}V \\ &= \int_V (\mathbf{BB}^{-1}\mathbf{DAL}\delta\mathbf{d})^T\boldsymbol{\sigma}\mathrm{d}V = \int_V (\mathbf{B}^{-1}\mathbf{DAL}\delta\mathbf{d})^T\mathbf{B}^T\boldsymbol{\sigma}\mathrm{d}V \\ &= \int_\Omega (\boldsymbol{\partial}\delta\mathbf{d})^T\int_S \mathbf{B}^T\boldsymbol{\sigma}\mathrm{d}S\mathrm{d}\Omega = \int_\Omega \delta\mathbf{e}^T\mathbf{s}\mathrm{d}\Omega\end{aligned} \tag{11}$$

where the following definitions for the generalized compatibility operator and the generalized stress vector emerge

$$\boldsymbol{\partial} = \mathbf{B}^{-1}\mathbf{DAL} \tag{12a}$$

$$\mathbf{s} = \int_S \mathbf{B}^T\boldsymbol{\sigma}\mathrm{d}S \tag{12b}$$

The external virtual work (second and third terms of the first member in (5)) reads

$$\begin{aligned}-\int_V \delta\mathbf{u}^T\bar{\mathbf{f}}\mathrm{d}V - \int_{\partial V_t}\delta\mathbf{u}^T\bar{\mathbf{t}}\mathrm{d}\mathcal{S}_t &= -\int_V \delta\mathbf{r}^T\mathbf{A}^T\bar{\mathbf{f}}\mathrm{d}V - \int_{\partial V_t}\delta\mathbf{r}^T\mathbf{A}^T\bar{\mathbf{t}}\mathrm{d}\partial V_t \\ &= -\int_\Omega \delta\mathbf{r}^T\left(\int_S \mathbf{A}^T\bar{\mathbf{f}}\mathrm{d}S + \int_{\partial S}\mathbf{A}^T\bar{\mathbf{t}}\mathrm{d}\partial S\right)\mathrm{d}\Omega \\ &\quad -\int_{\partial\Omega_t}\delta\mathbf{r}^T\int_S \mathbf{A}^T\bar{\mathbf{t}}\mathrm{d}S\mathrm{d}\partial\Omega_t \\ &= -\int_\Omega \delta\mathbf{r}^T\bar{\mathbf{q}}\mathrm{d}\Omega - \int_{\partial\Omega_t}\delta\mathbf{r}^T\bar{\mathbf{g}}\mathrm{d}\partial\Omega_t\end{aligned} \tag{13}$$

where the following definitions for the generalized pseudo-body force and pseudo-surface traction vectors for the domain and Neumann boundary emerges

$$\bar{\boldsymbol{q}} = \int_S \mathbf{A}^T\bar{\mathbf{f}}\mathrm{d}S + \int_{\partial S}\mathbf{A}^T\bar{\mathbf{t}}\mathrm{d}\partial S \qquad \text{in } \Omega \tag{14a}$$

$$\bar{\boldsymbol{g}} = \int_S \mathbf{A}^T\bar{\mathbf{t}}\mathrm{d}S \qquad \text{on } \partial\Omega_t \tag{14b}$$

Notice that the generalized pseudo-body force vector has a surface traction component. The generalized constitutive relation (6c) can be derived from (12b)

$$\mathbf{s} = \int_S \mathbf{B}^T\boldsymbol{\sigma}\mathrm{d}S = \int_S \mathbf{B}^T\mathbf{C}\boldsymbol{\varepsilon}\mathrm{d}S = \int_S \mathbf{B}^T\mathbf{CB}\mathrm{d}S\mathbf{e} = \mathbf{Ee} \tag{15}$$

with

$$\mathbf{E} = \int_S \mathbf{B}^T\mathbf{CB}\mathrm{d}S \tag{16}$$

Usually further assumptions are introduced when evaluating the previous relation, as particular stress/strain distributions or the introduction of extrinsic factors (e.g., shear factor).

Hence, the PVW for reads

$$\int_{\Omega} \delta \mathbf{e}^T \mathbf{s} \mathrm{d}\Omega - \int_{\Omega} \delta \mathbf{r}^T \overline{\mathbf{q}} \mathrm{d}\Omega - \int_{\partial\Omega_t} \delta \mathbf{r}^T \overline{\mathbf{g}} \mathrm{d}\partial\Omega_t = 0 \qquad \forall\, \delta \mathbf{d} \in \mathcal{W} \tag{17}$$

Notice that this equation holds irrespective of the variations $\delta\mathbf{d}$, thus, by virtue of (8b), also holds for arbitrary variations $\delta\mathbf{r}$. Integration by parts on the first term of the first member of (17) yields

$$\int_{\Omega} \delta \mathbf{d}^T \left(\boldsymbol{\partial}^* \mathbf{s} + \overline{\mathbf{p}}\right) \mathrm{d}\Omega - \int_{\partial\Omega_t} \delta \mathbf{r}^T \left(\mathbf{M}^T \mathbf{s} - \overline{\mathbf{b}}\right) \mathrm{d}\partial\Omega_t = 0 \tag{18}$$

Here $\overline{\mathbf{b}}$ is the generalized pseudo-boundary force vector

$$\overline{\mathbf{b}} = \int_S \mathbf{A}^T \overline{\mathbf{t}} \mathrm{d}S \qquad \text{on } \partial\Omega_t \tag{19}$$

Resorting to the fundamental lemma of the calculus of variations, the equilibrium conditions in the domain (6b) and the boundary conditions (7) are identified.

The generalized compatibility and equilibrium operators $\boldsymbol{\partial}$ and $\boldsymbol{\partial}^*$ possess an adjoint relationship. The pairs $\{\mathbf{s}, \mathbf{e}\}$, $\{\mathbf{d}, \overline{\mathbf{p}}\}$, $\{\mathbf{d}, \overline{\mathbf{b}}\}$, $\{\mathbf{r}, \overline{\mathbf{q}}\}$ and $\{\mathbf{r}, \overline{\mathbf{g}}\}$ are conjugated.

2.2 The Particular Cases

2.2.1 Beam Models

Consider the bending of a beam in the (x_1, x_3) plane. The domain problem is

$$V = \left\{(x_1, x_2, x_3) \in \mathbb{R}^3 | x_3 \in \Omega = [0, l] \subset \mathbb{R}, (x_1, x_2) \in S \subset \mathbb{R}^2\right\} \tag{20}$$

where Ω is the span, S is the cross section and l stands for the beam length. The boundary is given by (9b), where $\partial\Omega = \{0, l\}$ are the end points of the beam and ∂S is the contour of S. Hence, the first part of (9b) is the end surfaces, the second part being the lateral surface. Notice that in the last part only static boundary conditions are prescribed.

The standard models are expressed by the Euler–Bernoulli and Timoshenko theories. These are not derived here, but all the relevant variables and operators were expressed in the framework presented in the previous subsection and can be found in Tables 1 and 2, in Section 4 in conjunction with the DSPV model, described in Section 4.

In order to emphasize the common quantities and simplify the reading, in the case that an element of the table is common to the adjacent element(s), this is written only once.

2.2.2 Plate Models

Consider the bending of a plate in the (x_1, x_2) plane. In this case

$$V = \left\{ (x_1, x_2, x_3) \in \mathbb{R}^3 | (x_1, x_2) \in \Omega \subset \mathbb{R}^2, x_3 \in S = \left[-\frac{h}{2}, \frac{h}{2} \right] \subset \mathbb{R} \right\} \quad (21)$$

where Ω is the middle plane, S is the out-of-plane dimension and h is the plate thickness. The boundary is given by (9b), where $\partial\Omega$ is the middle surface contour and $\partial S = \{-\frac{h}{2}, \frac{h}{2}\}$ are the bottom and top surfaces. Notice that on the last part only static boundary conditions are prescribed.

The most simple models are the Kirchhoff–Love and Reissner–Mindlin. Tables 3 and 4 gather the variables and operators of these models. In order to shorten the notation some specific quantities were introduced. The tangential vector components is denoted by t_α. Note that $t_1 = -n_2$ and $t_2 = n_1$. Also $\partial_t = \partial_\alpha t_\alpha$, $\partial_n = \partial_\alpha n_\alpha$, $m_n = m_\alpha n_\alpha$, $m_t = m_\alpha t_\alpha$ and $m_\beta = m_{\alpha\beta} n_\alpha$.

The Kirchhoff–Love model has one particularity which was omitted in Table 4, namely, the fact that the boundary conditions are not only imposed along the edges but also at the points where the boundary is non-smooth. This information is not reproduced in the table as both shear deformable models do not posses this type of constraints. Thus, for the Kirchhoff–Love model

$$w_3 = 0 \text{ or } [\![-m_t + \overline{m}_t^{\partial\Omega}]\!] = 0 \quad (22)$$

3 A Consistency Approximation Approach Analysis

3.1 Reproducing the Classical Models Limit

In the following it is sufficient to address only the one-dimensional domain problem, the conclusion being later generalized. Consider the Timoshenko beam model. When the thickness of the beam tends to zero, $h \to 0$, the (engineering) shear strain of the beam axis, η_{13}, also tends to zero, $\eta_{13} \to 0$. Therefore, the result for $\boldsymbol{\partial}$ in Table 2 (Timoshenko model) and the last scalar equation of (6a) leads to $\frac{\partial w_1}{\partial x_3} - \theta_2 = 0$ or

$$\theta_2 = \frac{\partial w_1}{\partial x_3} \quad (23)$$

Equivalent consistency results can be obtained for curved beams, as been reported by Prathap [20] and Donning and Liu [8].

Notice that $\eta_{13} = \eta_{13}(x_3)$ is the (engineering) strain at the beam axis and should not be confused with the (engineering) strain $\gamma_{13} = \gamma_{13}(x_1, x_2, x_3)$ included in vector $\boldsymbol{\varepsilon} = \boldsymbol{\varepsilon}(x_1, x_2, x_3)$.

Observe that, contrary to the usual FEM interpolations, an *unequal order of approximation* does not provide consistent fields, as was shown by Kanok-Nukulchai et al. [13].[5]

For approximations based on linear combinations of functions

$$\theta_2 = \mathbf{\Phi}^{\theta_2}\boldsymbol{\theta}_2 \tag{24a}$$

$$w_1 = \mathbf{\Phi}^{w_1}\mathbf{w}_1 \tag{24b}$$

equation (23) is expressed as

$$\mathbf{\Phi}^{\theta_2}\boldsymbol{\theta}_2 = \frac{\partial \mathbf{\Phi}^{w_1}}{\partial x_3}\mathbf{w}_1 \tag{25}$$

Hence, in the classical limit it is sufficient that the two following conditions hold:

$$\mathbf{\Phi}^{\theta_2} = \frac{\partial \mathbf{\Phi}^{w_1}}{\partial x_3} \tag{26a}$$

$$\boldsymbol{\theta}_2 = \mathbf{w}_1 \tag{26b}$$

Thus, the first of these conditions can be imposed to construct the consistent approximation for the rotation field, the remainder of the EFG formulation (or for any other meshless method) being exactly the same as in the classical form.

The equivalent conditions for plates can be derived in the same way as (26a) and leads to

$$\mathbf{\Phi}^{\theta_1} = -\frac{\partial \mathbf{\Phi}^{w_3}}{\partial x_1} \tag{27a}$$

$$\mathbf{\Phi}^{\theta_2} = -\frac{\partial \mathbf{\Phi}^{w_3}}{\partial x_2} \tag{27b}$$

An awkward consequence of the approximations based on derivatives of the MLS functions is the fact that, under certain conditions, the value of the approximation at a given node may be *independent* of the nodal parameter associated to that node. The conditions are associated with symmetries (of the weight function, of the basis functions, the layout of the nodes and the non-intersection with the boundary) on the support of the node.

3.2 Completeness

A result that proved to be very useful to study the properties of the consistent approximation is lemma 3.3, namely the *m-consistency condition II*, due to Liu et al.

[5] The nomenclature employed in [13] is quite different from the one presented here. The consistency and unequal order of approximation in the present work are designated by matching and consistency approximation, respectively, in [13].

[16]. This lemma was proposed for the so-called *moving least-square reproducing kernel methods* and it establishes, for a complete m-order, ℓ component polynomial basis and a weight function $W(\boldsymbol{x}) \in C^m(\Omega)$, the following reproducing conditions

$$\sum_{I=1}^{NP} (\mathbf{x}_I - \mathbf{x})^{\boldsymbol{\alpha}} D_{\boldsymbol{\alpha}}^{\boldsymbol{\beta}} \boldsymbol{\Phi}_I(\mathbf{x}) = \boldsymbol{\alpha}!\delta_{\boldsymbol{\alpha}\boldsymbol{\beta}} \tag{28a}$$

$$\sum_{I=1}^{NP} \mathbf{x}_I^{\boldsymbol{\alpha}} D_{\boldsymbol{\alpha}}^{\boldsymbol{\beta}} \boldsymbol{\Phi}_I(\mathbf{x}) = \frac{\boldsymbol{\alpha}!}{(\boldsymbol{\alpha}-\boldsymbol{\beta})!}\mathbf{x}^{\boldsymbol{\alpha}-\boldsymbol{\beta}} \tag{28b}$$

where $0 \leq |\boldsymbol{\alpha}|, |\boldsymbol{\beta}| \leq m$ and multi-index notation is used.

A crucial aspect for the convergence of the use of some approximation in a Galerkin method is the completeness of the set of approximation functions. For the set of functions used to approximate the rotations this property can easily be revealed by the *m-consistency condition II*, equation (28b). As the interest here is only on the *first order derivative*, this equation is specialized in the following for that case. For one-dimensional space with $\boldsymbol{\alpha} = \alpha$ and $\boldsymbol{\beta} = 1$

$$\sum_{I=1}^{NP} x_{1I}^{\alpha} \frac{\partial \phi_I}{\partial x_1} = \frac{\alpha!}{(\alpha-1)!} x_1^{\alpha-1} \tag{29}$$

and for two-dimensional space with $\boldsymbol{\alpha} = (\alpha_1, \alpha_2)$ and $\boldsymbol{\beta} = (1, 0)$ and $\boldsymbol{\beta} = (0, 1)$

$$\sum_{I=1}^{NP} x_{1I}^{\alpha_1} x_{2I}^{\alpha_2} \frac{\partial \phi_I}{\partial x_1} = \frac{\alpha_1!}{(\alpha_1-1)!} x_1^{\alpha_1-1} x_2^{\alpha_2} \tag{30a}$$

$$\sum_{I=1}^{NP} x_{1I}^{\alpha_1} x_{2I}^{\alpha_2} \frac{\partial \phi_I}{\partial x_2} = \frac{\alpha_2!}{(\alpha_2-1)!} x_1^{\alpha_1} x_2^{\alpha_2-1} \tag{30b}$$

In both cases, the approximation for rotations will only reproduce exactly complete polynomials of degree $(m-1)$. Hence, contrary to the usual MLS approximation, the approximation based on the consistency conditions are not m complete. This should not be interpreted as a drawback, e.g., when the exact solution for the generalized displacements belongs to the space of polynomial of degree less or equal to m. In this case, the approximation can still attain the (exact) desired solution.

The minimum degree of the polynomial basis to ensure the reproduction of constant curvatures will be $m = 2$. This can readily be found by inspection of the maximum order of differentiation in $\boldsymbol{\partial}$ associated to the generalized rotations, see Tables 2 and 4. As this is equal to one and the approximations are evaluated by the derivative of the displacements, then $m \geq 2$, in both cases, will yield the required minimum reproducing requirement. This conclusion is similar to the *Kirchhoff mode reproducing conditions* presented by Wang and Chen [26].

3.3 Linear Dependence

However simple, the use of the approximations that fulfil *a priori* the consistency conditions give rise to a singular global system because the approximation functions are *linearly dependent*, i.e., given a set of n functions $\{u_1(\mathbf{x}), u_2(\mathbf{x}), \ldots, u_n(\mathbf{x})|\mathbf{x} \in \Omega\}$ there exists some set $\{c_i \in \mathbb{R}|i = 1, \ldots, n\}$ not all zero such that

$$\sum_{I=1}^{n} c_I u_I(\mathbf{x}) = 0, \quad \forall \mathbf{x} \in \Omega \tag{31}$$

This will be proved by rewriting the consistency condition (28a) as the linear dependency condition. For the one-dimensional case and $\boldsymbol{\beta} = 1$

$$\sum_{I=1}^{NP} (x_{1I} - x_1)^{\alpha} \frac{\partial \phi_I}{\partial x_1} = \alpha! \delta_{\alpha 1} \tag{32}$$

and for two-dimensional space with $\boldsymbol{\alpha} = (\alpha_1, \alpha_2)$ and $\boldsymbol{\beta} = (1, 0)$ and $\boldsymbol{\beta} = (0, 1)$

$$\sum_{I=1}^{NP} (x_{1I} - x_1)^{\alpha_1} (x_{2I} - x_2)^{\alpha_2} \frac{\partial \phi_I}{\partial x_1} = \alpha_1! \alpha_2! \delta_{\alpha_1 1} \delta_{\alpha_2 0} \tag{33a}$$

$$\sum_{I=1}^{NP} (x_{1I} - x_1)^{\alpha_1} (x_{2I} - x_2)^{\alpha_2} \frac{\partial \phi_I}{\partial x_2} = \alpha_1! \alpha_2! \delta_{\alpha_1 0} \delta_{\alpha_2 1} \tag{33b}$$

From equation (32) and $\alpha = 0$ the linear dependence condition arises with $c_I = 1$ and $u_I = \frac{\partial \phi_I}{\partial x}$. This is the only value of $\boldsymbol{\alpha}$ that fulfils the linear dependence condition, hence, in one-dimensional problems only one deficiency per generalized rotations field exists. Alternatively, these results could be revealed by using the partition of unity paradigm. For the one dimensional case, from the differentiation of

$$\sum_{I=1}^{NP} \phi_I = 1 \tag{34}$$

follows

$$\sum_{I=1}^{NP} \frac{\partial \phi_I}{\partial x} = 0, \quad \forall x \in \Omega \tag{35}$$

which agrees with (32) for $\alpha = 0$.

This means that for plane bending problems the rank deficiency of the global matrix is one (besides the two rigid body motions, (RBM's)) and for full three dimensional beams the rank deficiency of the global matrix is two (besides the six RBM's). Of course, the RBM's are suppressed by the imposition of the Dirichlet boundary conditions. Moreover, equation (35) shows that, in fact, a *partition of nullity* is used in this case for the approximation of the rotation fields.

From equation (33a) and $\alpha_1 = 0$ there are $\sum_{\alpha_2=0}^{m} \alpha_2 = m + 1$ possibilities that fulfil the linear dependence condition, $\{\forall \mathbf{x} | \mathbf{x} \in \Omega\}$. In this case, $c_I = (x_{2I} - x_2)^{\alpha_2}$ and $u_I = \frac{\partial \phi_I}{\partial x_1}$. Hence, contrary to the one-dimensional case, the number of deficiencies *depend* on the m-order of the polynomial basis. Similar conclusion can be drawn from equations (33b) and the total number of the linear dependencies for a plate is $2(m + 1)$.

Remarkably, these facts, that were numerically confirmed, passed unnoticed until now. This may be related to errors induced by the the non-exact integration of the weak form appearing in a Galerkin method. Hence, linearly dependent approximations will only induce very ill-conditioned system matrices. Nevertheless, if an exact integration procedure were to be carried out, the global system would be, in fact, singular.

This conclusion does not mean that the consistency approach should not be used. It only indicates that special care should be taken (i) in the solution algorithm of the system of equations and (ii) when the number of null eigenvalues of the system matrix plays an important role in the problem under analysis. An immediate application of the last result is the detection of a cross of a limit or bifurcation point in a geometrically non-linear analysis, where the rank deficiency caused by the approximation has to be taken into account. Notice that this kind of problems is especially pronounced in very slender elements, precisely where the shear locking phenomenon frequently occurs.

3.4 An Alternative Interpretation

The implementation of the consistency approach can be made by the embedding of the consistency conditions in the compatibility operators and by using the same approximation functions for all the generalized displacements. The resulting compatibility operators for beams and plates are

$$\boldsymbol{\partial} = \begin{bmatrix} \cdot & \partial_{33} \\ \partial_{33} & -1 \end{bmatrix} \qquad \text{and} \qquad \boldsymbol{\partial} = \begin{bmatrix} \cdot & -\partial_{11} & \cdot \\ \cdot & \cdot & -\partial_{22} \\ \cdot & -\partial_{12} & -\partial_{12} \\ \partial_1 & -\partial_1 & \cdot \\ \partial_2 & \cdot & -\partial_2 \end{bmatrix} \tag{36}$$

respectively, where "·" is equal zero.

It can be shown that this modification on the compatibility equations has its dual on the equilibrium equations. These new equilibrium equations can be obtained from the usual ones by the application of a certain differential operators.

From the previous expressions, it is clear the appearance of second order derivatives, which are typical of the classical theories.

4 Displacement-Strain Primary Variables (DSPV) Formulation

4.1 Introduction

As established in the last section, the use of consistent approximations has some drawbacks relatively to the usual Timoshenko formulation. In fact, the use of a linear dependent basis gives rise to the appearance of spurious kinematic modes. In order to circumvent this inconvenience, a first-order shear-deformation model for beams and plates is developed in this section. The main idea in this theory is to use the strains as primary variables *instead* of the rotations. An attempt to unify the concepts and the notation is made in order to identify similarities and point-out differences between the DSPV formulation and the traditional Euler–Bernoulli/Timoshenko (beams) and Kirchhoff–Love/Reissner–Mindlin (plate) models.

Usually, structural theories are derived departing directly from the domain Ω subjected to a certain loading. On the contrary, here the models are derived right from the 3D continuum, thus leading to the appearance of some loading terms which are not usually considered. Nevertheless, all of them have a simple physical interpretation.

4.2 The Plane Beam Theory

Consider a linear shear-deformable straight beam whose axis is along x_3 direction. To simplify the exposition it is assumed that the axis of the beam is constrained to lie in the (x_1, x_3) plane and not to twist or warp. In this case the domain V is given by (20).

Let the generalized pseudo-displacement vector, $\mathbf{r}$, the generalized displacement vector, $\mathbf{d}$, and the generalized strain vector, $\mathbf{e}$, be

$$\mathbf{r} = \begin{bmatrix} w_1 \\ \theta_3 \end{bmatrix} \qquad \mathbf{d} = \begin{bmatrix} w_1 \\ \eta_{13} \end{bmatrix} \qquad \mathbf{e} = \begin{bmatrix} \chi_2 \\ \eta_{13} \end{bmatrix} \tag{37}$$

where w_1 is the transversal displacement, θ_2 is the rotation along x_2 axis, κ_2 is the curvature along x_2 axis and η_{13} is the (engineering) strain.

Observe that the usual generalized displacement vector in the Timoshenko beam theory agrees with the pseudo-generalized displacement vector in DSPV.

The matrix, $\mathbf{A}$, the differential operator, $\mathbf{L}$, and the matrix, $\mathbf{B}$, read

$$\mathbf{A} = \begin{bmatrix} 1 & \cdot \\ \cdot & \cdot \\ \cdot & -x_1 \end{bmatrix} \qquad \mathbf{L} = \begin{bmatrix} 1 & \cdot \\ \partial_3 & -1 \end{bmatrix} \qquad \mathbf{B} = \begin{bmatrix} \cdot & \cdot \\ \cdot & \cdot \\ -x_1 & \cdot \\ \cdot & \cdot \\ \cdot & 1 \\ \cdot & \cdot \end{bmatrix} \tag{38}$$

The compatibility operator given by (12a) results in

$$\boldsymbol{\partial} = \begin{bmatrix} \partial_{33} & -\partial_3 \\ \cdot & 1 \end{bmatrix} \tag{39}$$

The generalized stress vector is given by (12b) and reads

$$\mathbf{s} = \begin{bmatrix} M_2 \\ V_1 \end{bmatrix} = \int_S \begin{bmatrix} -x_1\sigma_{33} \\ \sigma_{13} \end{bmatrix} \mathrm{d}S \tag{40}$$

where M_2 and V_1 are the bending moment and shear force, respectively.

The generalized pseudo-body force and generalized pseudo-surface traction vectors are given by (14) and read

$$\overline{\mathbf{q}} = \begin{bmatrix} \overline{p}_1^{\Omega} \\ \overline{m}_2^{\Omega} \end{bmatrix} = \int_S \begin{bmatrix} \overline{f}_1 \\ -x_1\overline{f}_3 \end{bmatrix} \mathrm{d}S + \int_{\partial S} \begin{bmatrix} \overline{t}_1 \\ -x_1\overline{t}_3 \end{bmatrix} \mathrm{d}\partial S \qquad \text{in } \Omega \tag{41a}$$

$$\overline{\mathbf{g}} = \begin{bmatrix} \overline{p}_1^{\partial\Omega} \\ \overline{m}_2^{\partial\Omega} \end{bmatrix} = \int_S \begin{bmatrix} \overline{t}_1 \\ -x_1\overline{t}_3 \end{bmatrix} \mathrm{d}S \qquad \text{on } \partial\Omega_t \tag{41b}$$

Integration by parts on the first term of the first member of (17) yields a set of equations with the format of (18) where

$$\boldsymbol{\partial}^* = \begin{bmatrix} -\partial_{33} & \cdot \\ -\partial_3 & -1 \end{bmatrix} \qquad \overline{\mathbf{p}} = \begin{bmatrix} \overline{p}_1^{\Omega} - \overline{m}_{2,3}^{\Omega} \\ -\overline{m}_2^{\Omega} \end{bmatrix} \tag{42}$$

and

$$\mathbf{M}^T = \begin{bmatrix} -n_3\partial_3 & \cdot \\ -n_3 & \cdot \end{bmatrix} \qquad \overline{\mathbf{b}} = \begin{bmatrix} \overline{p}_1^{\partial\Omega} + \overline{m}_2^{\Omega} n_3 \\ -\overline{m}_2^{\partial\Omega} \end{bmatrix} \tag{43}$$

After the plane stress assumptions, $\sigma_{11} = 0$ and $\sigma_{22} = 0$, are introduced in (3c) and after the introduction of the shear factor κ_x, the generalized constitutive operator resulting from (16) is

$$\mathbf{E} = \begin{bmatrix} EI_2 & \cdot \\ \cdot & \kappa_1 GA \end{bmatrix} \tag{44}$$

where $G = \frac{E}{2(1+\nu)}$ is the shear rigidity.

4.3 The Plate Theory

Consider a linear shear-deformable plate lying in the (x_1, x_2) plane. In this case the domain is given by (21).

Let the generalized pseudo-displacement vector, $\mathbf{r}$, the generalized displacement vector, $\mathbf{d}$, and the generalized strain vector, $\mathbf{e}$, be

$$\mathbf{r} = \begin{bmatrix} w_3 \\ \theta_2 \\ \theta_3 \end{bmatrix} \qquad \mathbf{d} = \begin{bmatrix} w_3 \\ \eta_{13} \\ \eta_{23} \end{bmatrix} \qquad \mathbf{e} = \begin{bmatrix} \chi_{11} \\ \chi_{22} \\ 2\chi_{12} \\ \eta_{13} \\ \eta_{23} \end{bmatrix} \tag{45}$$

The variable θ_α introduced in the definition of $\mathbf{r}$ is justified because it simplifies the subsequent expressions. Nevertheless, θ_α are *not* the rotations along x_α. These are denoted by $\hat{\theta}$ and the relations between them are $\hat{\theta}_1 = -\theta_2$ and $\hat{\theta}_2 = \theta_1$.

The matrix, $\mathbf{A}$, the differential operator, $\mathbf{L}$, and the matrix, $\mathbf{B}$, are

$$\mathbf{A} = \begin{bmatrix} \cdot & x_3 & \cdot \\ \cdot & \cdot & x_3 \\ 1 & \cdot & \cdot \end{bmatrix} \qquad \mathbf{L} = \begin{bmatrix} 1 & \cdot & \cdot \\ -\partial_1 & 1 & \cdot \\ -\partial_2 & \cdot & 1 \end{bmatrix} \qquad \mathbf{B} = \begin{bmatrix} x_3 & \cdot & \cdot & \cdot & \cdot \\ \cdot & x_3 & \cdot & \cdot & \cdot \\ \cdot & \cdot & \cdot & \cdot & \cdot \\ \cdot & \cdot & x_3 & \cdot & \cdot \\ \cdot & \cdot & \cdot & 1 & \cdot \\ \cdot & \cdot & \cdot & \cdot & 1 \end{bmatrix} \tag{46}$$

The compatibility operator is given by (12a) and results in

$$\boldsymbol{\partial} = \begin{bmatrix} -\partial_{11} & \partial_1 & \cdot \\ -\partial_{22} & \cdot & \partial_2 \\ -2\partial_{12} & \partial_2 & \partial_1 \\ \cdot & 1 & \cdot \\ \cdot & \cdot & 1 \end{bmatrix} \tag{47}$$

The generalized stress vector is given by (12b) and reads

$$\mathbf{s} = \begin{bmatrix} m_{11} \\ m_{22} \\ m_{12} \\ v_1 \\ v_2 \end{bmatrix} = \int_S \begin{bmatrix} x_3\sigma_{11} \\ x_3\sigma_{22} \\ x_3\sigma_{12} \\ \sigma_{13} \\ \sigma_{23} \end{bmatrix} \mathrm{d}S \tag{48}$$

The generalized pseudo-body force and pseudo-surface traction vectors are given by (14) and read

$$\overline{\mathbf{q}} = \begin{bmatrix} \overline{p}_3^{\Omega} \\ \overline{m}_1^{\Omega} \\ \overline{m}_2^{\Omega} \end{bmatrix} = \int_S \begin{bmatrix} \overline{f}_3 \\ -x_3\overline{f}_2 \\ x_3\overline{f}_1 \end{bmatrix} \mathrm{d}S + \int_{\partial S} \begin{bmatrix} \overline{t}_3 \\ -x_3\overline{t}_2 \\ x_3\overline{t}_1 \end{bmatrix} \mathrm{d}\partial S \qquad \text{in } \Omega \tag{49a}$$

$$\overline{\mathbf{g}} = \begin{bmatrix} \overline{p}_3^{\partial\Omega} \\ \overline{m}_1^{\partial\Omega} \\ \overline{m}_2^{\partial\Omega} \end{bmatrix} = \int_S \begin{bmatrix} \overline{t}_3 \\ -x_3\overline{t}_2 \\ x_3\overline{t}_1 \end{bmatrix} \mathrm{d}S \qquad \text{on } \partial\Omega_t \tag{49b}$$

Integration by parts on the first term of the first member of (17) yields equations with the format of (18) where

$$\boldsymbol{\partial}^* = \begin{bmatrix} \partial_{11} & \partial_{22} & 2\partial_{12} & \cdot & \cdot \\ \partial_1 & \cdot & \partial_2 & -1 & \cdot \\ \cdot & \partial_2 & \partial_1 & \cdot & -1 \end{bmatrix} \qquad \overline{\mathbf{p}} = \begin{bmatrix} \overline{p}_z^{\Omega} + \overline{m}_{\alpha,\alpha}^{\Omega} \\ \overline{m}_1^{\Omega} \\ \overline{m}_2^{\Omega} \end{bmatrix} \tag{50}$$

Table 1. Variables for the Euler–Bernoulli, DSPV and Timoshenko beam models.

Variable	Euler–Bernoulli	DSPV	Timoshenko
$\mathbf{r}$		$\begin{bmatrix} w_1 \\ \theta_2 \end{bmatrix}$	
$\mathbf{d}$	$\begin{bmatrix} w_1 \end{bmatrix}$	$\begin{bmatrix} w_1 \\ \eta_{13} \end{bmatrix}$	$\begin{bmatrix} w_1 \\ \theta_2 \end{bmatrix}$
$\mathbf{e}$	$\begin{bmatrix} \chi_2 \end{bmatrix}$	$\begin{bmatrix} \chi_2 \\ \eta_{13} \end{bmatrix}$	
$\mathbf{s}$	$\begin{bmatrix} M_2 \end{bmatrix}$	$\begin{bmatrix} M_2 \\ V_1 \end{bmatrix}$	
$\overline{\mathbf{q}}$	$\begin{bmatrix} \overline{p}_1^{\Omega} \\ \overline{m}_2^{\Omega} \end{bmatrix} = \int_S \begin{bmatrix} \overline{f}_1 \\ -x_1 \overline{f}_3 \end{bmatrix} \mathrm{d}S + \int_{\partial S} \begin{bmatrix} \overline{t}_1 \\ -x_1 \overline{t}_3 \end{bmatrix} \mathrm{d}\partial S$		
$\overline{\mathbf{g}}$		$\begin{bmatrix} \overline{p}_1^{\partial\Omega} \\ \overline{m}_2^{\partial\Omega} \end{bmatrix} = \int_S \begin{bmatrix} \overline{t}_1 \\ -x_1 \overline{t}_3 \end{bmatrix} \mathrm{d}S$	
$\overline{\mathbf{p}}$	$\begin{bmatrix} \overline{p}_1^{\Omega} - \overline{m}_{2,3}^{\Omega} \end{bmatrix}$	$\begin{bmatrix} \overline{p}_1^{\Omega} - \overline{m}_{2,3}^{\Omega} \\ -\overline{m}_2^{\Omega} \end{bmatrix}$	$\begin{bmatrix} \overline{p}_1^{\Omega} \\ -\overline{m}_2^{\Omega} \end{bmatrix}$
$\overline{\mathbf{b}}$	$\begin{bmatrix} \overline{p}_1^{\partial\Omega} + \overline{m}_2^{\Omega} n_3 \\ \overline{m}_2^{\partial\Omega} \end{bmatrix}$	$\begin{bmatrix} \overline{p}_1^{\partial\Omega} + \overline{m}_2^{\Omega} n_3 \\ -\overline{m}_2^{\partial\Omega} \end{bmatrix}$	$\begin{bmatrix} \overline{p}_1^{\partial\Omega} \\ \overline{m}_2^{\partial\Omega} \end{bmatrix}$

and

$$\mathbf{M}^T = \begin{bmatrix} n_1\partial_1 & n_2\partial_2 & n_\alpha\partial_\alpha & \cdot & \cdot \\ n_x & \cdot & n_y & \cdot & \cdot \\ \cdot & n_y & n_x & \cdot & \cdot \end{bmatrix} \qquad \overline{\mathbf{b}} = \begin{bmatrix} \overline{p}_z^{\partial\Omega} - \overline{m}_\alpha^{\Omega} n_\alpha \\ \overline{m}_1^{\partial\Omega} \\ \overline{m}_2^{\partial\Omega} \end{bmatrix} \tag{51}$$

After the plane stress assumption, $\sigma_{zz} = 0$, are introduced in (3c) and the inclusion of the shear factors, κ_1 and κ_2, the generalized constitutive operator resulting from (16) is

$$\mathbf{E} = \begin{bmatrix} D & \nu D & \cdot & \cdot & \cdot \\ \nu D & D & \cdot & \cdot & \cdot \\ \cdot & \cdot & D\frac{1-\nu}{2} & \cdot & \cdot \\ \cdot & \cdot & \cdot & \kappa_1 G h & \cdot \\ \cdot & \cdot & \cdot & \cdot & \kappa_2 G h \end{bmatrix} \tag{52}$$

where $D = \frac{Et^3}{12(1-\nu^2)}$ is the bending rigidity.

4.4 Comparison of DSPV with the Traditional Models

For an easy comparison between the DSPV and the traditional models the latter were rewritten in the notation of the present text and are confronted in Tables 1 and 2 for beams and Tables 3 and 4 for plates. These tables collect the variables or operators of each model.

Table 2. Operators for the Euler–Bernoulli, DSPV and Timoshenko beam models.

Operator	Euler–Bernoulli	DSPV	Timoshenko
$\mathbf{A}$		$\begin{bmatrix} 1 & \cdot \\ \cdot & \cdot \\ \cdot & -x_1 \end{bmatrix}$	
$\mathbf{L}$	$\begin{bmatrix} 1 \\ \partial_3 \end{bmatrix}$	$\begin{bmatrix} 1 & \cdot \\ \partial_3 & -1 \end{bmatrix}$	$\begin{bmatrix} 1 & \cdot \\ \cdot & 1 \end{bmatrix}$
$\mathbf{E}$	$\begin{bmatrix} EI \end{bmatrix}$	$\begin{bmatrix} EI & \cdot \\ \cdot & \kappa GA \end{bmatrix}$	
$\mathbf{B}^T$	$\begin{bmatrix} \cdot & \cdot & -x_1 & \cdot & \cdot & \cdot \end{bmatrix}$	$\begin{bmatrix} \cdot & \cdot & -x_1 & \cdot & \cdot & \cdot \\ \cdot & \cdot & \cdot & \cdot & 1 & \cdot \end{bmatrix}$	
$\boldsymbol{\partial}$	$\begin{bmatrix} \partial_{33} \end{bmatrix}$	$\begin{bmatrix} \partial_{33} & -\partial_3 \\ \cdot & 1 \end{bmatrix}$	$\begin{bmatrix} \cdot & \partial_3 \\ \partial_3 & -1 \end{bmatrix}$
$\boldsymbol{\partial}^*$	$\begin{bmatrix} -\partial_{33} \end{bmatrix}$	$\begin{bmatrix} -\partial_{33} & \cdot \\ -\partial_3 & -1 \end{bmatrix}$	$\begin{bmatrix} \cdot & \partial_3 \\ \partial_3 & 1 \end{bmatrix}$
$\mathbf{M}^T$	$\begin{bmatrix} -n_3\partial_3 \\ n_3 \end{bmatrix}$	$\begin{bmatrix} -n_3\partial_3 & \cdot \\ -n_3 & \cdot \end{bmatrix}$	$\begin{bmatrix} \cdot & n_3 \\ n_3 & \cdot \end{bmatrix}$

The analysis of the tables reveals the connection bridge character of the DSPV between the Kirchhoff–Love and the Reissner–Mindlin models. As a shear deformable model it shares the constitutive models and the number of boundary conditions with the Reissner–Mindlin formulation. As the shear strains can naturally tend to zero, it includes second-order differentials in the compatibility and equilibrium operators. Thus, it includes the governing equations of the Kirchhoff–Love model in the domain. In practice, this implies the use of, at least, C^1 approximations for the discretization of $w_1 (w_3)$ for the transversal displacement and C^0 approximations for η_{13} (η_{12} and η_{13}) in beam (plate) models. Thus, this model is not suitable for FEM discretization, but can easily be implemented using meshfree approximations. In fact, there is no inconvenient in employing the *same* approximation for all fields, thus speeding up the numerical computations.

Hence, the DSPV theory can be regarded as a generalization of the thin models to include shear deformation.

In the DSVP theory the Dirichlet boundary conditions are imposed over a specific set of pseudo generalized displacements, $\mathbf{r}$, which are different from the generalized displacements $\mathbf{d}$. This also happens in the classical models, where the Dirichlet boundary conditions are imposed in terms of transversal displacement and rotations, but the generalized displacements is only the transversal displacement. On the contrary, for the first-order traditional models, $\mathbf{r} = \mathbf{d}$.

Table 3. Variables for the Kirchhoff–Love, DSPV and Reissner–Mindlin plate models.

Variable	Kirchhoff–Love	DSPV	Reissner–Mindlin
$\mathbf{r}$		$\begin{bmatrix} w_3 \\ \theta_1 \\ \theta_2 \end{bmatrix}$	
$\mathbf{d}$	$\begin{bmatrix} w_3 \end{bmatrix}$	$\begin{bmatrix} w_3 \\ \eta_{13} \\ \eta_{23} \end{bmatrix}$	$\begin{bmatrix} w_3 \\ \theta_1 \\ \theta_2 \end{bmatrix}$
$\mathbf{e}$	$\begin{bmatrix} \chi_{11} \\ \chi_{22} \\ 2\chi_{12} \end{bmatrix}$	$\begin{bmatrix} \chi_{11} \\ \chi_{22} \\ 2\chi_{12} \\ \eta_{13} \\ \eta_{23} \end{bmatrix}$	
$\mathbf{s}$	$\begin{bmatrix} m_{11} \\ m_{22} \\ m_{12} \end{bmatrix}$	$\begin{bmatrix} m_{11} \\ m_{22} \\ m_{12} \\ v_1 \\ v_2 \end{bmatrix}$	
$\overline{\mathbf{q}}$	$\begin{bmatrix} \overline{p}_3^{\Omega} \\ \overline{m}_1^{\Omega} \\ \overline{m}_2^{\Omega} \end{bmatrix} = \int_S \begin{bmatrix} \overline{f}_3 \\ x_3 \overline{f}_2 \\ x_3 \overline{f}_1 \end{bmatrix} \mathrm{d}S + \int_{\partial S} \begin{bmatrix} \overline{t}_3 \\ x_3 \overline{t}_2 \\ x_3 \overline{t}_1 \end{bmatrix} \mathrm{d}\partial S$		
$\overline{\mathbf{g}}$		$\begin{bmatrix} \overline{p}_3^{\partial\Omega} \\ \overline{m}_1^{\partial\Omega} \\ \overline{m}_2^{\partial\Omega} \end{bmatrix} = \int_S \begin{bmatrix} \overline{t}_1 \\ x_3 \overline{t}_2 \\ x_3 \overline{t}_1 \end{bmatrix} \mathrm{d}S$	
$\overline{\mathbf{p}}$	$\begin{bmatrix} \overline{p}_3^{\Omega} + \overline{m}_{\alpha,\alpha}^{\Omega} \end{bmatrix}$	$\begin{bmatrix} \overline{p}_3^{\Omega} + \overline{m}_{\alpha,\alpha}^{\Omega} \\ \overline{m}_1^{\Omega} \\ \overline{m}_2^{\Omega} \end{bmatrix}$	$\begin{bmatrix} \overline{p}_3^{\Omega} \\ \overline{m}_1^{\Omega} \\ \overline{m}_2^{\Omega} \end{bmatrix}$
$\overline{\mathbf{b}}$	$\begin{bmatrix} \overline{p}_3^{\partial\Omega} - \overline{m}_\alpha^{\Omega} n_\alpha + \overline{m}_{t,t}^{\partial\Omega} \\ \overline{m}_\alpha^{\partial\Omega} n_\alpha \end{bmatrix}$	$\begin{bmatrix} \overline{p}_3^{\partial\Omega} - \overline{m}_\alpha^{\Omega} n_\alpha \\ \overline{m}_1^{\partial\Omega} \\ \overline{m}_2^{\partial\Omega} \end{bmatrix}$	$\begin{bmatrix} \overline{p}_3^{\partial\Omega} \\ \overline{m}_1^{\partial\Omega} \\ \overline{m}_2^{\partial\Omega} \end{bmatrix}$

5 Conclusions

The shear locking phenomenon is undoubtedly present in the numerical solutions of first order theories based on meshfree approximations. The traditional method used for alleviating the problem in the FEM, based on the equivalence theorem, are not appropriate for meshless methods because there is no exact integration rule for the non-scattered data approximation.

The consistency approach is analyzed and (i) it is proved that it leads to a linear dependent approximation (which gives rise to a singular system of equations within

Table 4. Operators for the Kirchhoff–Love, DSPV and Reissner–Mindlin plate models.

Operator	Kirchhoff–Love	DSPV	Reissner–Mindlin
$\mathbf{A}$		$\begin{bmatrix} \cdot & x_3 & \cdot \\ \cdot & \cdot & x_3 \\ 1 & \cdot & \cdot \end{bmatrix}$	
$\mathbf{L}$	$\begin{bmatrix} 1 \\ -\partial_1 \\ -\partial_2 \end{bmatrix}$	$\begin{bmatrix} 1 & \cdot & \cdot \\ -\partial_1 & 1 & \cdot \\ -\partial_2 & \cdot & 1 \end{bmatrix}$	$\begin{bmatrix} 1 & \cdot & \cdot \\ \cdot & 1 & \cdot \\ \cdot & \cdot & 1 \end{bmatrix}$
$\mathbf{E}$	$D \begin{bmatrix} 1 & \nu & \cdot \\ \nu & 1 & \cdot \\ \cdot & \cdot & \frac{1-\nu}{2} \end{bmatrix}$	$\begin{bmatrix} D & \nu D & \cdot & \cdot & \cdot \\ \nu D & D & \cdot & \cdot & \cdot \\ \cdot & \cdot & D\frac{1-\nu}{2} & \cdot & \cdot \\ \cdot & \cdot & \cdot & \kappa_1 Gh & \cdot \\ \cdot & \cdot & \cdot & \cdot & \kappa_2 Gh \end{bmatrix}$	
$\mathbf{B}^T$	$\begin{bmatrix} x_3 & \cdot & \cdot & \cdot & \cdot & \cdot \\ \cdot & x_3 & \cdot & \cdot & \cdot & \cdot \\ \cdot & \cdot & \cdot & x_3 & \cdot & \cdot \end{bmatrix}$	$\begin{bmatrix} x_3 & \cdot & \cdot & \cdot & \cdot & \cdot \\ \cdot & x_3 & \cdot & \cdot & \cdot & \cdot \\ \cdot & \cdot & \cdot & x_3 & \cdot & \cdot \\ \cdot & \cdot & \cdot & \cdot & 1 & \cdot \\ \cdot & \cdot & \cdot & \cdot & \cdot & 1 \end{bmatrix}$	
$\boldsymbol{\partial}$	$\begin{bmatrix} -\partial_{11} \\ -\partial_{22} \\ -2\partial_{12} \end{bmatrix}$	$\begin{bmatrix} -\partial_{11} & \partial_1 & \cdot \\ -\partial_{22} & \cdot & \partial_2 \\ -2\partial_{12} & \partial_2 & \partial_1 \\ \cdot & 1 & \cdot \\ \cdot & \cdot & 1 \end{bmatrix}$	$\begin{bmatrix} \cdot & \partial_1 & \cdot \\ \cdot & \cdot & \partial_2 \\ \cdot & \partial_2 & \partial_1 \\ \partial_1 & 1 & \cdot \\ \partial_2 & \cdot & 1 \end{bmatrix}$
$\boldsymbol{\partial}^*$	$\begin{bmatrix} \partial_{11} & \partial_{22} & 2\partial_{12} \end{bmatrix}$	$\begin{bmatrix} \partial_{11} & \partial_{22} & 2\partial_{12} & \cdot & \cdot \\ \partial_1 & \cdot & \partial_2 & -1 & \cdot \\ \cdot & \partial_2 & \partial_1 & \cdot & -1 \end{bmatrix}$	$\begin{bmatrix} \cdot & \cdot & \cdot & \partial_1 & \partial_2 \\ \partial_1 & \cdot & \partial_2 & -1 & \cdot \\ \cdot & \partial_2 & \partial_1 & \cdot & -1 \end{bmatrix}$
$\mathbf{M}^T$	$\begin{bmatrix} -n_1 n_2 \partial_t + n_1 \partial_1 & n_1 n_2 \partial_t + n_2 \partial_2 & -\left(n_2^2 - n_1^2\right)\partial_t + n_1 \partial_2 + n_2 \partial_1 \\ n_1^2 & n_2^2 & 2n_1 n_2 \end{bmatrix}$	$\begin{bmatrix} n_1\partial_1 & n_2\partial_2 & \partial_n & \cdot & \cdot \\ n_x & \cdot & n_y & \cdot & \cdot \\ \cdot & n_y & n_x & \cdot & \cdot \end{bmatrix}$	$\begin{bmatrix} \cdot & \cdot & \cdot & n_1 & n_2 \\ n_1 & \cdot & n_2 & \cdot & \cdot \\ \cdot & n_2 & n_1 & \cdot & \cdot \end{bmatrix}$

a Galerkin method), (ii) the number of dependencies are[6] one in a plane beam, two in a spacial beam and $2(m+1)$ in a plate and (iii) the reproducing properties are reduced by one order.

Departing from the concept of change of variables, the DSPV first order model for beams and plates is presented. As C^1 continuity is required, (most of the) meshfree approximations are natural candidates for its numerical implementation, thus taking advantage of a specific property of meshless approximations: the (arbitrary) continuity. On the contrary, for the plates case the FEM is naturally excluded.

[6] Besides the RBM's, which are removed by the correct imposition of the essential boundary conditions.

The analysis of the consistency and the DSPV models seams to indicate a common and unavoidable path for the shear-locking free analysis of structures: the evaluation of second order derivatives of the approximation. This should be neither a surprise nor a disappointment, since this is the case in the classical theories, which we are trying to reproduce in the first place.

References

1. S. N. Atluri, H.-G. Kim, and J. Y. Cho. A critical assessment of the truly Meshless Local Petrov–Galerkin (MLPG), and Local Boundary Integral Equation (LBIE) methods. *Computational Mechanics*, 24(5):348–372, 1999.
2. S. Beissel and T. Belytschko. Nodal integration of the element-free Galerkin method. *Computer Methods in Applied Mechanics and Engineering*, 139(1–4):49–74, 1996.
3. T. Belytschko, Y. Y. Lu, and L. Gu. Element-Free Galerkin Methods. *International Journal for Numerical Methods in Engineering*, 37(2):229–256, 1994.
4. J. Y. Cho and S. N. Atluri. Analysis of shear flexible beams, using the meshless local Petrov–Galerking method, based on a locking-free formulation. *Engineering Computations*, 18(1–2):215–240, 2001.
5. S. De and K.-J. Bathe. Displacement/pressure mixed interpolation in the method of finite spheres. *International Journal for Numerical Methods in Engineering*, 51(3):275–292, 2001.
6. P. de T. R. Mendonça, C. S. de Barcellos, and A. Duarte. Investigations on the hp-cloud method by solving Timoshenko beam problems. *Computational Mechanics*, 25(2–3):286–295, 2000.
7. J. Dolbow and T. Belytschko. Volumetric locking in the element free Galerkin method. *International Journal for Numerical Methods in Engineering*, 46(6):925–942, 1999.
8. B. M. Donning and W. K. Liu. Meshless methods for shear-deformable beams and plates. *Computer Methods in Applied Mechanics and Engineering*, 152(1–2):47–71, 1998.
9. C. A. Duarte and J. Tinsley Oden. *H-p* clouds – An *h-p* meshless method. *Numerical Methods for Partial Differential Equations*, 12(6):673–705, 1996.
10. O. Garcia, E. A. Fancello, C. S. de Barcellos, and C. A. Duarte. *hp*-Clouds in Mindlin's thick plate model. *International Journal for Numerical Methods in Engineering*, 47(8):1381–1400, 2000.
11. T. J. R. Hughes. *The Finite Element Method: Linear Static and Dynamic Finite Element Analysis*. Dover Publications, Mineola, NY, 2000.
12. T. J. R. Hughes and T. E. Tezduyar. Finite elements based upon mindlin plate theory with particular reference to the four-node isoparametric element. *Journal of Applied Mechanics, Transactions of the ASME*, 48(3):587–596, 1981.
13. W. Kanok-Nukulchai, W. Barry, K. S.-Yasoontorn, and P. H. Bouillard. On elimination of shear locking in the element-free Galerkin method. *International Journal for Numerical Methods in Engineering*, 52(7):705–725, 2001.
14. P. Lancaster and K. Šalkauskas. Surfaces genarated by moving least squares methods. *Mathematics of Computation*, 37(155):141–158, 1981.
15. W. K. Liu, S. Jun, and Y. F. Zhang. Reproducing kernel particle methods. *International Journal for Numerical Methods in Engineering*, 20(8–9):1081–1106, 1995.

16. W.-K. Liu, S. Li, and T. Belytschko. Moving least-square reproducing kernel methods (I): Methodology and convergence. *Computer Methods in Applied Mechanics and Engineering*, 143(1–2):113–154, 1997.
17. Y. Y. Lu, T. Belytschko, and L. Gu. A new implementation of the element free Galerkin method. *Computer Methods in Applied Mechanics and Engineering*, 113(3–4):397–414, 1994.
18. D. S. Malkus and T. J. R. Hughes. Mixed finite element methods-reduced and selective integration techniques: A unification of concepts. *Computer Methods in Applied Mechanics and Engineering*, 15(1):63–81, 1978.
19. W. D. Pilkey and W. Wunderlich. *Mechanics of Structures: Variational and Computational Methods*, CRC Press, second edition, 2003.
20. G. Prathap. *The Finite Element Method in Structural Mechanics: Principles and Practice of Design of Field-Consistent Elements for Structural and Solid Mechanics*, Solid Mechanics and Its Applications, Vol. 24, Kluwer Academic Publishers, 1993.
21. J. C. Simo and M. S. Rifai. A class of mixed assumed strain methods and the method of incompatible modes. *International Journal for Numerical Methods in Engineering*, 29(8):1595–1638, 1990.
22. N. Sukumar, B. Moran, and T. Belytschko. The natural element method in solid mechanics. *International Journal for Numerical Methods in Engineering*, 43(5):839–887, 1998.
23. C. Tiago and V. Leitão. On the procedures to eliminate shear locking in meshless methods. In V. Leitão, C. Alves, and C. Duarte (Eds), *ECCOMAS Thematic Conference on Meshless Methods*, Lisbon, Portugal, 2005, pp. A14.1–A14.8.
24. C. Tiago and P. Pimenta. Geometrically exact analysis of space frames by a meshless method. In V. Leitão, C. Alves, and C. Duarte (Eds), *ECCOMAS Thematic Conference on Meshless Methods*, Lisbon, Portugal, 2005, pp. C44.1–C44.8
25. C. Tiago and P. M. Pimenta. Geometrically exact analysis of shells by a meshless approach. In C. A. Mota Soares (Ed.), *Third European Conference on Computational Mechanics – Solids, Structures and Coupled Problems in Engineering*, Lisbon, Portugal, June 2006.
26. D. Wang and J.-S. Chen. Locking-free stabilized conforming nodal integration for meshfree Mindlin–Reissner plate formulation. *Computer Methods in Applied Mechanics and Engineering*, 193(12–14):1065–1083, 2004.
27. O. C. Zienkiewicz, R. L. Taylor, and J. M. Too. Reduced integration technique in general analysis of plates and shells. *International Journal for Numerical Methods in Engineering*, 3(2):275–290, 1971.

FEM/SPH Coupling Technique for High Velocity Impact Simulations

Levent Aktay* and Alastair F. Johnson

Institute of Structures and Design, German Aerospace Center (DLR), Pfaffenwaldring 38–40, 70569, Stuttgart, Germany; E-mail: alastair.johnson@dlr.de
**Currently at Tecosim Technische Simulation GmbH, Heidenheimer Str. 5, 71229, Leonberg, Germany; E-mail: levent.aktay@de.tecosim.com*

Abstract. In this work, impact simulations using both meshfree Smooth Particle Hydrodynamics (SPH) and combined FEM/SPH Method were carried out for a sandwich composite panel with carbon fibre fabric/epoxy face skins and polyetherimide (PEI)foam and hybrid (Nomex/PEI foam) core. A numerical model was developed using the dynamic explicit finite element (FE) structure analysis program PAM-CRASH. The carbon fibre/epoxy facings were modelled with layered shell elements, whilst SPH particles replaced solid elements in the core. The efficiency and the advantages of pure meshfree SPH and combined FEM/SPH methods were demonstrated by comparing the core deformation modes and impact force pulses measured in the experiments to predicted structural impact response for a range of impact velocities.

Key words: High Velocity Impact (HVI), impact damage, sandwich composite, numerical modelling, Finite Element Method (FEM), Smooth Particle Hydrodynamics (SPH).

1 Introduction

Composite sandwich construction consists of a lightweight core material sandwiched between two stiff facings. There are essentially two different classes of cores, namely foams and honeycombs, with a wide range of materials and properties within each type. Composite facings are commonly made of laminated fibre reinforced plastics. In terms of structural efficiency, with a small additional weight in the core it is possible to produce an improved shell structure compared with a monolithic composite laminate, particularly under transverse and bending loads.

The application of polymeric composite sandwich structures in the aerospace industry has been continuously increasing as new fibre types, resin systems, adhesives, new lightweight core materials and advanced manufacturing techniques have been developed and introduced into the market. In aircraft structures, sandwich materials may be used in ailerons, spoilers, passenger floors and numerous nacelles and

V.M.A. Leitão et al. (eds), Advances in Meshfree Techniques, 147–167.

fairings. The latest commercial aeroplane projects of Airbus and Boeing, the A380 and the 787, respectively, show the evolutionary growth in the use of composite materials. The main structure of Airbus A380 includes 25% of carbon fibre reinforced plastic (CFRP) structures, compared to 15% in the Airbus A340 launched in 2002.

The development of a software tool that can be used to predict the resulting damage from impact which is directly related to the structural integrity and safety requirements of engineering structures is of interest to aircraft manufacturers. Despite extensive research and development of sandwich structures, their impact response is still not fully understood [1, 2]. Although experimental tests under impact conditions provide considerable information about the tested specimen and their characteristic parameters, the dynamic properties and failure behaviour of composite sandwich structures are complex and test programmes are destructive, time consuming and consequently expensive for industry [3, 4]. In order to reduce the development and certification costs, computational methods can be used to predict structural integrity under crash and impact loads. The development of sandwich models and their validation with experimental tests allow prediction of impact response of sandwich structures under different impact scenarios without extensive impact testing. Additionally, geometry and material parameters can be varied in numerical models providing output for engineers to design improved concept structures.

A drawback of structural sandwich components is their relatively low resistance to impact damage due to the thin outer composite skins. When composite skin laminates are subjected to impact by a projectile, many fracture processes, with their associated energy absorbing capacities, can occur [5] such as matrix cracking, fibre matrix debonding, delamination and fibre failure. Interaction of failure modes and the effects of fibre type and lay-up, the matrix fibre resin bond and environmental effects result in complex failure modes. After fracture of the skin an impacting projectile may damage and penetrate into the core. For relatively low velocity projectiles, a sandwich panel may respond by bending and no damage will occur if the energy of the projectile can be accommodated by the elastic strain energy in the panel. At higher impact velocities a critical condition is reached when a local contact stress exceeds a local material strength, which may be a laminate bending strength, core compression strength or interface delamination strength. Thus in order to improve durability and damage tolerance of sandwich structures, a structural analysis of skin, core and interface damage should be addressed within the design process.

Modelling of high velocity impact (HVI) and crash scenarios involving material failure and large deformation using classical FEM is complex. Although the FEM is established as an effective tool for predicting structural behaviour in different loading conditions, FEM is less reliable for large deformations and structural failure modelling under dynamic loads. Additionally it is difficult to simulate structural failure containing the fracture of material into fragments since FEM is based on a continuum mechanics formulation requiring element connectivity. The enhancement of existing numerical methods based on FEM is a current research theme. Proposed adaptive remeshing seems promising however it is computationally expensive since the procedure determining the error estimation for remeshing criteria requires long computational time.

As an alternative to FEM, meshfree methods have been developed and applied to numerical simulations involving material failure and damage. Meshfree methods replace finite elements by a set of nodes or particles within the problem domain and its boundaries. This feature makes meshfree methods very effective since mesh connectivity is not as critical as in FEM. There are several meshfree methods and new meshfree methods are being developed in current research. Among them SPH is one of the earliest particle methods in computational mechanics. SPH was developed by Lucy [6] to solve astrophysical problems in 3D open space. Since its invention SPH has been extensively studied and extended to dynamic response with material strength, fracture and impact simulations, failure of brittle solids and metal forming simulations. The study presented here proposes the SPH Method and its combination with FEM to overcome the limitations of explicit FEM such as too small time steps, hour-glassing, mesh size dependency and element distortion during impact simulations of composite structures.

2 High Velocity Impact Tests

DLR carries out high velocity gas gun impact tests, aiming to determine failure modes, impact resistance, contact force-time curves and total energy dissipation of impact events. In these tests various projectiles are fired against a test structure or against a plate specimen mounted on a load cell. Hard projectiles, representative of runway and engine debris impactors and soft projectiles, representative of bird strike or burst tyre rubber fragments, are possible impact scenarios for an aircraft structure [7, 8]. Impact velocity is varied in gas gun tests by changing the pressure in the pressure vessel and calibration curves for vessel pressure and projectile velocity are used to achieve the required impact velocity. The most important result of a HVI test is to define modes of failure and failure thresholds, such as kinetic energy to cause damage on skin and core. Impact loads as functions of time are recorded in a dynamic impact test, if the impacted structure is mounted on a load cell.

A HVI test program was carried out at the DLR in which several sandwich plates with three different core configurations were tested with two different projectile geometries, representative of stones and runway debris. Sandwich plates were manufactured with PEI or Nomex core, 33 mm thick, with $[0/45/0/45]_2$ carbon fibre fabric/epoxy facings with a nominal thickness of 1.6 mm. As a third configuration called demonstrator a hybrid cored sandwich plate consisting of PEI (1/3 of the thickness) and Nomex (2/3 of the thickness) core was investigated under different impact velocities (Figures 1 and 2). In this paper the impact test results are based on a projectile 26 mm in diameter, with length 37 mm and mass of 36.4 g impacting on PEI cored

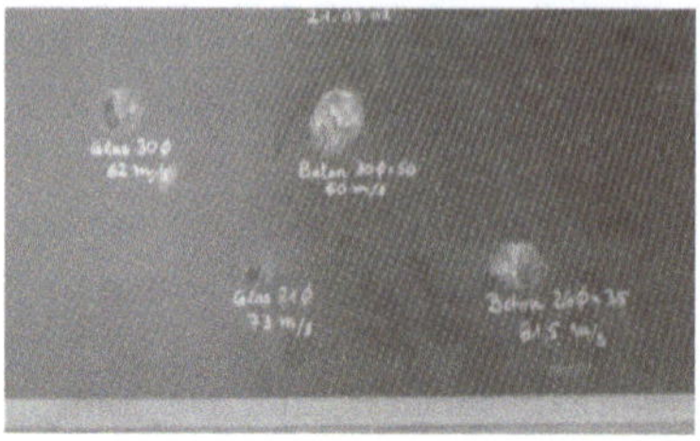

Fig. 1. PEI core sandwich plate.

Fig. 2. Hybrid core.

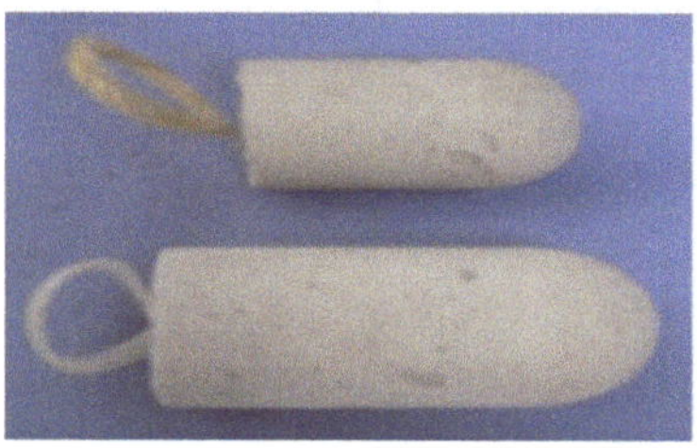

Fig. 3. Concrete impactors.

Fig. 4. Glass impactors.

sandwich plates and glass impactors 21 and 29 mm in diameter with mass of 13 and 31 g, respectively impacting on hybrid cored sandwich panels (Figures 3 and 4).

A nominal impact velocity of 60 m/s was chosen, which corresponds to the maximum expected impact speed of runway debris on an aircraft structure during start and landing. The actual impact velocity is measured optically using two light beams set 100 mm apart, which are broken by the passage of the projectile. Quantitative test data are difficult to obtain in HVI gas gun tests, since it is not possible to instrument the projectile which is the procedure adopted in low velocity drop tower tests. The DLR has developed a procedure which measures the resultant force pulse on the panel specimen by mounting it on a ring load cell.

3 Theoretical Fundamentals

This work aims at demonstrating the efficiency and the advantages of meshfree SPH and combined FEM/SPH methods by comparing the core deformation modes and impact force pulses measured in the experiments to predict the structural impact response. Therefore in this section an overview about the SPH Method is given.

SPH is a Lagrangian technique which allows the numerical grid to be embedded in the material as it deforms which reduces some of the material interface problems associated with Eulerian codes. Furthermore, the ability to handle severe distortions allows the SPH technique to be applied to problems with large deformations or severe element distortions. The SPH Method is generally not as efficient as FEM for structural response applications due to the high computational costs for particle methods. On the other hand the advantage of SPH is that it is possible for both severe distortions and normal structural responses to be performed with a single Lagrangian code.

First point to be considered is that the foundation of SPH is interpolation theory. Second point is that the conservation laws of continuum dynamics, in the form of partial differential equations, are transformed into integral equations in a weak formulation of the field equations. Computational information is known only at discrete points, so that the integrals are evaluated as sums over neighbouring particles.

SPH is based on two interpolation approximations: Kernel approximation and particle approximation. Considering the function $f(x)$ in Equation (1), value at a point of $f(x)$ over domain Ω could be extracted from its integral using the delta function (δ) as a filter,

$$\langle f(\overline{x})\rangle = \int_{\Omega} f(\overline{x'})\delta(\overline{x}-\overline{x'})d\overline{x'}. \tag{1}$$

One can define delta function as follows:

$$\int_{\Omega} \delta(\overline{x}-\overline{x'})d\overline{x'} = 1. \tag{2}$$

As $h \to 0$, $\delta(\overline{x}-\overline{x'})$ can be replaced by with a kernel function $W(\overline{x}-\overline{x'}, h)$ which has a support domain determined by the parameter h,

$$\lim_{h\to 0} W(\overline{x}-\overline{x'}, h) = \delta(\overline{x}-\overline{x'}). \tag{3}$$

Therefore Equation (3) yields to the following:

$$\langle f(\overline{x})\rangle = \int_{\Omega} f(\overline{x'})W(\overline{x}-\overline{x'}, h)d\overline{x'}. \tag{4}$$

Since the domain is represented by discrete particles, the summation of the contributions of each discrete particle within the kernel approximation range results the smoothed value of $f(x)$ at a point (particle approximation), as

$$\langle f(\overline{x})\rangle = \sum_{j=1}^{N}\left(\frac{m_j}{\rho_j}\right) f_j W(|\overline{x}-\overline{x'}|, h) \tag{5}$$

in which N represents the number of discrete particles, m_j and ρ_j stand for mass and the density of the particle j, respectively.

3.1 Approximation of Smoothing Function

Any numerical approximation should represent the physical equations such as differential conservation equations, as the number of particles tends to infinity and as the smoothing length tends to zero. By using Taylor series expansion it can be understood how accurate is an SPH function and how well it represents the real behaviour of the material.

Suppose that $f(x)$ is a smooth function then applying the Taylor series expansion:

$$f(x') = f(x) + f'(x)(x'-x) + \frac{1}{2}f''(x)(x'-x)^2 + \cdots \tag{6}$$

$$f(x') = \sum_{k=0}^{n} \frac{(-1)^k h^k f^k(x)}{k!}\left(\frac{x'-x}{h}\right)^k + r_n\left(\frac{x'-x}{h}\right) \tag{7}$$

where r is the reminder after approximating the derivative to order k. Substituting Equation (7) into Equation (4) leads to:

$$f(x') = \sum_{k=0}^{n} A^k f^k(x) + r_n\left(\frac{x'-x}{h}\right) \tag{8}$$

where

$$A^k = \frac{(-1)^k h^k}{k!}\int\left(\frac{x'-x}{h}\right)^k W(x'-x,h)dx' \tag{9}$$

Finally, comparing the left and right hand sides of Equation (8), one can obtain the coefficients A as

$$\begin{aligned} A^0 &= \int W(x'-x,h)dx' = 1 \\ A^1 &= h\int (\frac{x'-x}{h})W(x'-x,h)dx' = 0 \\ &\vdots \\ A^n &= \frac{(-1)^n h^n}{k!}\int\left(\frac{x'-x}{h}\right)^n W(x'-x,h)dx' = 0 \end{aligned} \tag{10}$$

or alternatively dividing the coefficients by smoothing length h one obtains

$$\begin{aligned} M^0 &= \int W(x'-x,h)dx' = 1 \\ M^1 &= \int (x'-x)W(x'-x,h)dx' = 0 \\ &\vdots \\ M^n &= \int (x'-x)^n W(x'-x,h)dx' = 0 \end{aligned} \tag{11}$$

3.2 Approximation of Derivatives

The approximation of the first derivative can be obtained by replacing the function $f(x)$ with its derivative

$$f'(x) = \int_{\Omega} f'(x')W(x'-x,h)dx' \tag{12}$$

Reformulating Equation (12) using integration by parts one obtains

$$f'(x) = \int_{S} f'(x')W(x'-x,h)\cdot\mathbf{n}\cdot ds - \int f'(x')W'(x'-x,h)dx' \tag{13}$$

n is the unit vector of the surface and the first integral is carried on the surface s of the computational domain. Using again the Taylor series expansion yields

$$f'(x) = \int_{S} f'(x')W(x'-x,h)\cdot\mathbf{n}\cdot ds + \sum_{k=0}^{n} A'^k f^k(x) + r_n\left(\frac{x'-x}{h}\right) \tag{14}$$

$$A'^k = \frac{(-1)^{k+1}h^k}{k!}\int\left(\frac{x'-x}{h}\right)^k W(x'-x,h)dx' \tag{15}$$

And finally the simplified coefficients can be calculated as

$$\begin{aligned} M'^0 &= \int W'(x'-x,h)dx' = 0 \\ M'^1 &= \int (x'-x)W'(x'-x,h)dx' = 1 \\ &\vdots \\ M'^n &= \int (x'-x)^n W'(x'-x,h)dx' = 0 \end{aligned} \tag{16}$$

$$W_s(x'-x,h) = 0 \tag{17}$$

Equation (17) defines the smoothing function value on the surface to be zero, which determines that the surface integration vanish for arbitrarily selected function $f(x)$. The first expression in Equation (16) is actually another representation of Equation (17) as can be observed from the following expression:

$$\begin{aligned} \int W'(x'-x,h)dx' &= \int_{s} 1.W(x'-x,h)\mathbf{n}\cdot ds - \int (1)'W(x'-x,h)dx' \\ &= \int_{0}^{s} W(x'-x,h)\mathbf{n}\cdot ds = 0 \end{aligned} \tag{18}$$

The approximation of the second derivative can be obtained similarly to the first derivative and the simplified coefficients can be derived as

$$
\begin{aligned}
M''^0 &= \int W''(x'-x,h)dx' = 0 \\
M''^1 &= \int (x'-x)W''(x'-x,h)dx' = 0 \\
M''^2 &= \int (x'-x)^2 W''(x'-x,h)dx' = 2 \\
&\vdots \\
M'^n &= \int (x'-x)^n W''(x'-x,h)dx' = 0
\end{aligned}
\tag{19}
$$

$$W_s(x'-x,h) = 0 \tag{20}$$

$$W_s'(x'-x,h) = 0 \tag{21}$$

The last two equations determine the surface term to vanish for an arbitrary selected function $f(x)$ and its first derivative. In other words

$$
\begin{aligned}
\int W''(x'-x,h)dx' &= \int 1.W'(x'-x,h)\mathbf{n}\cdot ds - \int (1)'W'(x'-x,h)dx' \\
&= \int_0^s W'(x'-x,h)\mathbf{n}\cdot ds = 0
\end{aligned}
\tag{22}
$$

So it can be seen that the derivatives in SPH Method are computed via analytic differentiation of smoothing function and there is no need for grid generation. It is also clear that the previously discussed requirements on the smoothing function are the representations of SPH approximations for a function and its derivatives. As an illustration the compact supportness property of the smoothing function is also a constituent of the surface Equation (21). Consequently approximating the smoothing function and its derivatives allows estimations of accelerations, strain rates, etc. in continuum equations and plays a very important role in forming the basis of SPH.

4 Modelling Composite Properties

For composite materials dynamic failure behaviour is very complex due to the different fibres and matrices available, the different fibre reinforcement types such as unidirectional (UD) fibres and fabrics, the possibility of both fibre dominated or matrix dominated failure modes, and the rate dependency of the polymer resin properties. Thus at present there are no universally accepted materials laws for crash and impact simulations with composites. It was considered that a homogeneous orthotropic elastic damaging material was an appropriate model for UD and fabric laminates, as this is applicable to brittle materials whose properties are degraded by micro cracking. Constitutive laws for orthotropic elastic materials with internal damage parameters are described in [9], and take the general form

$$\epsilon = S\sigma \tag{23}$$

where σ and ϵ are vectors of stress and strain and S is the elastic compliance matrix. In the plane stress case required here to characterise the properties of composite plies or shell elements with orthotropic symmetry axes (x_1, x_2), the in-plane stress and strain components are

$$\sigma = \begin{pmatrix} \sigma_{11} \\ \sigma_{22} \\ \sigma_{12} \end{pmatrix} \qquad \epsilon = \begin{pmatrix} \epsilon_{11} \\ \epsilon_{22} \\ 2\epsilon_{12} \end{pmatrix} \tag{24}$$

Using a strain equivalent damage mechanics formulation, the elastic compliance matrix S may then be written as

$$\mathbf{S} = \begin{pmatrix} \frac{1}{E_1(1-d_1)} & \frac{-\nu_{12}}{E_1} & 0 \\ \frac{-\nu_{12}}{E_1} & \frac{1}{E_2(1-d_2)} & 0 \\ 0 & 0 & \frac{1}{G_{12}(1-d_{12})} \end{pmatrix} \tag{25}$$

where ν_{12} is the principal Poisson's ratio, which for simplicity is assumed not to be degraded. This general plane stress form for an orthotropic elastic material with damage has 3 scalar damage parameters d_1, d_2, d_{12} and 4 'undamaged' elastic constants: the Young's moduli in the principal orthotropy directions E_1, E_2, the in-plane shear modulus G_{12}, and the principal Poisson's ratio ν_{12}. The damage parameters have values $0 \leq \mathrm{d}_i \leq 1$ and represent modulus reductions under different loading conditions due to progressive damage in the material. Thus for unidirectional (UD) plies with fibres in the x_1 direction, d_1 is associated with damage or failure in the fibres, d_2 transverse to the fibres, and d_{12} with in-plane shear failure. For fabric reinforcements then d_2 is associated with the second fibre direction. In order to proceed further evolution equations are required which relate the damage parameters to other state variables. Several different models have been proposed in the literature. In [9, 10] it is postulated that the damage parameters are functions of strain energy release rates in the material, and this method has been developed further, to include plasticity and rate effects in damage evolution. The damage mechanics formulation outlined above is currently being implemented and tested along with a new approach to delamination modelling described in [10]. The crash simulations of composite aircraft components reported in [9, 11] are based on alternative models in which it is supposed that damage evolution is dependent on the strain invariants, which can be determined from measured stress-strain curves. The elastic damaging materials law for fabric reinforcements may be modelled in PAM-CRASH as a 'degenerate bi-phase' model in

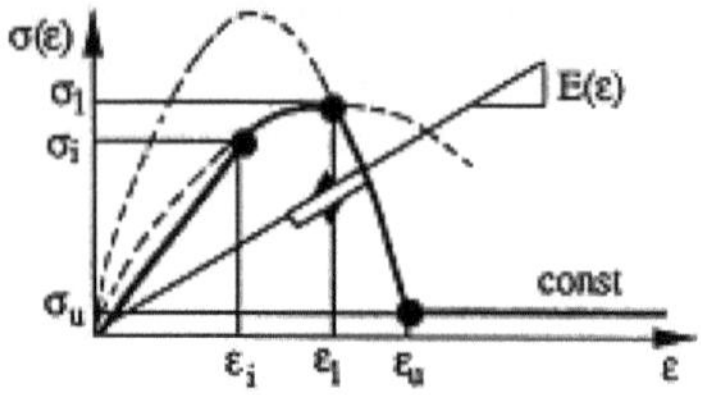

Fig. 5. Stress-strain diagram.

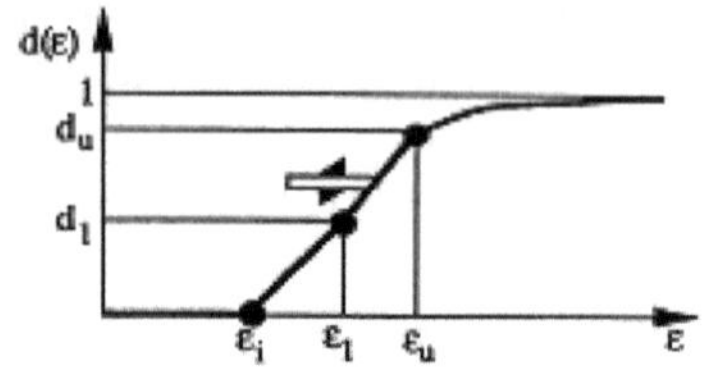

Fig. 6. Corresponding fracturing damage function.

which the UD fibre phase is omitted, and the 'matrix' phase is assumed to be orthotropic. If the simplifying assumption is made that $d_1 = d_2 = d_{12} = d$, the composite fabric ply has orthotropic stiffness properties, but a single 'isotropic' damage function d which degrades all the stiffness constants equally. The code does however allow different damage functions in tension and compression. This model has been found to be easy to apply and appropriate for quasi-isotropic laminates, which are commonly used in aircraft structures. An improved bi-phase model in [11] has been developed and implemented, which allows two independent fibre directions in plies and may be used when orthotropic damage effects are more significant.

It is necessary to determine parameters for the chosen composites model from measured stress-strain curves. Uniaxial stress-strain curves are assumed to have a damage evolution equation which is a bilinear function of strain, in which there are two damage constants d_1 and d_u to be determined (note d_1 here should not be confused with the damage parameter d_1). Typical uniaxial stress-strain curves have the general form shown in Figure 5 where ϵ_i is strain at the onset of initial damage, ϵ_1 is the strain at the peak failure stress, and ϵ_u is a limiting strain above which the stress is assumed to take a constant value σ_u. Measured test data for fabric composites are used to calibrate the materials model and to determine the damage parameters d_1 and d_u for the analysis. The parameter d_1 is related to the departure from linearity at the first 'knee' in the stress-strain curves, and is thus small in tension, whilst the parameter d_u determines the residual stress value σ_u shown in Figure 6. For the FE analysis it is not good practice to reduce the material stresses directly to zero at material fracture, as this may lead to numerical instabilities. Thus under tensile stresses typically $\mathrm{d}_u \approx 0.95$, indicating that the element is nearly fully damaged, whilst in compression $\mathrm{d}_u \approx 0.5$ to model the residual compression crushing stresses. In this way it is possible to attribute measured crush stresses to shell elements under axial compression load so that energy absorption in composite structures may be simulated with shell models. Materials rate dependence is not included in the modelling presented here.

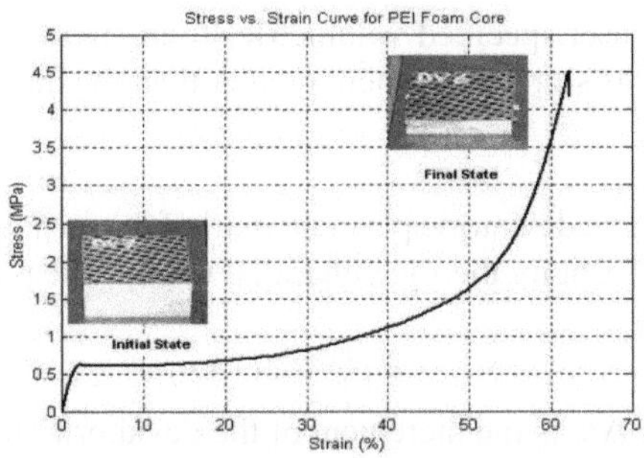

Fig. 7. Stress-strain plot of PEI foam core.

5 Modelling of Sandwich Core Materials

5.1 Modelling of PEI Foam Core

Foam core materials are widely used in automotive and aerospace industry such as energy absorbers and comfort enhancers. Because of high energy absorbing capability of foams, they became very important in vehicle crashworthiness. However they are susceptible to in-plane shear and core compression failure, buckling instability and face sheet-to-core debonding. The core must be capable of taking a compressive loading without premature failure. This helps to prevent the thin skins from wrinkling, and failing in a buckling mode. The DLR carried out compression tests on PEI foam core and a typical stress-strain curve is shown in Figure 7. To model this behaviour in PAM-CRASH a crushable foam solid was used. The elastic behaviour of this material is described by the shear modulus and the initial tangent modulus. The inelastic behaviour exhibits both, volumetric (bulk) plasticity and deviatoric (shear) plasticity. The coupling between both parts of the material response is established via a pressure (p) dependent von Mises (J_2 plasticity) yield surface.

$$\phi_s = J_2 - (a_0 + a_1 p + a_2 p^2) = 0 \tag{26}$$

where

$$J_2 = \frac{1}{2} S_{ij} S_{ij} = \frac{1}{3} {\sigma_Y}^2 \tag{27}$$

is the second invariant of the deviatoric stress tensor on the yield surface. a_0, a_1 and a_2 are user specified material parameters, which are linked to the user specified pressure cutoff for tensile fracture.

Volumetric Plasticity. The volumetric plasticity is computed first. It is governed by the volumetric yield function

$$\phi_v(p) = p - f_p(\epsilon_v) \tag{28}$$

where the $f_p(\epsilon_v)$ is a user specified volumetric strain-stress curve. The curve must be specified such that its slope is nowhere greater than the unloading bulk modulus, K_o.

Deviatoric Plasticity. The deviatoric plasticity part of the response is calculated next, using a conventional plasticity theory with radial return. The program evaluates trial deviatoric stresses as

$$S_{trial} = S_n + 2G\Delta e_n. \tag{29}$$

where $\Delta[e_{ij}]_n = [\dot{e_{ij}}]_n \Delta t_n$ is the increment of the deviatoric strain tensor. Using the trial stress, the program calculates $J_{2,trial}$ and compares it with the present yield level of $J_2 = a_0 + a_1 p + a_2 p^2$. If the trial value exceeds the yield value of J_2, a simple radial return is performed to obtain the stresses

$$S_{n+1} = (J_2 / J_{2,trial})^{1/2} S_{trial} \tag{30}$$

at the end of this time step. The total stress is calculated from

$$\sigma_{n+1} = S_{n+1} - p_{n+1}\delta. \tag{31}$$

Here, J_2 is the second invariant of the deviatoric stress tensor on the von Mises yield surface, a_0, a_1, a_2 are user specified material parameters, p and σ_Y stand for pressure and effective yield stress, respectively. In order to overcome the high mesh distortion which causes numerical problems with the timestep assignment in explicit codes, the FE meshes in highly distorted local damage zone has been replaced by interacting particles. The material response assigned for discrete SPH particles is an isotropic elastic-plastic-hydrodynamics solid material model in which the pressure-volume relation is modelled by an equation of state (EOS). This material model was originally developed for ballistic impact in metals and describes an isotropic elastic-plastic material at low pressure, whose properties are defined by the shear modulus and tangent modulus or effective plastic stresses and effective plastic strains. Additionally with EOS describes the 'hydrodynamic' pressure-volume behaviour at high pressures. In this case it is given as follows:

$$p = C_0 + C_1\mu + C_2\mu^2 + C_3\mu^3 \quad \text{where} \quad \mu = \frac{\rho}{\rho_0} - 1. \tag{32}$$

C_0, C_1, C_2, C_3 are material constants, μ is a dimensionless compressibility parameter defined in terms of the ratio current density, ρ, to initial density, ρ_0. The polynomial form is an established approximation of the observed EOS for many materials (for examples, see [12]), with the feature that it reduces to a dilatational elastic materials law with bulk modulus C_1 when C_0=C_2=C_3=0.

5.2 Modelling of Aramid Paper Honeycomb (Nomex) Core

Nomex honeycomb is made from Nomex paper, a form of paper based on aramid, rather than cellulose fibres. The initial paper honeycomb is usually dipped in a phen-

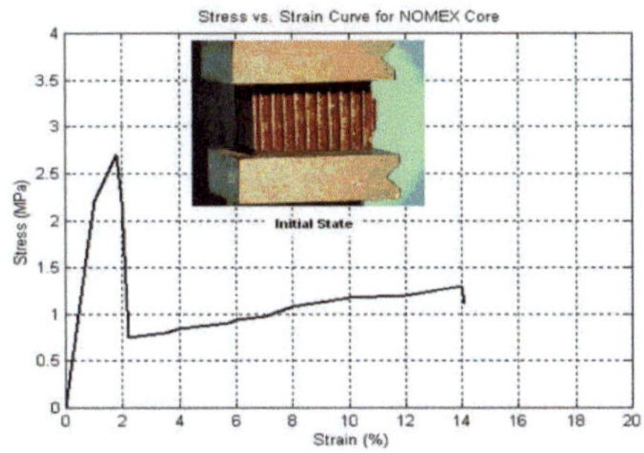

Fig. 8. Stress-strain plot of Nomex core.

olic resin to produce a honeycomb core with high compression strength and good fire resistance.

The honeycomb core properties depend on the size of the cells and the thickness and strength of the web material. They can give stiff and very light sandwich laminates with composite skins and high-performance resin systems such as epoxy provided that the necessary adhesion to the laminate skins can be achieved. Figure 8 shows the compression behaviour of the Nomex honeycomb as measured in tests at the DLR. In PAM-CRASH a nonlinear fibre bi-phase solid, was chosen to simulate the material response. Material Model 31 corresponds to highly anisotropic and nonlinear honeycomb cores which the through thickness compression properties are modelled as elastic-plastic.

6 Numerical Analysis using Combined FEM/SPH Method

The proposed finite element mesh based model, described in [13], can be used for prediction of failure modes in sandwich panels. However topological element connectivity in FEM can lead to numerical instabilities (for higher velocities) and further enhancement is needed for better quantitative correlations. Following this idea, carbon fibre/epoxy facings were modelled with standard layered shell elements, whilst SPH particles are used for the PEI core where extensive crushing and fracture by the rigid impactor occurs. Use of solid elements here leads to aforementioned difficulties with excessive distortion. Beside the advantages of its meshfree nature, as a result of the dynamic neighbouring search algorithm, the SPH Method is computationally expensive. One alternative numerical solution technique that is commonly used is coupling. Since SPH uses a Lagrangian formulation, a possible coupling between SPH and standard Lagrangian FEM is straightforward using contact interfaces. This means (for impact problems) a coupling between discrete smoothed particles for the parts where large deformation occurs and finite elements for the parts where small deformation takes place is possible. Such coupling would exploit the potential of

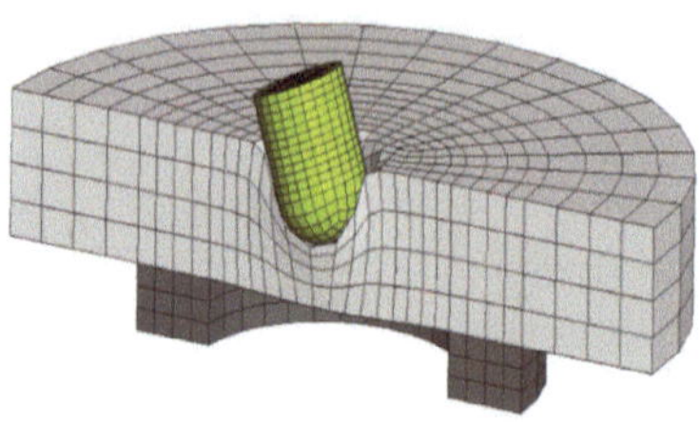

Fig. 9. Numerical impact (FEM) model.

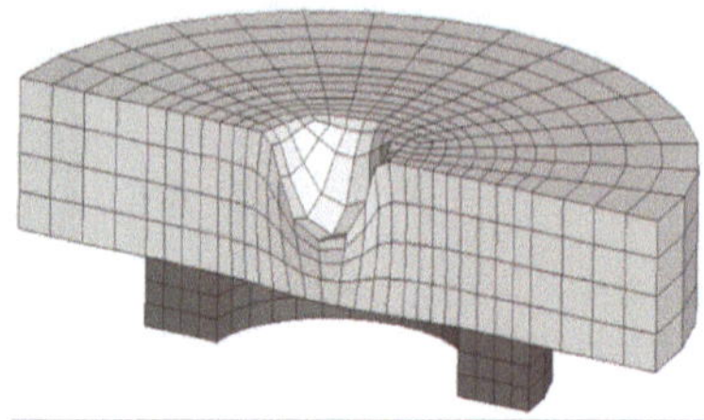

Fig. 10. Maximum penetration.

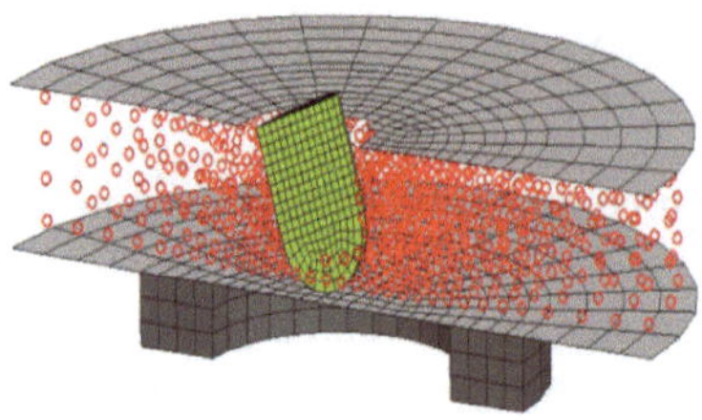

Fig. 11. Numerical impact (SPH) model.

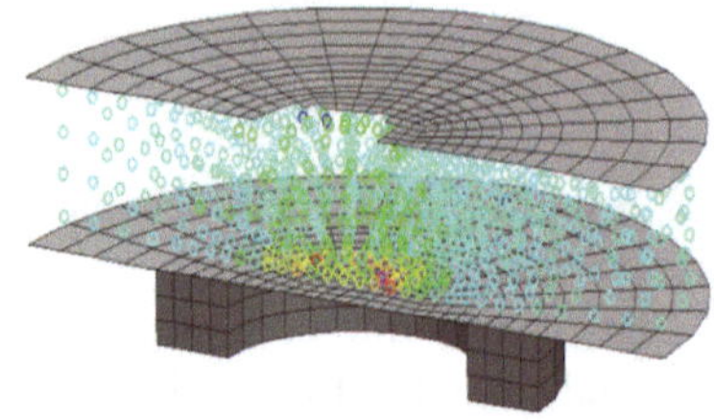

Fig. 12. Maximum penetration.

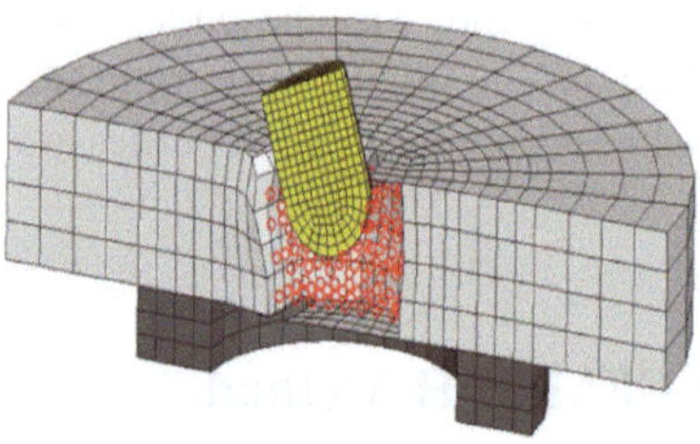

Fig. 13. Numerical impact (FEM/SPH) model.

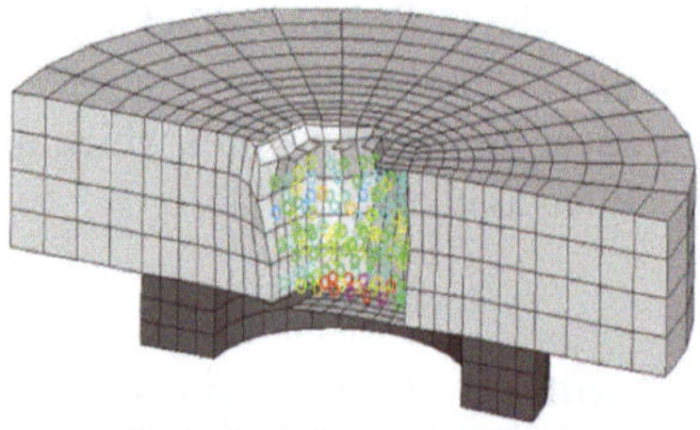

Fig. 14. Maximum penetration.

each method while avoiding their deficiencies. In this work, coupling was applied through a sliding interface condition. The mesh patterns or both discretizations can be combined with a tied kinematic constraint type contact that connects two contact interfaces defined on two meshed parts of a structure that are close to each other but whose respective discretization grids are not necessarily matching. Discrete SPH particles are generated with a transformation of finite element mesh into masspoints.

To make a qualitative comparison between FEM, SPH and combined FEM/SPH, impact simulations were carried out for a sandwich composite panel mounted on a ring load cell and impact at the centre. Normal impact from a rigid impactor was considered at nominal impact velocity of 60 m/s, which corresponds to the typical impact speed of runway debris on an aircraft structure during start and landing.

Here three different approaches have been used to model the impact damage on PEI core sandwich panels. In all three cases impactors first damage the upper skin and penetrate into the sandwich core. Figures 9 and 10 show the maximum penetration and post-impact core deformation with pure FEM. One can observe the large deformation on the PEI core. The core material under impact load shows classical elastic-plastic response. FEM approximations are continuum-based approximations and the topological connectivity in FEM is very critical. The modelling of core deformation with higher impact velocities can cause element distortion problems resulting in error termination. Therefore the core has been modelled using discrete SPH particles. Figures 11 and 12 show the maximum penetration and core deformation for the numerical impact simulation with SPH Method. In SPH the topological connectivity is not as critical as in FEM and the core deformation can be modelled without having element distortion problem. Since in SPH the number of discrete particles and the domain of influence of each particle are decisive for the CPU times, combined FEM/SPH Method, which combines the faster computation nature of FEM and accuracy of SPH, is proposed. Figures 13 and 14 show the maximum penetration and post-impact core deformation with proposed combined approach. Here one can observe the interaction between the FEM-based and SPH-based modelled impact damage zone. After damaging the upper skin the impactor penetrates into the sandwich core. In both SPH and combined FEM/SPH approaches the impactor compresses the particles in damage zone and its kinetic energy is absorbed by the particles in the damage zone. However one can observe that the penetration of the impactor in the pure SPH approach is larger that the pure FEM and combined FEM/SPH cases. In experimental investigation the maximum penetration value has been measured as 25 mm. The maximum penetration values are 20, 32 and 19.5 mm, respectively, in the example computed here.

Contact force history comparison between experimental analysis and FEM simulation in Figure 15 shows that the proposed FEM model provides reasonable accurate contact force history of HVI on sandwich panel. As one can observe from the experimental curve, the first peak load is about 7 kN and the second peak is about 4 kN. Depending on the energy absorption mechanism of the sandwich plate, the projectile loses its kinetic energy and this results in a lower second peak. Proposed FEM model estimates especially the first peak value accurately, which is very critical value for impact scenarios. However as Figures 9 and 10 show, FEM model produces very severe deformation on core which can lead to numerical errors, inaccurate results and numerical instabilities for higher velocities. In the SPH Method depending on the deformation mechanism neighbouring particles can separate and as Figure 16 illustrates core deformation is much more realistic than that of FEM.

Figure 16 shows that the proposed SPH model for the sandwich core approximates the experimental impact force pulse quite accurately. However the initiation of the peak force is later than the experimental one and the pulse is more oscillatory than the FEM results and the experimental pulse. Figure 17 shows the comparison

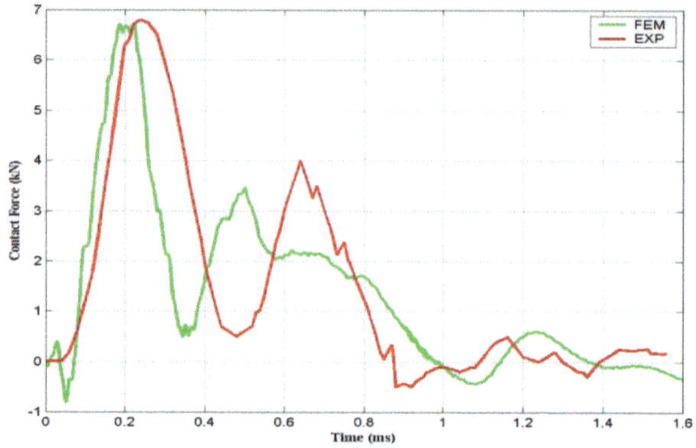

Fig. 15. Contact force history (FEM vs. experiment).

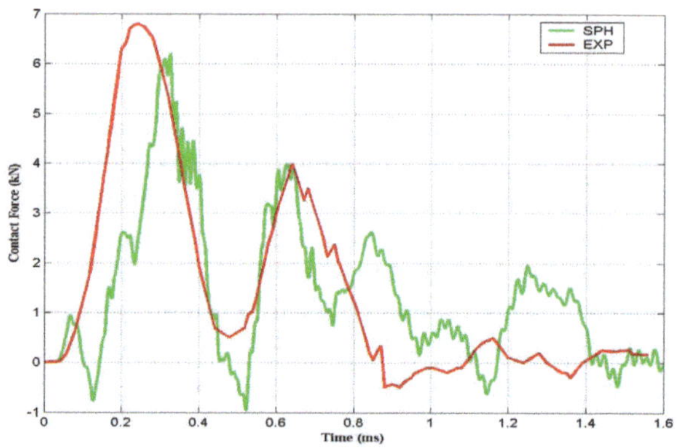

Fig. 16. Contact force history (SPH vs. experiment).

of the contact force histories between experimental and numerical simulation with combined FEM/SPH. Here once can observe that the numerical pulse is not as oscillatory as in the SPH case since the topological connectivity is still in use outside the damage zone. The combined FEM/SPH Method approximates the impact pulse more accurately. The pulse duration which is important as well as the peak force value is very similar to the experimental one and better than the approximations produced by the pure SPH Method.

Additional to the PEI core sandwich panels the accuracy of combined FEM/SPH Method has been tested in three different impact cases using a demonstrator hybrid cored sandwich panel. Three different impact scenarios in which glass impactors with 21 and 29 mm in diameter with the impact velocities of 65.8, 95.1 and 108 m/s have been chosen. Therefore the efficiency of the combined FEM/SPH Method can be tested in a range of impact velocities showing different damage modes.

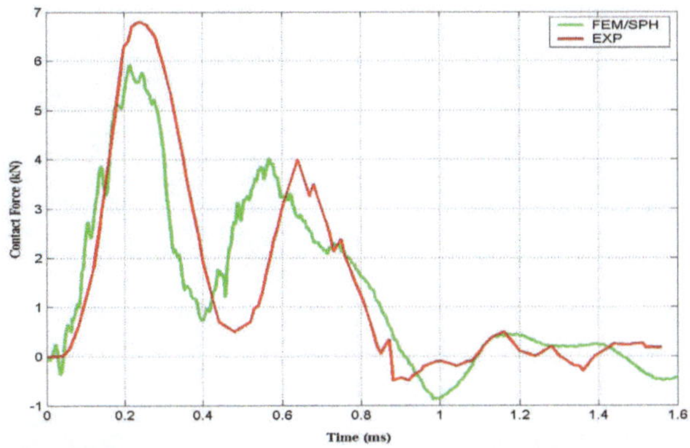

Fig. 17. Contact force history (FEM/SPH vs. experiment).

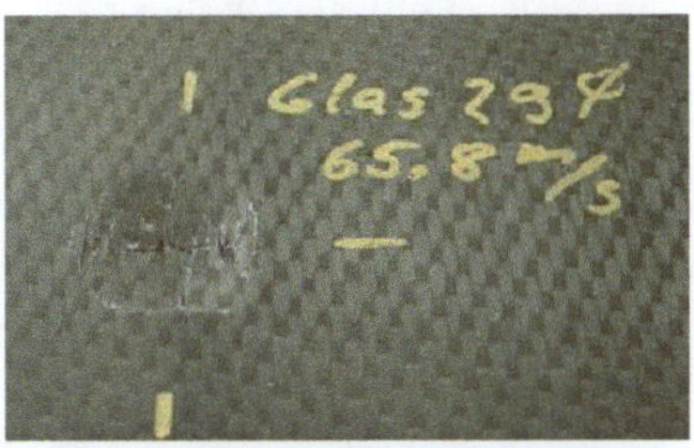

Fig. 18. Impact damage on upper skin.

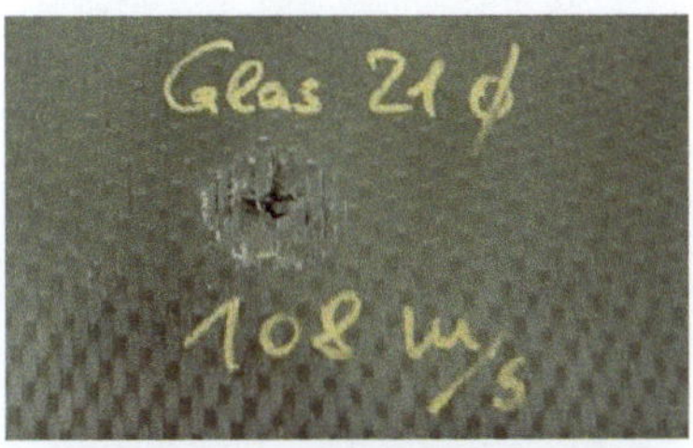

Fig. 19. Damage on upper skin and core.

For the first comparison an impact case with a glass ball impactor 29 mm in diameter with an impact velocity of 65.8 m/s has been chosen. The impactor damages the upper skin introduces an impact load on the core and bounces back (Figure 18). In the second scenario an impactor 21 mm in diameter has been impacted on the plate with an impact velocity of 108 m/s. The impactor damages the upper skin, penetrates into the core and since its remaining kinetic energy is absorbed by the core it resides inside the sandwich plate. The damage zone is smaller that the previous case since the diameter of the impactor is smaller (Figure 19). In the last case a glass ball impactor 29 mm in diameter and with an impact velocity of 95.1 m/s has been tested. The impactor damages the upper skin, penetrates into the core and since its kinetic energy was not absorbed completely by the core it continues its movement and damages the lower skin (Figures 20 and 21). Since the hybrid core consists of both Nomex and PEI foam, the Nomex core has been modelled using classical finite elements with discrete particles used for the PEI layer. For the Nomex layer a non-linear fiber bi-phase material model with element elimination technique has been used, as discussed above. This technique eliminates the elements which reach a prescribed element strain threshold value. Note that Nomex is orthotropic

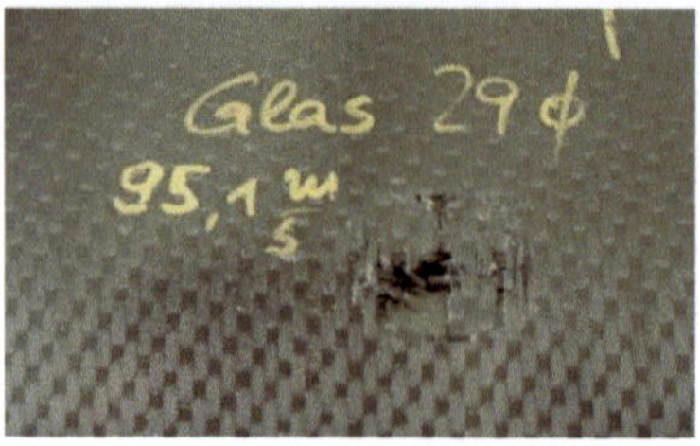

Fig. 20. Damage on upper skin and core.

Fig. 21. Backside damage.

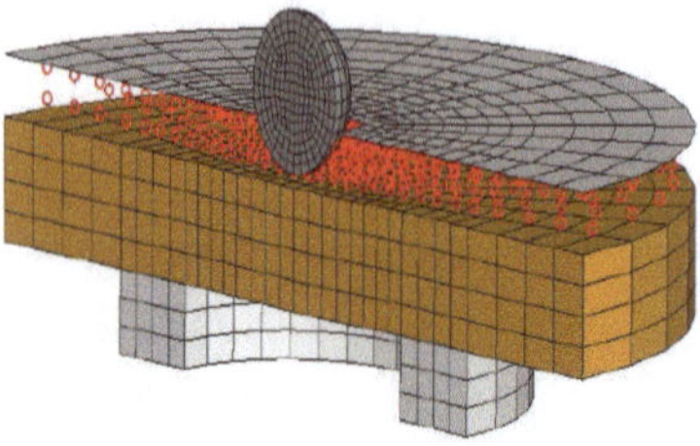

Fig. 22. Impact state, $t = 0.16$ ms.

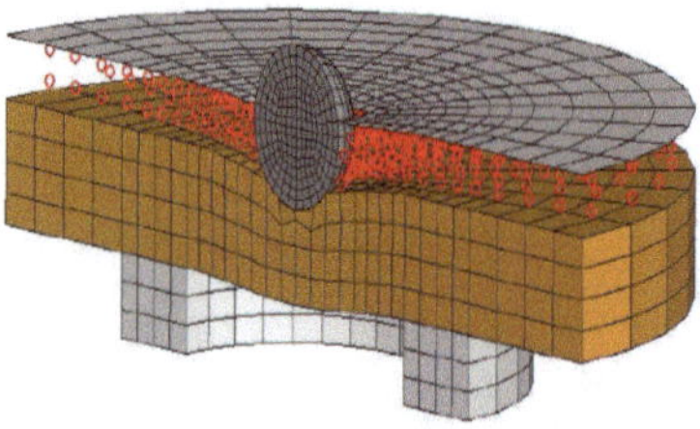

Fig. 23. Impact state, $t = 0.36$ ms.

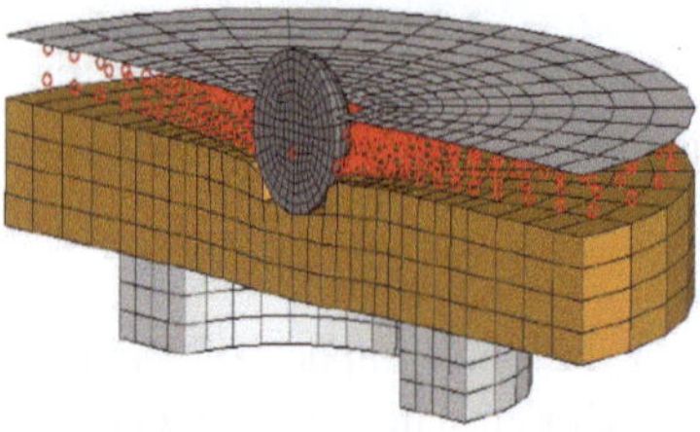

Fig. 24. Impact state, $t = 0.72$ ms.

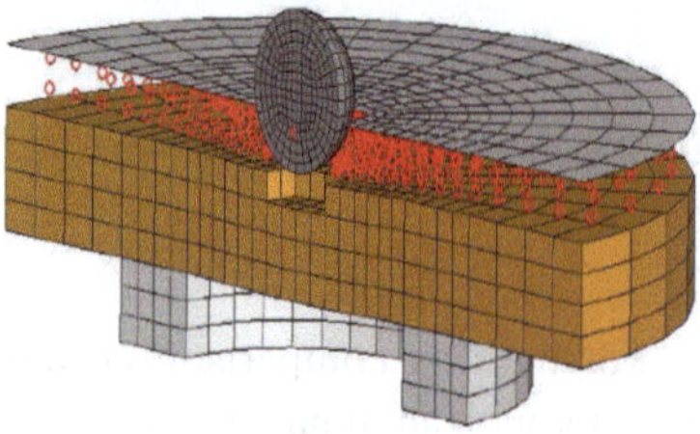

Fig. 25. Impact state, $t = 1.2$ ms.

in its mechanical properties and is thus less suitable for a particle model which is intrinsically isotropic in the SPH formulation used here.

Figures 22–25 show the states of impact for the impact case with impactor 29 mm in diameter and 65.8 m/s impact velocity. The impactor damages the upper skin and compressed the PEI core. Discrete SPH particles compressed together and impactor damages the Nomex core. After crushing the small area in Nomex core, the impactor bounces back. The damage area in the numerical investigation is larger than the experimental cases because of the element elimination technique used for modelling the damage of the composite skins. In real cases the damaged composite skin fragments and it continues to carry reduced loads.

Figures 26–29 show the states of numerical impact simulation in which an impactor 21 mm in diameter and with an impact velocity of 108 m/s. The impactor

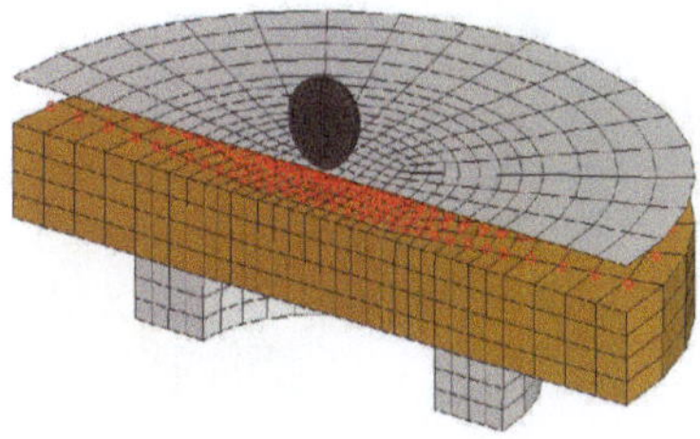

Fig. 26. Impact state, $t = 0.16$ ms.

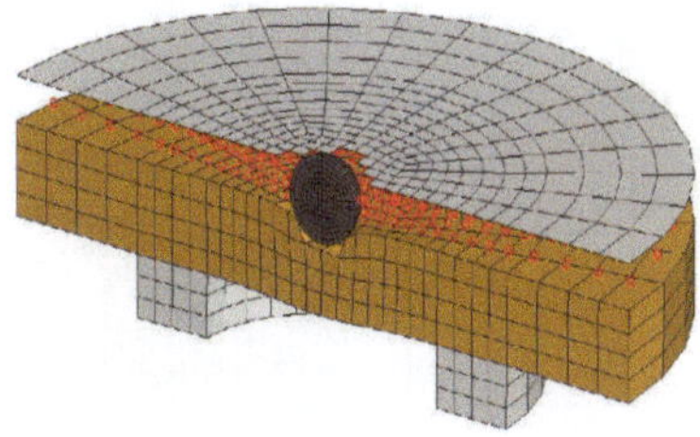

Fig. 27. Impact state, $t = 0.2$ ms.

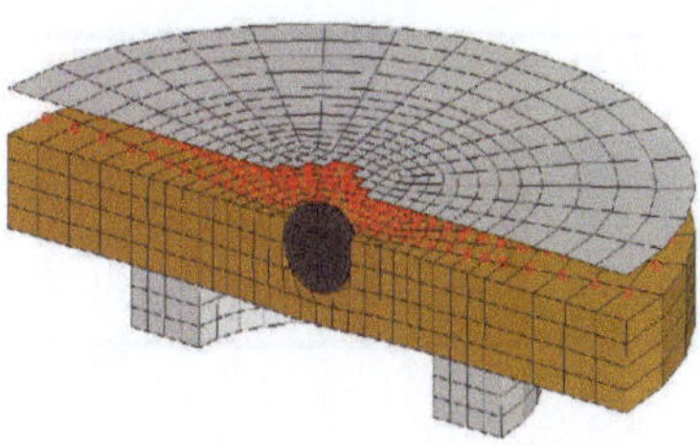

Fig. 28. Impact state, $t = 0.36$ ms.

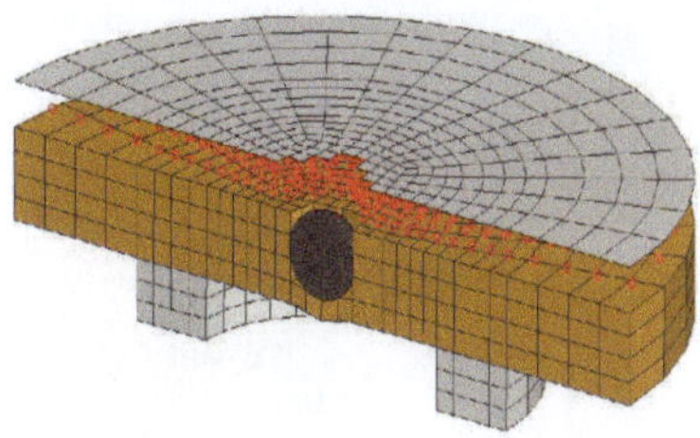

Fig. 29. Impact state, $t = 0.8$ ms.

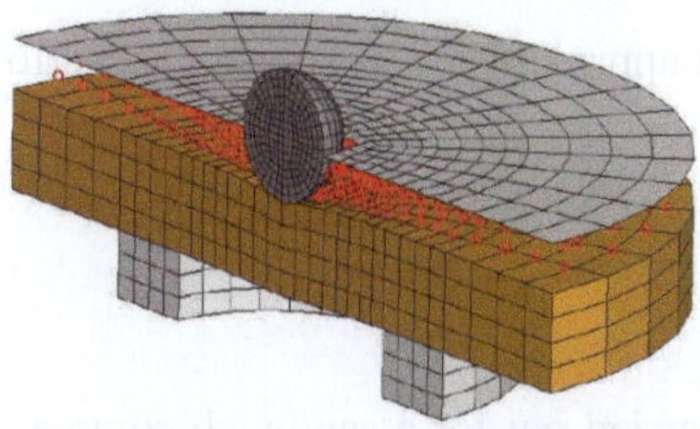

Fig. 30. Impact state, $t = 0.16$ ms.

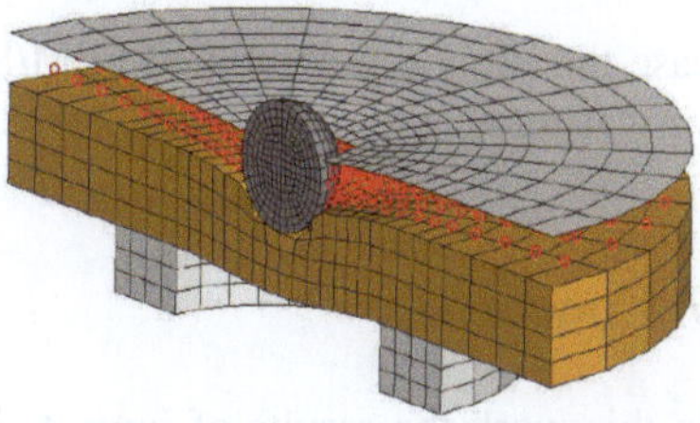

Fig. 31. Impact state, $t = 0.28$ ms.

damages the skin and firstly penetrates into the PEI core. The discrete particles are compressed together and since the kinetic energy of the impactor is not yet zero, it continues it movement. It penetrates into the Nomex core and causes a large deformation. During this deformation process the residual kinetic energy of the impactor absorbed by the Nomex core and impactor resides inside the sandwich plate. The same impact scenario observed in the experimental impact test.

Figures 30–34 show the last impact case in which an impactor 29 mm in diameter with 95.1 m/s impact velocity impact onto a hybrid core sandwich plate. The same impact stages can be also observed in this case. However the residual kinetic energy of the impactor is not absorbed by the Nomex core and the impactor continues its movement. The impactor damages the lower skin and rear face damage occurs. Figure 35 shows the side view of the damage zone. Here one can also observe the deformation in the damage zone modelled with discrete particles. Again in this impact

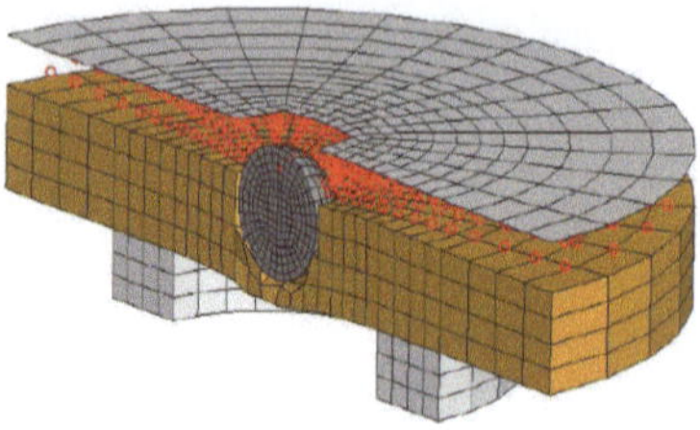

Fig. 32. Impact state, $t = 0.52$ ms.

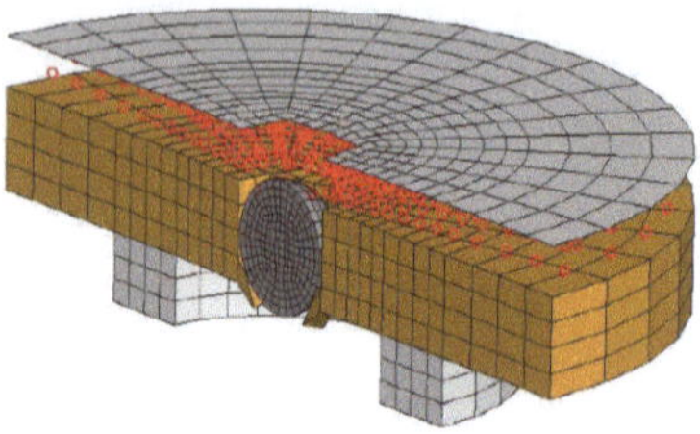

Fig. 33. Impact state, $t = 0.88$ ms.

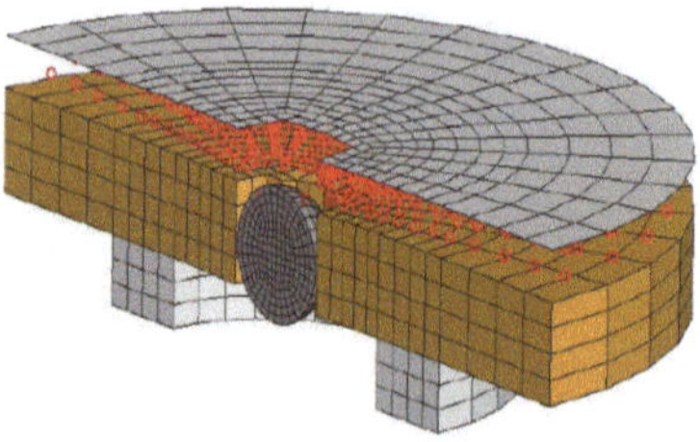

Fig. 34. Impact state, $t = 1$ ms.

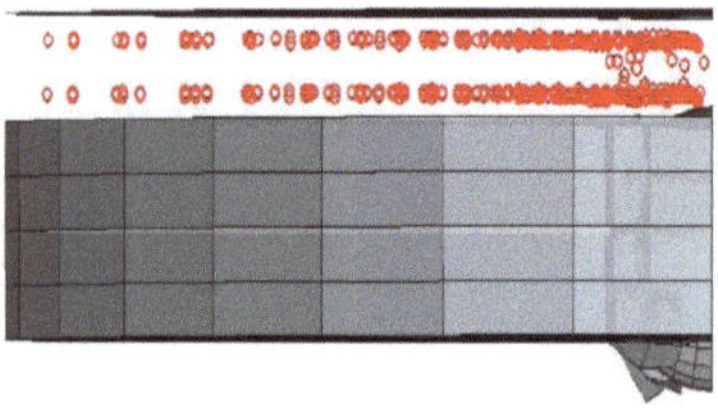

Fig. 35. Backside damage, side view.

case the proposed combined FEM/SPH Method approximates the impact scenario very accurate and realistically.

7 Conclusion

In this work the results of impact simulations carried out for a sandwich composite plate with carbon fibre fabric/epoxy face skins and PEI foam core using SPH and FEM/SPH Methods have been presented. The deformation pattern and the contact force histories obtained with pure SPH and combined FEM/SPH Methods have been compared with that of experimental and pure FEM-based numerical results. The both approaches give quite accurate results. However since the computational time depends on the domain modelled by discrete particles, a combined FEM/SPH Method has been shown to be a good alternative for the numerical simulations where large deformation and element distortion are critical. Later to show the efficiency of the combined FEM/SPH Method in different velocity ranges three impact scenarios in which a hybrid core consisting of PEI and Nomex sandwich plate impacted by spherical glass impactors have been modelled. The combined FEM/SPH Method reproduces the impact failure modes and observed damage on the sandwich plates to a good approximation.The paper demonstrated the capability of meshfree SPH Method and its combination with FEM to be an effective candidate to overcome the drawbacks of FEM for modelling fragmentation failures in impact. Future work is studying extensions of the method using a semi-adaptive coupling of SPH and FEM and applying this to study crush response in composite energy absorbing structural elements.

References

1. Arias A., Zaera R., Lopez-Puente J. and Navarro C., Numerical modelling of the impact behaviour of new particulate-loaded composite materials. *Composite Structures*, 61:151–159, 2003.
2. Shim V. P. W. and Yap K. Y., Modelling impact deformation of foam-plate sandwich systems. *Int. J. Impact Engng.*, 19:615–636, 1997.
3. Wada A., Kawasaki T., Minoda Y., Kataoka A., Tashiro S. and Fukuda H., A method to measure shearing modulus of the foamed core for sandwich plates. *Composite Structures*, 60:385–390, 2003.
4. Lopatnikov S. L., Gama B. A., Haque M. J., Krauthauser C., Gillespie J. W., Güden M. and Hall I. W., Dynamics of metal foam deformation during Taylor cylinder-Hopkinson bar impact experiment. *Composite Structures*, 61:61–71, 2003.
5. Abrate S., *Impact on Composite Structures*. Cambridge University Press, 1998.
6. Lucy L. B., A numerical approach to the testing of the fission hypothesis. *The Astronomical Journal*, 82:1013–1024, 1997.
7. Johnson A. F. and Holzapfel M., Models for ice and stone projectiles, CRAHVI D1.4.6. DLR Report IB 435-200/14, 2002.
8. Johnson A. F. and Holzapfel M., Modelling soft body impact on composite structures. In: *14th US National Congress of Theoretical and Applied Mechanics, Special Symposium: Impact on Composites 2002*, Blacksburg VA, 23–28 June 2002.
9. Ladeveze P. and Le Dantec E., Damage modelling of the elementary ply for laminated composites. *Composites Science and Technology*, 43:257–267, 1992.
10. Johnson A. F. and Pickett A. K., Impact and crash modelling of composite structures: A challenge for damage mechanics. In: *European Conference on Computational Mechanics, ECMM'99*, Munich, 1999.
11. Johnson A. F. and Kohlgrüber D., Design and performance of energy absorbing subfloor structures in aerospace applications. In: *IMechE Seminar S672*, London, 9 May 2000.
12. Hiermaier S. and Thoma K., Computational simulation of high velocity impact situations using smoothed particle hydrodynamics. In: *9th DYMAT Technical Conference on Materials and Structural Modelling in Collosion Research*, TU Munich, Germany, 1995.
13. Aktay L., Johnson A. F. and Holzapfel M., Prediction of impact damage on sandwich composite panels. *Comp. Mat. Sci.*, 32:252–260, 2005.

On the Construction of Mass Conservative and Meshless Adaptive Particle Advection Methods

Armin Iske

Department of Mathematics, University of Hamburg, D-20146 Hamburg, Germany; E-mail: iske@math.uni-hamburg.de

Abstract. This contribution concerns the construction of meshless Lagrangian particle methods for the numerical simulation of multiscale phenomena in linear transport problems, where mass conservative discretization methods are essentially required. The proposed discretization scheme works with a finite set of unstructured nodes, each corresponding to one flow particle. In this method, corresponding particle average values are maintained during the simulation. The discrete nodes are subject to adaptive modifications, leading to semi-Lagrangian particle simulations, whose adaption rules rely on the insertion (refinement) and removal (coarsening) of the nodes at each time step. The resulting meshless particle method is mass conservative by construction. The required algebraic rules for the downstream particle advection and the local redistribution of the particle masses are developed. Moreover, the implementation of boundary conditions is addressed. The efficacy of the proposed conservative and meshless adaptive particle method is finally shown by using one numerical simulation concerning the slotted cylinder, a popular standard test case for passive advection.

Key words: Hyperbolic conservation problem, finite volume particle method, semi-Lagrangian particle advection, mass conservative particle advection, mass conservative adaption rules, slotted cylinder, passive advection, implementation of boundary conditions.

1 Introduction

Many physical phenomena in time-dependent evolution processes are modelled by hyperbolic conservation laws, where relevant applications, e.g. from fluid flow simulation, essentially require *conservative* discretization schemes. The *finite volume method* (FVM) is a standard conservative method to construct numerical approximations for solutions of hyperbolic conservation problems. The FVM relies on a suitable partitioning of the computational domain into small cells, the control volumes, where each control volume of the partitioning bears a cell average value. To enforce

V.M.A. Leitão et al. (eds), Advances in Meshfree Techniques, 169–186.

mass conservation, fluxes are described at the cell interfaces, resulting in a explicit one-step Eulerian discretization scheme. For a comprehensive treatment of finite volume methods, see [10].

Therefore, the classical FVM is *mesh-based* by the domain partitioning, although its construction relies merely on (mesh-independent) geometrical coefficients: cell volumes, cell surfaces, and cell normal vectors [8]. This observation has recently motivated the construction of a meshless finite volume particle method (FVPM) [6].

Another approach for solving linear hyperbolic equations is given by the *method of characteristics*. This Lagrangian method relies on downstream advection of particles along their streamlines, see [4, 11]. In contrast to Eulerian schemes, Lagrangian particle methods are better suited to construct *meshless* methods, leading to highly flexible discretizations, which are particularly useful for problems with complicated domain geometries, moving boundaries, and large-scale deformations of the solution.

On the down side, Lagrangian methods are usually not conservative, although considerable effort has been made recently in order to construct conservative Lagrangian advection methods, see [9, 12–14]. In our previous paper [7], a conservative and adaptive semi-Lagrangian method for passive advection has finally been proposed. Yet it seems to be difficult to combine the basic concepts of conservative schemes with those of meshless methods.

This contribution combines three desirable requirements for the construction of effective Lagrangian particle methods in linear hyperbolic conservation problems:

- mass conservation;
- adaptivity;
- mesh-independence.

To this end, algebraic rules for mass conservation are developed. The algebraic rules, given by local mass balance equations, are concerning the mass transfer for the advection step, the particle adaption (i.e., adaptive insertion and removal of particles), and the implementation of boundary conditions. The proposed method can be viewed as a generalization of our previous paper [7], which is conservative and adaptive but not meshless.

The outline of this work is as follows. In the following Section 2, the governing equations for passive advection are briefly reviewed, before some basic ingredients of finite volume particle methods are introduced in Section 3. Then, in Section 4, the mass conservative and meshless advection of *particle averages* is discussed. Section 5 is concerning the adaption of the particles. The implementation of boundary conditions is then explained in Section 6. The efficacy of the proposed conservative and meshless adaptive particle advection is finally illustrated in Section 7 by one numerical simulation concerning the slotted cylinder, a popular standard test case for passive advection.

2 Passive Advection and Mass Conservation

The numerical simulation of linear transport processes, *passive advection*, is governed by scalar time-dependent *hyperbolic conservation laws* of the form

$$\frac{\partial u}{\partial t} + \nabla \cdot (\mathbf{a}u) = 0, \tag{1}$$

where for a compact time interval $I = [0, T] \subset \mathbb{R}$, $T > 0$, and the computational domain $\Omega = \mathbb{R}^2$, the *velocity field*

$$\mathbf{a} = \mathbf{a}(t, \mathbf{x}), \qquad t \in I, \ \mathbf{x} \in \Omega,$$

is assumed to be given. The scalar solution $u : I \times \Omega \to \mathbb{R}$ of (1) is the density (or concentration) of a physical quantity, which is subject to a conservation law. We consider solving (1) numerically under given initial condition

$$u(0, \mathbf{x}) = u_0(\mathbf{x}), \qquad \text{for } \mathbf{x} \in \Omega. \tag{2}$$

Later in Section 6, boundary conditions are added to the Cauchy problem (1), (2), where the computational domain $\Omega \subset \mathbb{R}^2$ is assumed to be bounded.

In many relevant applications, *mass conservation*,

$$\frac{d}{dt} \int_{\Omega} u(t, \mathbf{x})\, d\mathbf{x} = 0, \tag{3}$$

is an important requirement. The aim of this work is to construct mass conservative particle advection methods satisfying (3).

3 Finite Volume Particle Methods

This section discusses selected features of finite volume particle methods, where we combine basic ingredients from the classical Eulerian FVM with recent concepts from particle methods. To this end, let us first briefly recall some basic details of the (mesh-based) FVM, before we turn to a mesh-independent (re)formulation to design meshless particle advection schemes.

3.1 Mesh-Based Formulation

According to the classical FVM discretization, the computational domain Ω is first decomposed into a partition $\mathcal{V} = \{V\}_{V \in \mathcal{V}} \subset \Omega$ of finitely many pairwise disjoint cells V, *control volumes*, so that

$$\Omega = \bigcup_{V \in \mathcal{V}} V.$$

In order to establish mass conservation, corresponding *cell average values*

$$\bar{u}_V(t) = \frac{1}{|V|}\int_V u(t,\mathbf{x})\,d\mathbf{x}, \qquad \text{for } V \in \mathcal{V},$$

of the numerical solution $u \equiv u(t,\mathbf{x})$ are maintained during the simulation. Here, for any $V \in \mathcal{V}$, $|V|$ denotes the volume of the cell $V \subset \Omega$, and so

$$m_V(t) = |V| \cdot \bar{u}_V(t), \qquad \text{for } V \in \mathcal{V},$$

is the *total mass* contained in the cell V at time t. Therefore,

$$M(t) = \int_\Omega u(t,\mathbf{x})\,d\mathbf{x} = \sum_{V\in\mathcal{V}}\int_V u(t,\mathbf{x})\,d\mathbf{x} = \sum_{V\in\mathcal{V}} m_V(t)$$

yields the total mass over the domain Ω at time t.

Note that a finite volume discretization of the above form cannot be meshless. Indeed, as soon as the computational domain Ω is partitioned into finite control volumes, this essentially requires a mesh.

3.2 Meshless Formulation

Now let us work with a more general (generic) formulation of the FVM to obtain a particle advection method, which does not necessarily rely on a mesh, but which is mass conservative by construction. This needs a few conceptual preparations and some reformulations of the FVM, to be explained in the remainder of this subsection.

Meshless finite volume particle methods work with a finite set $\mathbf{V} = \{\mathbf{v}\}_{\mathbf{v}\in\mathbf{V}} \subset \Omega$ of nodes, each of which corresponds at a time t to one flow particle. The basic idea of the meshless method, proposed in this work, is to assign, to any node $\mathbf{v} \in \mathbf{V}$, a *volume* $|\mathbf{v}|$ and a *particle average value*

$$\bar{u}_\mathbf{v}(t) \approx u(t,\mathbf{v}), \qquad \text{for } \mathbf{v} \in \mathbf{V},$$

so that

$$m_\mathbf{v}(t) = |\mathbf{v}| \cdot \bar{u}_\mathbf{v}(t), \qquad \text{for } \mathbf{v} \in \mathbf{V},$$

is the total mass which is, at time t, attached to the node $\mathbf{v}$.

The choice of this particular construction is motivated by the mean value theorem, which states that for an "influence area" $V \in \Omega$ surrounding node $\mathbf{v} \in \mathbf{V}$ we have the identity

$$\bar{u}_\mathbf{v}(t) \approx \bar{u}_V(t) = u(t,\xi), \qquad \text{for some } \xi \in V,$$

which establishes a one-to-one correspondence between the control volumes $\mathcal{V} = \{V\}_{V\in\mathcal{V}}$ of the previous subsection and the node set $\mathbf{V} = \{\mathbf{v}\}_{\mathbf{v}\in\mathbf{V}}$ of this subsection.

This way, the meshless concept replaces the *cell average values* $\bar{u}_V(t)$ and the *cell volumes* $|V|$ of the mesh-based FVM by *particle average values* $\bar{u}_{\mathbf{v}}(t)$ and *particle volumes* $|\mathbf{v}|$. In this modified concept, the volume $|\mathbf{v}|$ is for any $\mathbf{v} \in \mathbf{V}$ regarded as a geometric quantity that should depend on the local geometry of the nodes $\mathbf{V}$ around $\mathbf{v}$, rather than on any connectivities between the nodes. Moreover, $|\mathbf{v}|$ should reflect the density of the nodes $\mathbf{V}$ in a local neighbourhood of $\mathbf{v}$, and the sum of the node volumes is required to make up the total volume $|\Omega|$ of the domain Ω, i.e.,

$$|\Omega| = \sum_{\mathbf{v}\in\mathbf{V}} |\mathbf{v}|.$$

Therefore, for any node $\mathbf{v} \in \mathbf{V}$, its corresponding volume $|\mathbf{v}|$ can be viewed as the *weight* of $\mathbf{v}$ in $\mathbf{V}$.

Initially, for a suitable node set $\mathbf{V} = \{\mathbf{v}\}_{\mathbf{v}\in\mathbf{V}} \subset \Omega$, the particle average values are given by the initial condition (2), so that we let

$$\bar{u}_{\mathbf{v}}(0) = u_0(\mathbf{v}), \qquad \text{for all } \mathbf{v} \in \mathbf{V}. \tag{4}$$

This in turn yields the total mass

$$M \equiv M(0) = \sum_{\mathbf{v}\in\mathbf{V}} |\mathbf{v}| \cdot \bar{u}_{\mathbf{v}}(0) = \sum_{\mathbf{v}\in\mathbf{V}} m_{\mathbf{v}}(0)$$

at initial time $t = 0$.

The aim of the subsequent construction is to solve the Cauchy problem (1), (2) numerically, such that the total mass $M \equiv M(0)$ is constant during the simulation. To be more precise, at any time step $t \to t + \tau$, with given time step size $\tau > 0$, we wish to establish algebraic rules for the required identity $M(t) = M(t + \tau)$, i.e.,

$$M(t) = \sum_{\mathbf{v}\in\mathbf{V}} m_{\mathbf{v}}(t) = \sum_{\mathbf{v}\in\mathbf{V}} m_{\mathbf{v}}(t+\tau) = M(t+\tau), \qquad \text{for } t, t+\tau \in I, \tag{5}$$

so that $M \equiv M(t)$ holds for all $t \in I$ and all $\mathbf{V} \equiv \mathbf{V}(t)$. Later in Section 6, where the implementation of boundary conditions is addressed, we replace the algebraic conditions accordingly, by considering incoming and outgoing flow across the boundary $\partial\Omega$ of a bounded domain $\Omega \subset \mathbb{R}^2$.

4 Mass Conservation by Construction

This section explains the key ingredients for mass conservation in meshless particle advection methods. To this end, we wish to compute, at any time step $t \to t + \tau$ and for a current node set $\mathbf{V} \equiv \mathbf{V}(t)$, updated mass values $\{m_{\mathbf{v}}(t + \tau) : \mathbf{v} \in \mathbf{V}\}$ from the current masses $\{m_{\mathbf{v}}(t) : \mathbf{v} \in \mathbf{V}\}$, which are assumed to be known, where the initial masses $\{m_{\mathbf{v}}(0) : \mathbf{v} \in \mathbf{V}\}$ are given by (4). According to our mass conservative construction (to be explained below), the updates on the masses are done, such that the identity (5) holds.

To this end, we follow our previous paper [7], where a mesh-based conservative advection method is proposed. The Lagrangian approach in [7] works with downstream mass advection from a finite set $\mathcal{U}$ of *upstream cells* onto a current set $\mathcal{V}$ of (downstream) control cells. To be more precise, any *upstream cell* $U \in \mathcal{U}$ contains, at time t, those particles, which by traversing along their *streamlines* arrive at time $t + \tau$ in the corresponding control cell $V \in \mathcal{V}$. By this duality relation, there is a one-to-one correspondence between the upstream cells and the control cells.

4.1 Upstream Nodes

In what follows, we wish to establish a similar duality relation between current nodes in $\mathbf{V}$ and *upstream nodes*, to be collected in a point set $\mathbf{U}$. But this requires a few comments concerning the *streamlines* of the particle flow in the context of Lagrangian advection schemes. For any node $\mathbf{v} \in \mathbf{V}$, its corresponding *upstream node* is given by $\mathbf{u} = \mathbf{x}(t)$, where $\mathbf{x}$ denotes the unique solution of the *ordinary differential equation* (ODE)

$$\dot{\mathbf{x}} = \frac{d\mathbf{x}}{dt} = \mathbf{a}(t, \mathbf{x}), \tag{6}$$

with initial condition $\mathbf{x}(t + \tau) = \mathbf{v}$.

Therefore, the upstream node $\mathbf{u} \equiv \mathbf{u}(\mathbf{v})$ of $\mathbf{v}$ can be viewed as the unique location of a flow particle at time t, which by traversing along its trajectory arrives at node $\mathbf{v}$ at time $t+\tau$, see Figure 1. In case of passive advection, governed by the linear transport equation (1), the shape of the particles' flow trajectories, termed *streamlines*, are entirely and uniquely determined by the given velocity field $\mathbf{a} = \mathbf{a}(t, \mathbf{x})$. Moreover, the solution u of (1) is constant along these streamlines, and so the streamlines are the characteristic curves of the hyperbolic equation (1).

Adopting some standard notation from dynamical systems [3], we express the upstream node $\mathbf{u}$ of $\mathbf{v}$ as

$$\mathbf{u} = \Phi^{t,t+\tau}\mathbf{v}, \tag{7}$$

where $\Phi^{t,t+\tau} : \Omega \to \Omega$ denotes the *continuous evolution* of the (backward) flow of (6). An equivalent formulation for (7) is given by $\mathbf{v} = \Phi^{t+\tau,t}\mathbf{u}$, since $\Phi^{t+\tau,t}$ is the inverse of $\Phi^{t,t+\tau}$.

Likewise, for the sake of notational simplicity, it is convenient to express any approximation to upstream node $\mathbf{u}$ as

$$\Psi^{t,t+\tau}\mathbf{v} \approx \mathbf{u}, \qquad \text{for } \mathbf{v} \in \mathbf{V},$$

where $\Psi^{t,t+\tau} : \Omega \to \Omega$ is the *discrete evolution* of the (backward) flow, and where the operator $\Psi^{t,t+\tau}$ is given by any suitable numerical method for solving the above ODE (6). Note that the implementation of $\Psi^{t,t+\tau}$, and thus the numerical approximation to $\mathbf{u}$, does *not* need a mesh.

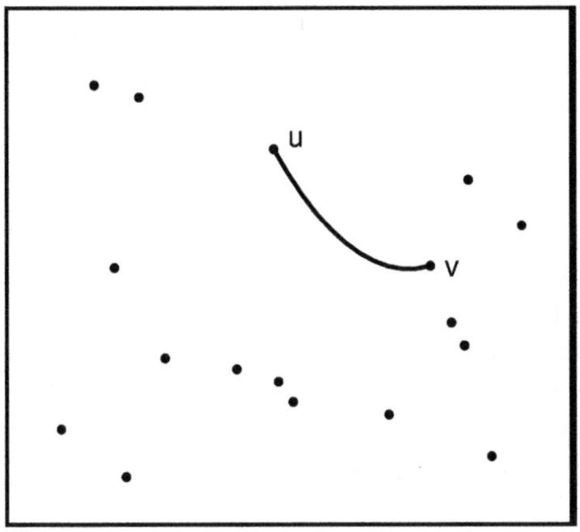

Fig. 1. Upstream node **u** of $\mathbf{v} \in \mathbf{V}$.

For details concerning the implementation of $\Psi^{t,t+\tau}$ we refer to our previous paper [7], where the duality relation for the pair $(\mathbf{U}, \mathbf{V})$ is expressed as

$$\mathbf{U} = \Psi^{t,t+\tau}\mathbf{V}, \qquad \text{for } \mathbf{V} \in \mathcal{V}. \tag{8}$$

Moreover, in the mesh-based setting of [7], global mass conservation is obtained by establishing *local mass conservation* through the balance equation

$$\int_{\mathbf{U}} u_h(t, \mathbf{x})\, dx = \int_{\mathbf{V}} u_h(t+\tau, \mathbf{x})\, d\mathbf{x}, \qquad \text{for every } \mathbf{V} \in \mathcal{V},$$

where u_h denotes a piecewise linear approximation to the solution u of (1),(2).

In the meshless particle advection scheme proposed in this work, a corresponding identity for local mass conservation is accommodated by the mass balance equation

$$m_{\mathbf{u}}(t) = m_{\mathbf{v}}(t+\tau), \qquad \text{for every } \mathbf{v} \in \mathbf{V}, \tag{9}$$

where $m_{\mathbf{u}}(t)$ denotes the portion of mass which is, at time step $t \to t+\tau$, advected from **u** onto **v**. Note that (9) is equivalent to

$$\bar{u}(t+\tau, \mathbf{v}) = \frac{|\mathbf{u}|}{|\mathbf{v}|}\,\bar{u}(t, \mathbf{u}), \qquad \text{for every } \mathbf{v} \in \mathbf{V},$$

where $|\mathbf{u}|$ is the volume of **u** w.r.t. **U**. But neither the mass $m_{\mathbf{u}}(t)$ nor the particle average value $\bar{u}(t, \mathbf{u})$ is known at time t, unless **u** is a current node in **V**.

4.2 Conservative Mass Transfer

For the sake of mass conservation, we determine the masses $\{m_{\mathbf{u}}(t) : \mathbf{u} \in \mathbf{U}\}$ from the current masses $\{m_{\mathbf{v}}(t) : \mathbf{v} \in \mathbf{V}\}$, such that the *local* mass balance equations (9) are satisfied. This then yields the desired *global* mass conservation (5) by construction.

To establish suitable algebraic conditions for local mass conservation (9), we transfer for any node $\mathbf{v} \in \mathbf{V}$ its entire mass $m_{\mathbf{v}}(t)$ to upstream nodes $\{\mathbf{u}\}_{\mathbf{u}\in\mathbf{U}}$ lying in the local neighbourhood of $\mathbf{v}$. This is accomplished as follows. Let $\gamma_{\mathbf{v}\to\mathbf{u}} \in [0, 1]$ denote the fraction of mass which is being transferred from $\mathbf{v} \in \mathbf{V}$ to any $\mathbf{u} \in \mathbf{U}$, so that

$$m_{\mathbf{v}\to\mathbf{u}}(t) = \gamma_{\mathbf{v}\to\mathbf{u}} \cdot m_{\mathbf{v}}(t), \qquad \text{for } \mathbf{v} \in \mathbf{V}, \mathbf{u} \in \mathbf{U},$$

is the portion of mass being transferred from $\mathbf{v}$ to $\mathbf{u}$. Hence, if we let $\gamma_{\mathbf{v}\to\mathbf{u}} = 0$, then no mass is transferred from $\mathbf{v}$ to $\mathbf{u}$. This is the case for upstream nodes $\mathbf{u} \in \mathbf{U}$ which are *not* lying in the local neighbourhood of $\mathbf{v}$. In contrast, we select a positive value $\gamma_{\mathbf{v}\to\mathbf{u}} > 0$, whenever $\mathbf{u} \in \mathbf{U}$ lies in the local neighbourhood of $\mathbf{v}$.

Therefore, for any $\mathbf{u} \in \mathbf{U}$, the total mass $m_{\mathbf{u}}$, which is transferred from the nodes in $\mathbf{V}$ to $\mathbf{u}$, is given by the weighted sum

$$m_{\mathbf{u}}(t) = \sum_{\mathbf{v}\in\mathbf{V}} m_{\mathbf{v}\to\mathbf{u}}(t) = \sum_{\mathbf{v}\in\mathbf{V}} \gamma_{\mathbf{v}\to\mathbf{u}} \cdot m_{\mathbf{v}}(t). \tag{10}$$

Now for the sake of mass conservation, we essentially require the property

$$\sum_{\mathbf{u}\in\mathbf{U}} \gamma_{\mathbf{v}\to\mathbf{u}} = 1, \qquad \text{for all } \mathbf{v} \in \mathbf{V}, \tag{11}$$

i.e., the non-negative multipliers $\gamma_{\mathbf{v}\to\mathbf{u}}$ are required to form a *partition of unity*.

Note that the condition in (11) states that the total mass $m_{\mathbf{v}}$ of any $\mathbf{v} \in \mathbf{V}$ is completely transferred to the nodes in $\mathbf{U}$. But this immediately establishes the desired identity (5) by the following simple calculations, where we use (9) (10), and (11).

$$\begin{aligned}
M(t+\tau) &= \sum_{\mathbf{v}\in\mathbf{V}} m_{\mathbf{v}}(t+\tau) \\
&= \sum_{\mathbf{u}\in\mathbf{U}} m_{\mathbf{u}}(t) \\
&= \sum_{\mathbf{u}\in\mathbf{U}} \sum_{\mathbf{v}\in\mathbf{V}} \gamma_{\mathbf{v}\to\mathbf{u}} \cdot m_{\mathbf{v}}(t) \\
&= \sum_{\mathbf{v}\in\mathbf{V}} m_{\mathbf{v}}(t) \sum_{\mathbf{u}\in\mathbf{U}} \gamma_{\mathbf{v}\to\mathbf{u}} \\
&= \sum_{\mathbf{v}\in\mathbf{V}} m_{\mathbf{v}}(t) \\
&= M(t).
\end{aligned}$$

4.3 Barycentric Coordinates and Voronoi Coefficients

Yet it remains to determine the coefficients $\gamma_{\mathbf{v}\to\mathbf{u}}$ satisfying (11), whose construction should depend on the node sets $\mathbf{V}$ and $\mathbf{U}$, and so by the duality (8) also on the solution u of (1), (2).

One possible option for the construction of the coefficients $\gamma_{\mathbf{v}\to\mathbf{u}}$ is to work with *generalized barycentric coordinates* [5]. In this case, non-negative coordinates $\gamma_{\mathbf{v}\to\mathbf{u}} \in [0, 1]$ are computed subject to conditions

$$\mathbf{v} = \sum_{\mathbf{u}\in\mathbf{U}} \gamma_{\mathbf{v}\to\mathbf{u}} \cdot \mathbf{u} \quad \text{with} \sum_{\mathbf{u}\in\mathbf{U}} \gamma_{\mathbf{v}\to\mathbf{u}} = 1, \qquad \text{for } \mathbf{v} \in \mathbf{V}. \tag{12}$$

Note that the linear system for the coordinates $\gamma_{\mathbf{v}\to\mathbf{u}}$ in (12) is overdetermined, whenever $\mathbf{U}$ contains more than three points. In this case, the solution of (12) is not unique. For different options and characterizations of generalized barycentric coordinates, we refer to [5].

In the proposed particle advection method, however, we prefer to work with *Voronoi coefficients*, yielding another possibility for computing the non-negative multipliers $\gamma_{\mathbf{v}\to\mathbf{u}}$ subject to constraints (11). As supported by our numerical experiments, this leads to a very flexible mass conservative discretization with superior accuracy, which moreover works particularly well in combination with particle adaption and the implementation of boundary conditions.

In this method, we regard for any node $\mathbf{v} \in \mathbf{V}$, the Voronoi tile

$$\mathrm{Vor}_{\mathbf{V}}(\mathbf{v}) = \{\mathbf{x} \in \Omega : \|\mathbf{x} - \mathbf{v}\| \leq \|\mathbf{x} - \mathbf{w}\| \text{ for all } \mathbf{w} \in \mathbf{V}\},$$

of $\mathbf{v}$, and we let $|\mathbf{v}| = |\mathrm{Vor}_{\mathbf{V}}(\mathbf{v})|$ for the volume of $\mathbf{v}$. Accordingly, for any $\mathbf{u} \in \mathbf{U}$, we let $|\mathbf{u}| = |\mathrm{Vor}_{\mathbf{U}}(\mathbf{u})|$ for the particle volume of $\mathbf{u}$, where

$$\mathrm{Vor}_{\mathbf{U}}(\mathbf{u}) = \{\mathbf{x} \in \Omega : \|\mathbf{x} - \mathbf{u}\| \leq \|\mathbf{x} - \mathbf{w}\| \text{ for all } \mathbf{w} \in \mathbf{U}\}.$$

Then we let

$$\gamma_{\mathbf{v}\to\mathbf{u}} = \frac{|\mathrm{Vor}_{\mathbf{V}}(\mathbf{v}) \cap \mathrm{Vor}_{\mathbf{U}}(\mathbf{u})|}{|\mathbf{v}|}, \qquad \text{for all } \mathbf{u} \in \mathbf{U}, \mathbf{v} \in \mathbf{V}.$$

It is easy to see that for any $\mathbf{v} \in \mathbf{V}$ the non-negative Voronoi coefficients $\gamma_{\mathbf{v}\to\mathbf{u}}$ form a partition of unity, so that (11) holds.

Now the proposed utilization of Voronoi coefficients deserves a rather philosophical comment concerning *meshless* versus *mesh-based* methods. Note that the Voronoi coefficients are based on a very natural geometric concept which mainly relies on the distribution of the nodes in $\mathbf{U}$ and $\mathbf{V}$, rather than being dominated by the topology of a mesh. Although the proposed construction is strictly speaking – through the required Voronoi diagrams of $\mathbf{U}$ and $\mathbf{V}$ – not entirely mesh-independent, it does not introduce any restriction to the spatial distribution of the moving particles in $\mathbf{V}$. This is particularly important for the construction of flexible adaption rules (as discussed in the following section), where the (unconnected) moving particles are subject to dynamic modifications during the simulation. Finally, it seems to be common practice in meshless methods to work with background Voronoi diagrams (or similar geometric data structures), e.g. to handle geometric queries (such as nearest neighbour search etc.), efficiently.

4.4 Mass Conservative Particle Advection

We close this section by providing the following algorithm, which returns, at any time step $t \to t + \tau$, on given masses $\{m_{\mathbf{v}}(t) : \mathbf{v} \in \mathbf{V}\}$, updated mass values $\{m_{\mathbf{v}}(t + \tau) : \mathbf{v} \in \mathbf{V}\}$ satisfying (5).

Algorithm 1 (Mass Conservative Particle Advection).

INPUT: *Time step* τ, *node set* $\mathbf{V} \subset \Omega$, *mass values* $\{m_{\mathbf{v}}(t) : \mathbf{v} \in \mathbf{V}\}$.

(1) *Compute upstream nodes* $\mathbf{U} = \{\mathbf{u} : \mathbf{u} = \Psi^{t,t+\tau}\mathbf{v}\}$ *from node set* $\mathbf{V}$.
(2) *Compute coordinates* $\{\gamma_{\mathbf{v}\to\mathbf{u}} \in [0, 1] : \mathbf{v} \in \mathbf{V}, \mathbf{u} \in \mathbf{U}\}$, *satisfying (11);*

FOR *each* $\mathbf{v} \in \mathbf{V}$ **DO**

(3a) *Compute total mass* $m_{\mathbf{u}}(t)$ *of* $\mathbf{u} = \Psi^{t,t+\tau}\mathbf{v}$ *via (10);*
(3b) *Let* $m_{\mathbf{v}}(t + \tau) = m_{\mathbf{u}}(t)$.

OUTPUT: *Mass values* $\{m_{\mathbf{v}}(t + \tau) : \mathbf{v} \in \mathbf{V}\}$, *satisfying (5).*

5 Mass Conservative Adaption Rules

In order to balance the two conflicting requirements of good approximation quality and small computational costs, we need to combine the proposed particle advection scheme with a suitable strategy for particle adaption. Adaptivity requires customized rules for the modification of the node set $\mathbf{V}$ after each time step $t \to t + \tau$ of Algorithm 1. Indeed, for the sake of reducing the computational complexity we wish to reduce the size of the node set $\mathbf{V}$, whereas for the sake of good approximation quality we prefer to increase the density (and thus the size) of the node set $\mathbf{V}$ in Ω.

This section combines the robust and effective adaption strategy of our previous work [1, 2, 7] with the basic requirements of mass conservation. To this end, we explain how the conservative distribution of mass is accomplished during the adaptive modification of the node set $\mathbf{V}$. But we do not intend to explain all details of the utilized node adaption scheme. For the purposes of this work, it is sufficient to say that the node adaption is done by the removal of current nodes from $\mathbf{V}$, *coarsening*, and by the insertion of new nodes to $\mathbf{V}$, *refinement.* The decision on the node removal and insertion rely on a customized a posteriori error indicator. For details on this, we refer to the previous work [1].

In the remainder of this section, it is explained how these two operations, coarsening and refinement, are accomplished, such that the total mass is conserved.

5.1 Coarsening (Node Removal)

A node $\mathbf{v}^* \in \mathbf{V}$ is *coarsened* by its removal from the current node set $\mathbf{V}$, i.e., in this case we let $\mathbf{V} = \mathbf{V} \setminus \mathbf{v}^*$. Moreover, the portion of mass $m_{\mathbf{v}^*}$, which is currently attached to $\mathbf{v}^*$ is distributed to remaining nodes in the neighbourhood of $\mathbf{v}^*$. In order to further explain this, let $\mathbf{V}_* \subset \mathbf{V} \setminus \mathbf{v}^*$ denote a set of neighbouring nodes to $\mathbf{v}^*$.

For the purpose of mass (re)distribution, we first compute coefficients $\gamma_{\mathbf{v}^* \to \mathbf{v}} \in [0, 1]$ satisfying

$$\sum_{\mathbf{v} \in \mathbf{V}_*} \gamma_{\mathbf{v}^* \to \mathbf{v}} = 1.$$

Next, we update for any node $\mathbf{v} \in \mathbf{V} \setminus \mathbf{v}^*$ its current mass by letting

$$m_{\mathbf{v}}^* = \begin{cases} m_{\mathbf{v}} + \gamma_{\mathbf{v}^* \to \mathbf{v}} \cdot m_{\mathbf{v}^*}, & \text{for } \mathbf{v} \in \mathbf{V}_*, \\ m_{\mathbf{v}}, & \text{for } \mathbf{v} \in \mathbf{V} \setminus (\mathbf{V}_* \cup \mathbf{v}^*), \end{cases}$$

so that the mass $m_{\mathbf{v}}$, attached to $\mathbf{v}$, is updated by $m_{\mathbf{v}}^*$.

This leads to a *mass conservative* removal of the node $\mathbf{v}^*$ from $\mathbf{V}$ by

$$\begin{aligned} M &= \sum_{\mathbf{v} \in \mathbf{V}} m_{\mathbf{v}} \\ &= \sum_{\mathbf{v} \in \mathbf{V} \setminus \mathbf{v}^*} m_{\mathbf{v}} + m_{\mathbf{v}^*} \\ &= \sum_{\mathbf{v} \in \mathbf{V} \setminus \mathbf{v}^*} m_{\mathbf{v}} + \sum_{\mathbf{v} \in \mathbf{V}_*} \gamma_{\mathbf{v}^* \to \mathbf{v}} \cdot m_{\mathbf{v}^*} \\ &= \sum_{\mathbf{v} \in \mathbf{V} \setminus (\mathbf{V}_* \cup \mathbf{v}^*)} m_{\mathbf{v}} + \sum_{\mathbf{v} \in \mathbf{V}_*} (m_{\mathbf{v}} + \gamma_{\mathbf{v}^* \to \mathbf{v}} \cdot m_{\mathbf{v}^*}) \\ &= \sum_{\mathbf{v} \in \mathbf{V} \setminus \mathbf{v}^*} m_{\mathbf{v}}^*. \end{aligned}$$

5.2 Refinement (Node Insertion)

A node $\mathbf{v}^* \in \mathbf{V}$ is *refined* by the insertion of new nodes $\widehat{\mathbf{V}}$ in the neighbourhood of $\mathbf{v}^*$, so that $\mathbf{V}$ is updated accordingly by letting $\mathbf{V} = \mathbf{V} \cup \widehat{\mathbf{V}}$.

This modification requires (re)distributing the current masses of neighbouring nodes $\mathbf{V}_*$ around $\mathbf{v}^*$ to the new nodes in $\widehat{\mathbf{V}}$. This is done by working with coefficients $\gamma_{\mathbf{v} \to \widehat{\mathbf{v}}} \in (0, 1)$ satisfying

$$0 < \sum_{\widehat{\mathbf{v}} \in \widehat{\mathbf{V}}} \gamma_{\mathbf{v} \to \widehat{\mathbf{v}}} < 1, \qquad \text{for } \mathbf{v} \in \mathbf{V}_*,$$

where each $\gamma_{\mathbf{v} \to \widehat{\mathbf{v}}}$ yields the fraction of mass being distributed from $\mathbf{v} \in \mathbf{V}_*$ to $\widehat{\mathbf{v}} \in \widehat{\mathbf{V}}$, so that by $0 < m_{\mathbf{v}}^* \le m_{\mathbf{v}}$, for $\mathbf{v} \in \mathbf{V}_*$, the updated mass $m_{\mathbf{v}}^*$ of $\mathbf{v} \in \mathbf{V}_*$ is positive. For notational convenience, we let $\gamma_{\mathbf{v} \to \widehat{\mathbf{v}}} = 0$ for all $\mathbf{v} \in \mathbf{V} \setminus \mathbf{V}_*$.

Hence, for any $\mathbf{v} \in \mathbf{V}$, its current mass $m_{\mathbf{v}}$ is updated by letting

$$m_{\mathbf{v}}^* = \left(1 - \sum_{\widehat{\mathbf{v}} \in \widehat{\mathbf{V}}} \gamma_{\mathbf{v} \to \widehat{\mathbf{v}}}\right) \cdot m_{\mathbf{v}}, \qquad \text{for all } \mathbf{v} \in \mathbf{V}.$$

Moreover, the mass $m_{\widehat{\mathbf{v}}}^*$ of any new node $\widehat{\mathbf{v}} \in \widehat{\mathbf{V}}$ (to be inserted) is given by

$$m_{\widehat{\mathbf{v}}}^* = \sum_{\mathbf{v} \in \mathbf{V}_*} \gamma_{\mathbf{v} \to \widehat{\mathbf{v}}} \cdot m_{\mathbf{v}} = \sum_{\mathbf{v} \in \mathbf{V}} \gamma_{\mathbf{v} \to \widehat{\mathbf{v}}} \cdot m_{\mathbf{v}} \qquad \text{for } \widehat{\mathbf{v}} \in \widehat{\mathbf{V}}.$$

This leads to a *mass conservative* insertion of the new nodes $\mathbf{V}_*$ by

$$\begin{aligned}
M &= \sum_{\mathbf{v} \in \mathbf{V}} m_{\mathbf{v}} \\
&= \sum_{\mathbf{v} \in \mathbf{V}} \left[\sum_{\widehat{\mathbf{v}} \in \widehat{\mathbf{V}}} \gamma_{\mathbf{v} \to \widehat{\mathbf{v}}} \cdot m_{\mathbf{v}} + \left(1 - \sum_{\widehat{\mathbf{v}} \in \widehat{\mathbf{V}}} \gamma_{\mathbf{v} \to \widehat{\mathbf{v}}}\right) \cdot m_{\mathbf{v}} \right] \\
&= \sum_{\mathbf{v} \in \mathbf{V}} \sum_{\widehat{\mathbf{v}} \in \widehat{\mathbf{V}}} \gamma_{\mathbf{v} \to \widehat{\mathbf{v}}} \cdot m_{\mathbf{v}} + \sum_{\mathbf{v} \in \mathbf{V}} \left(1 - \sum_{\widehat{\mathbf{v}} \in \widehat{\mathbf{V}}} \gamma_{\mathbf{v} \to \widehat{\mathbf{v}}}\right) \cdot m_{\mathbf{v}} \\
&= \sum_{\widehat{\mathbf{v}} \in \widehat{\mathbf{V}}} \sum_{\mathbf{v} \in \mathbf{V}} \gamma_{\mathbf{v} \to \widehat{\mathbf{v}}} \cdot m_{\mathbf{v}} + \sum_{\mathbf{v} \in \mathbf{V}} m_{\mathbf{v}}^* \\
&= \sum_{\widehat{\mathbf{v}} \in \widehat{\mathbf{V}}} m_{\widehat{\mathbf{v}}}^* + \sum_{\mathbf{v} \in \mathbf{V}} m_{\mathbf{v}}^* \\
&= \sum_{\mathbf{v} \in \mathbf{V} \cup \widehat{\mathbf{V}}} m_{\mathbf{v}}^*.
\end{aligned}$$

6 Implementation of Boundary Conditions

Now let us finally turn to the implementation of boundary conditions. In the above discussion until now, we have considered the special case where the computational domain Ω is the whole plane, i.e., $\Omega = \mathbb{R}^2$, so that $(\mathbf{U} \cup \mathbf{V}) \subset \Omega$ at any time t. However, in specific applications of interest, Ω is bounded, and, moreover, boundary conditions are of relevance. Therefore, suppose from now that Ω is bounded.

Recall that our proposed scheme works with a set $\mathbf{V} = \{\mathbf{v}\}_{\mathbf{v} \in \mathbf{V}}$ of finite nodes, and corresponding upstream nodes $\mathbf{U} = \{\mathbf{u}\}_{\mathbf{u} \in \mathbf{U}}$, satisfying $\mathbf{u} = \Psi^{t,t+\tau}\mathbf{v}$ for all $\mathbf{v} \in \mathbf{V}$.

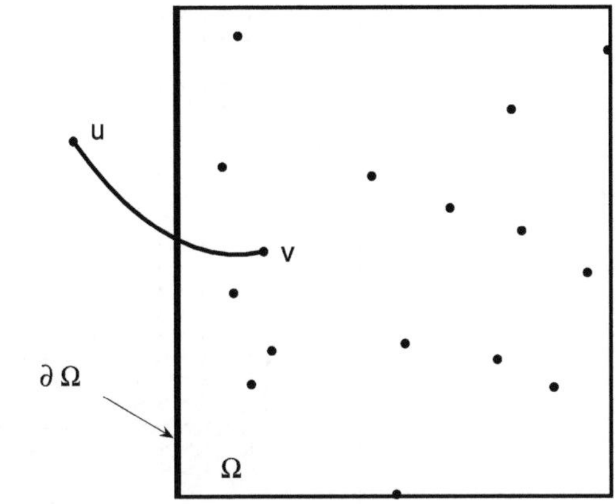

Fig. 2. Incoming flow from upstream node $\mathbf{u} \in \mathbf{U} \setminus \Omega$, to corresponding downstream node $\mathbf{v} = \Psi^{t+\tau,t}\mathbf{u} \in \Omega$ across the domain boundary $\partial\Omega$.

6.1 Incoming Flow

The following discussion is relevant, when an upstream node $\mathbf{u} \in \mathbf{U}$ lies outside the domain Ω, i.e., $\mathbf{u} \in \mathbf{U} \setminus \Omega$, see Figure 2 for illustration. In this case, a fraction of mass, $\gamma_{\partial\Omega\to\mathbf{u}} \in [0, 1]$, is advected from $\mathbf{u}$ to $\mathbf{v} = \Psi^{t+\tau,t}\mathbf{u} \in \Omega$ across the boundary $\partial\Omega$, where the coefficients $\gamma_{\partial\Omega\to\mathbf{u}}$ are required to form a partion of unity, i.e.,

$$\sum_{\mathbf{u}\in\mathbf{U}\setminus\Omega} \gamma_{\partial\Omega\to\mathbf{u}} = 1.$$

In order to implement boundary conditions concerning the *incoming flow*, we assign a boundary value $\gamma_{\partial\Omega\to\mathbf{u}} \cdot m_{\partial\Omega}$ to each $\mathbf{u} \in \mathbf{U} \setminus \Omega$, giving the portion of mass which is advected from $\mathbf{u}$ into Ω across the boundary $\partial\Omega$. Hence,

$$M_{\text{in}} = \sum_{\mathbf{u}\in\mathbf{U}\setminus\Omega} \gamma_{\partial\Omega\to\mathbf{u}} \cdot m_{\partial\Omega} \equiv m_{\partial\Omega}$$

is the total amount of incoming mass across the boundary $\partial\Omega$.

Moreover, for any $\mathbf{u} \in \mathbf{U} \setminus \Omega$, portions of mass from current nodes $\mathbf{v} \in \mathbf{V}$ may also be distributed to $\mathbf{u}$. In this case, we obtain

$$m_{\mathbf{u}} = \gamma_{\partial\Omega\to\mathbf{u}} \cdot m_{\partial\Omega} + \sum_{\mathbf{v}\in\mathbf{V}} \gamma_{\mathbf{v}\to\mathbf{u}} \cdot m_{\mathbf{v}}, \qquad \text{for } \mathbf{u} \in \mathbf{U} \setminus \Omega,$$

for the total mass which is assigned to the upstream node $\mathbf{u}$.

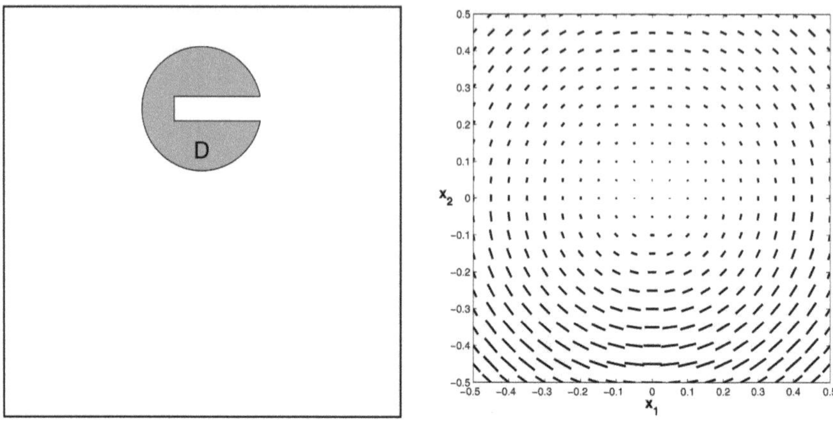

(a) (b)

Fig. 3. The slotted cylinder. (a) Initial condition and (b) velocity field.

6.2 Outgoing Flow

Let us regard the situation, where for $\mathbf{v} \in \mathbf{V}$ all its coordinates $\gamma_{\mathbf{v}\to\mathbf{u}}$ vanish, i.e., $\gamma_{\mathbf{v}\to\mathbf{u}} = 0$ for all $\mathbf{u} \in \mathbf{U}$. In this case, *no* mass is distributed from $\mathbf{v}$ to any point in $\mathbf{U}$. The amount of mass $m_{\mathbf{v}}$, attached to $\mathbf{v}$, is then being advected from $\mathbf{v} \in \Omega$ to the exterior of Ω across the domain boundary $\partial\Omega$.

In consequence, the total outgoing mass is given by

$$M_{\text{out}} = \sum_{\mathbf{v}\in\mathbf{V}_{\text{out}}} m_{\mathbf{v}},$$

where we let

$$\mathbf{V}_{\text{out}} = \{\mathbf{v} \in \mathbf{V} : \gamma_{\mathbf{v}\to\mathbf{u}} = 0 \text{ for all } \mathbf{u} \in \mathbf{U}\} \subset \mathbf{V}$$

for the set of nodes whose masses are advected across $\partial\Omega$ to the exterior of Ω.

7 The Slotted Cylinder – A Test for Passive Advection

In this section, the performance of our advection scheme is evaluated by using one numerical experiment. In this experiment, taken from [7, Section 5], we consider the rotating *slotted cylinder*, a popular test case suggested by Zalesak [15].

Here, $\Omega = [-0.5, 0.5]^2 \subset \mathbb{R}^2$ and the initial condition is given by

$$u(0, \mathbf{x}) = \begin{cases} 1, & \text{for } \mathbf{x} \in D, \\ 0, & \text{otherwise,} \end{cases} \tag{13}$$

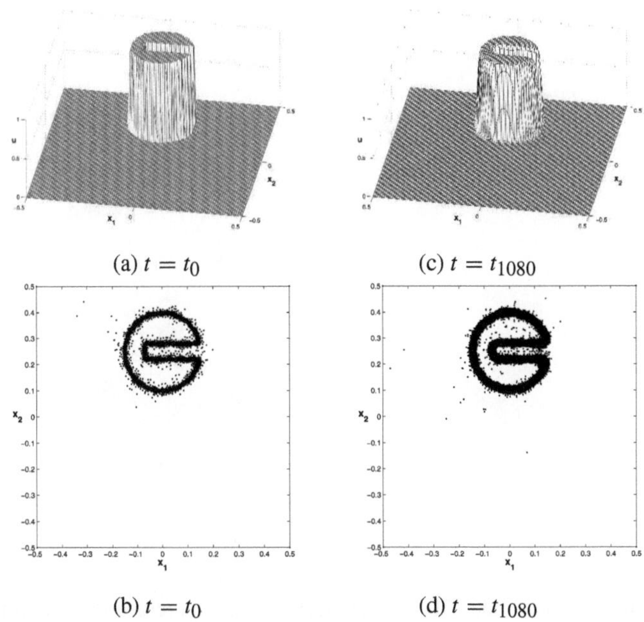

(a) $t = t_0$ (c) $t = t_{1080}$

(b) $t = t_0$ (d) $t = t_{1080}$

Fig. 4. The slotted cylinder. (a) 3D view, (b) node distribution, of the initial condition (left column), and after six revolutions (right column), (c), (d).

where $D \subset \Omega$ is the slotted disc of radius $r = 0.15$, centered at $(0, 0.25)$ with slot width 0.06 and length 0.22, see Figure 3(a).

In the original test case of Zalesak, the slotted cylinder is rotated by a steady flow field $\mathbf{a}(\mathbf{x}) \sim (x_2, -x_1)$, where $\mathbf{x} = (x_1, x_2)$. We decided to replace the velocity field in [15] by the somewhat more complicated velocity field

$$\mathbf{a}(\mathbf{x}) = (x_2, -x_1) \begin{cases} \frac{1}{2}\sin(2\varphi(\mathbf{x}) - \frac{\pi}{2}) + \frac{3}{2}, & \text{for } x_2 < 0, \\ 1, & \text{for } x_2 \geq 0, \end{cases}$$

whose azimuth angle is given by

$$\varphi(\mathbf{x}) = \begin{cases} \arctan(-x_2/x_1), & \text{for } x_1 > 0, \\ \arctan(x_1/x_2) + \frac{\pi}{2}, & \text{for } x_1 \leq 0. \end{cases}$$

This velocity field rotates the slotted cylinder clockwise with constant angular velocity in the first and second quadrant, whereas the cylinder is accelerated in the fourth quadrant, and decelerated in the third quadrant, see Figure 5. The maximum angular velocity $\omega = 2$ is attained in the lower half of the coordinate system, namely at the points on the vertical line

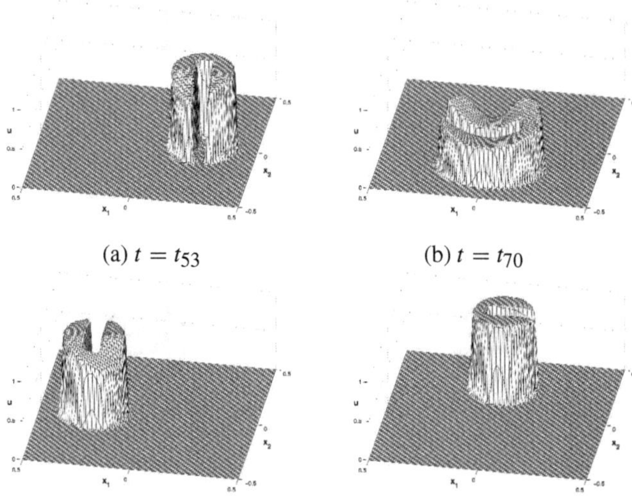

(a) $t = t_{53}$ (b) $t = t_{70}$

(c) $t = t_{102}$ (d) $t = t_{180}$

Fig. 5. The slotted cylinder. 3D view on $\bar{u}_{\mathbf{v}}(t)$ during first revolution at four times.

$$\{\mathbf{x} = (x_1, x_2) : x_1 = 0, x_2 < 0\}.$$

The slotted cylinder is *stretched* when passing through the *acceleration part* of the velocity field in the fourth quadrant, whereas it is *squashed* in the *deceleration part* of the third quadrant in order to recover its original shape of the initial condition at each full revolution.

Initially, a set $\mathbf{V} \subset \Omega$ of 1500 randomly distributed nodes is chosen. The initial condition (13) is used in order to assign a particle average value $\bar{u}_{\mathbf{v}}(0)$ in (4) to each node $\mathbf{v} \in \mathbf{V}$ at time $t = t_0$. The nodes in $\mathbf{V}$ are automatically adapted to the discontinuities of the initial condition u_0, see Figure 4(b).

At each revolution of the slotted cylinder, the particle average values $\bar{u}_{\mathbf{v}}$ are decreasing, as soon as the cylinder enters the acceleration part of the velocity field, see Figure 5. This behaviour is due to the mass conservation of the scheme. In contrast to this, in the deceleration part, the particle average values $\bar{u}_{\mathbf{v}}$ are increasing. Moreover, in this region, the initial shape of slotted cylinder is gradually recovered, see Figures 5(b)–(d).

Our simulation of this model problem comprises six full revolutions of the slotted cylinder. Figure 4(a) shows the 3D view of the particle averages $\bar{u}_{\mathbf{v}}$, and Figure 4(b) shows the node distribution for the initial condition (13). In comparison, Figures 4(c)–(d) shows to the corresponding numerical result after six full revolutions. Observe that the shape of the cylinder is accurately maintained during the sim-

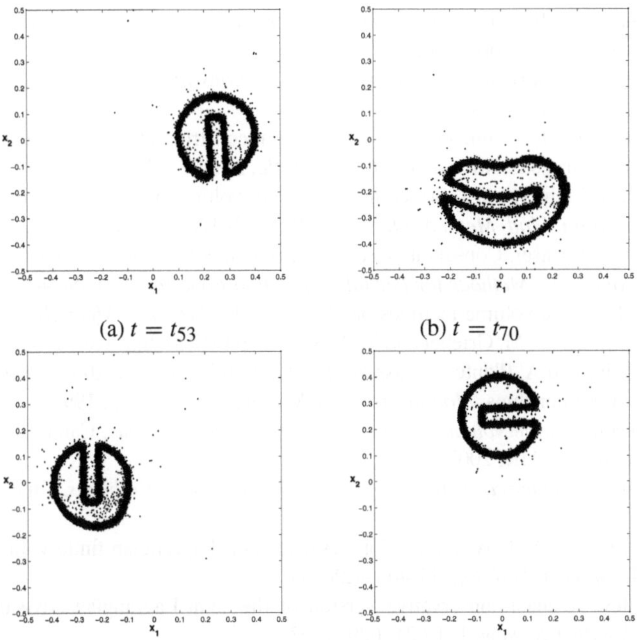

(a) $t = t_{53}$ (b) $t = t_{70}$

(c) $t = t_{102}$ (d) $t = t_{180}$

Fig. 6. The slotted cylinder. Node distribution during first revolution at four times.

ulation, and numerical diffusion is widely suppressed. For more details concerning the model problem and the discussion of the numerical results, we refer to [7].

Acknowledgement

The fruitful joint collaboration with Martin Käser in previous work on related particle methods is greatly appreciated.

References

1. J. Behrens, A. Iske, and S. Pöhn. Effective node adaption for grid-free semi-Lagrangian advection. In: *Discrete Modelling and Discrete Algorithms in Continuum Mechanics*, T. Sonar and I. Thomas (Eds), Logos, Berlin, 2001, pp. 110–119.
2. J. Behrens, A. Iske, and M. Käser. Adaptive meshfree method of backward characteristics for nonlinear transport equations. In: *Meshfree Methods for Partial Differential Equations*, M. Griebel and M.A. Schweitzer (Eds), Springer-Verlag, Heidelberg, 2002, pp. 21–36.

3. P. Deuflhard and F. Bornemann. *Scientific Computing with Ordinary Differential Equations*. Springer, New York, 2002.
4. D.R. Durran. *Numerical Methods for Wave Equations in Geophysical Fluid Dynamics*. Springer, New York, 1999.
5. M.S. Floater, K. Hormann, and G. Kós. A general construction of barycentric coordinates over convex polygons. *Advances in Comp. Math.*, 24:311–331, 2006.
6. D. Hietel, K. Steiner, and J. Struckmeier: A finite-volume particle method for compressible flows. *Math. Mod. Meth. Appl. Sci.*, 10(9):1363–1382, 2000.
7. A. Iske and M. Käser. Conservative semi-Lagrangian advection on adaptive unstructured meshes. *Numerical Methods for Partial Differential Equations*, 20:388–411, 2004.
8. M. Junk. Do finite volume methods need a mesh? In: *Meshfree Methods for Partial Differential Equations*, M. Griebel and M.A. Schweitzer (Eds), Springer, 2003, pp. 223–238.
9. J.P.R. Laprise and A. Plante. A class of semi-Lagrangian integrated-mass (SLIM) numerical transport algorithms. *Monthly Weather Review*, 123:553–565, 1995.
10. R.L. LeVeque. *Finite Volume Methods for Hyperbolic Problems*. Cambridge University Press, Cambridge, UK, 2002.
11. K.W. Morton. *Numerical Solution of Convection-Diffusion Problems*. Chapman & Hall, London, 1996.
12. T.N. Phillips and A. J. Williams. Conservative semi-Lagrangian finite volume schemes. *Numer. Meth. Part. Diff. Eq.*, 17:403–425, 2001.
13. A. Priestley. A quasi-conservative version of the semi-Lagrangian advection scheme. *Monthly Weather Review*, 121:621–629, 1993.
14. J.S. Scroggs and F.H.M. Semazzi. A conservative semi-Lagrangian method for multidimensional fluid dynamics applications. *Numer. Meth. Part. Diff. Eq.*, 11:445–452, 1995.
15. S. T. Zalesak. Fully multidimensional flux-corrected transport algorithms for fluids. *J. Comput. Phys.*, 31:335–362, 1979.

Spectral-Like Accuracy in Space of a Meshless Vortex Method

L.A. Barba

Department of Mathematics, University of Bristol, University Walk, Clifton, Bristol BS8 1TW, U.K.; E-mail: l.a.barba@bristol.ac.uk

Abstract. The convergence of a meshless vortex method is studied numerically. The method uses core spreading for diffusion and radial basis function interpolation for spatial adaption of the Lagrangian particles. Spectral accuracy in space is observed in the absence of convection error, and second order of convergence is obtained it its presence.

Key words: Vortex method, radial basis functions, convergence.

1 Introduction

We consider the accuracy and convergence of a meshless method for fluid dynamics based on vortex particles. The vortex method has a long history, beginning with the use of point vortices to study the instability of a vortex sheet [14]. Although many important achievements have been produced since, there continue to be some frustrations. In fact, the vortex method is still viewed in some circles as a modelling approach and not as direct simulation. The following opinion, expressed some years ago, is relevant today: "*There has been and perhaps there will always be some skepticism about the use of vortices for flow simulation*" [18]. For this reason, it is important to make contributions to the validation and verification of vortex method codes, and this work is a step in that direction.

The vortex method solves the Navier–Stokes equation in vorticity formulation,

$$\frac{\partial \boldsymbol{\omega}}{\partial t} + \mathbf{u} \cdot \nabla \boldsymbol{\omega} = \boldsymbol{\omega} \cdot \nabla \mathbf{u} + \nu \Delta \boldsymbol{\omega} \tag{1}$$

by discretizing the vorticity field into smooth Lagrangian particles:

$$\boldsymbol{\omega}(\mathbf{x}, t) \approx \boldsymbol{\omega}^h(\mathbf{x}, t) = \sum_{i=1}^{N} \boldsymbol{\Gamma}_i(t) \zeta_\sigma \left(\mathbf{x} - \mathbf{x}_i(t)\right). \tag{2}$$

V.M.A. Leitão et al. (eds), *Advances in Meshfree Techniques*, 187–197.

Here, $\mathbf{\Gamma}_i$ is the vector circulation strength (a scalar in 2D) of a vortex particle located at $\mathbf{x}_i$. The particle's vorticity distribution function, ζ_σ, often called the cutoff function, can be for example a Gaussian; in two dimensions:

$$\zeta_\sigma(\mathbf{x}) = \frac{1}{2\pi\sigma^2} \exp\left(\frac{-|\mathbf{x}|^2}{2\sigma^2}\right). \tag{3}$$

The particles are evolved by integrating their trajectories with the velocity at their center, evaluated using the Biot–Savart law, which in 2D is:

$$\mathbf{u}(\mathbf{x}, t) = \frac{-1}{2\pi} \int \frac{(\mathbf{x} - \mathbf{x}') \times \omega(\mathbf{x}', t)\hat{\mathbf{k}}}{|\mathbf{x} - \mathbf{x}'|^2} d\mathbf{x}'. \tag{4}$$

Using the discretized form of the vorticity, Equation (2), in the Biot–Savart law, the discretized velocity is obtained. The Gaussian cutoff function allows the integral in (4) to be evaluated analytically, after which the velocity at one particle's location is obtained by summing the influence over all others. Clearly, the direct evaluation of the discrete velocity is an N-body problem, for which reason the computational efficiency can be greatly improved using a fast multipole method [10]. Once the velocity is computed, the vortex method is expressed in the following system of equations:

$$\frac{d\mathbf{x}_i}{dt} = \mathbf{u}(\mathbf{x}_i, t) \tag{5}$$

$$\frac{d\omega}{dt} = \nu\nabla^2\omega + \text{ B.C.} \tag{6}$$

The above equations express the fact that the vorticity transport is solved by moving the vortex-particle elements with the local velocity of the fluid. This is the only part of the vortex method that is required for ideal flow. For viscous flow, the method has been extended by providing a change in vorticity at the vortex particles due to the viscous effects. This is generally accomplished in a split-step formulation. In addition, the presence of boundaries in the flow can also be translated into changes in the vorticity, which is expressed by the term "B.C." in Equation (6).

The variety of schemes used in vortex methods to provide viscous effects are reviewed in [6]. In this work, we have used the core spreading method, in which particle core sizes are grown to exactly solve the diffusion part of the equation [12]. The representative length scale of the computational vortex elements, σ, is made to grow according to the following equation:

$$\frac{d\sigma^2}{dt} = 2\nu, \tag{7}$$

Note that this scheme is exact in its representation of the viscous effects, due to the fact that the Gaussian function is an exact solution of the heat equation. It is important to realize that this method is formulated specifically for the Gaussian cutoff in two dimensions. Some authors have neglected this fact, resulting in applications of questionable consistency [3].

The core spreading method is fully localized and grid-free; however, it requires core size control to tackle consistency issues brought up in [9]. Simply stated, the characteristic length scale of the computational particles cannot be allowed to grow uncontrolled, as they represent the smallest resolved scales. The consistency problem arises due to the fact that the Gaussian elements are convected without deformation. In the present vortex method, control of the particle sizes is provided in a spatial adaption algorithm using radial basis function interpolation, as described later.

The rest of the paper will present a discussion of the discretization accuracy of the vortex method, and in particular the so-called "convection error" of the Lagrangian formulation (Section 2), and subsequently a report of numerical convergence studies performed using a parallel implementation of the method described in [6]. We use both axisymmetric test flows (for which an analytic solution allows precise error measurements) and non-axisymmetric tests, which are more challenging.

The vortex method with core spreading for diffusion, and a spatial adaption process that uses radial basis function (RBF) interpolation, was implemented in parallel using the PETSc library [1] in a C++/MPI code. In this implementation, the RBF interpolations are solved using the built-in pre-conditioners and GMRES solver of PETSc. More details of the parallel implementation are given in [2].

2 Discretization Accuracy of the Vortex Method

The Lagrangian approach described above is devoid of numerically diffusive truncation errors, which makes the vortex method especially suited for the study of high-Reynolds number vortical flows.

This often praised feature of Lagrangian particle methods too frequently remains unexplained. The low numerical dissipation of the vortex method arises from the fact that the nonlinear term in the Navier–Stokes equation is replaced by a set of ordinary differential equations for the particle locations. As a result, the equivalent of a local truncation error for the vortex method is what is called "convection error" in the specialized literature.

Convection error in vortex methods refers to the error which originates from convecting the vortex particles without deformation with the velocity at their centers. A derivation of the estimate for this error when using Gaussian particles is presented in [12]. The estimate is obtained by subtracting the discrete representation of the velocity to the exact velocity in the nonlinear convective term, as follows:

$$\varepsilon(\mathbf{x}) = \nabla \cdot \left[\sum_{i=1}^{N} \Gamma_i \, \zeta_\sigma(\mathbf{x} - \mathbf{x}_i) \left(\mathbf{u}(\mathbf{x}) - \frac{d\mathbf{x}_i}{dt} \right) \right] \tag{8}$$

which, after a Taylor expansion and some algebra results in the following estimate:

$$\varepsilon(\mathbf{x}) = \sigma^2 \frac{\partial^2 \omega}{\partial x_j \partial x_k} \frac{\partial u_j}{\partial x_k}. \tag{9}$$

Thus, one sees that this error is second order in σ and that it does not diffuse vorticity. The result in (9) also indicates that axisymmetric flows do not suffer from convection error, as only the symmetric part of the velocity gradient tensor has a contribution, hence the result of (9) in principal coordinates is zero for axisymmetric flow. This observation is important to have in mind, as the use of axisymmetric test problems for validation of vortex method codes is widespread practice.

In addition to convection error, the vortex method suffers from an increase of the spatial discretization error over time, due to the vortex particles becoming disordered. In the sense of function approximation, the spatial discretization using particles depends on a measure of the particle density: smooth particles need to overlap at all times to be able to represent a continuous field. As particles follow the flow map, there is a chance of gaps appearing in the particle distribution, or particles clustering together in other areas. This problem must be corrected with a scheme that adapts the particle in such a way that overlap is always maintained, and clustering avoided.

The standard solution to this situation is applying a remeshing scheme, as introduced in [11]. This has allowed long-time calculations with the vortex method by effectively controlling the growth of discretization error, but at the cost of introducing some numerical dissipation and grid-dependency. Remeshing schemes are based on tensor product formulations, and are subject to interpolation errors.

An alternative to the standard remeshing schemes is using radial basis function (RBF) interpolation techniques in a fully meshless spatial adaption process; the method is introduced and described in detail in [6]. It is basically a way to rediscretize the vorticity field with a set of new vortex particles that are well-overlapped and represent accurately the current flow field.

When interpreting the spatial discretization of the vortex method as an approximation problem using RBFs, one can see the potential for spectral-like accuracy in space. It has been established that RBF interpolation has spectral accuracy when using Gaussian bases [19].

In this paper, we show numerical experiments that confirm spectral accuracy in a vortex method in the absence of convection error. It is noteworthy that this result is obtained in a *viscous* calculation, showing that core spreading does not limit the accuracy that can be obtained from the vortex method (unlike other viscous approaches; for example, the particle strength exchange method [8] is limited to second order accuracy, and deteriorates as particles become disordered).

Convection error, however, will be present in the general case. Further numerical experiments presented here verify *observed* second order accuracy in the presence of convection error when using Gaussian bases. Hence, the possibility of increasing the convergence rate now lies in reducing convection error, perhaps by using deformable basis functions or some other approach.

3 Numerical Convergence Study

3.1 Axisymmetric Test Flow

A first convergence study is performed using an axially symmetric viscous flow, for which an analytic solution is available. This is the Lamb–Oseen vortex, for which the vorticity evolves as follows, with $r = x^2 + y^2$:

$$\omega(r,t) = \frac{\Gamma_0}{4\pi\nu t}\exp\left(-\frac{r^2}{4\nu t}\right). \tag{10}$$

Here, Γ_0 represents the total circulation of the vortex, and ν is the fluid viscosity. The Lamb–Oseen vortex simply spreads over time, while the maximum vorticity decreases, in a self-similar way.

To minimize time-stepping errors in these numerical tests, a very small time step was used of $\Delta t = 0.002$ in a 4th order Runge–Kutta scheme. The calculations were advanced for 200 time steps, and spatial adaption was performed every 5 steps. Although no regular particle arrangement is required by the formulation, the particles were placed in a triangular lattice, covering the vorticity support to a minimum circulation level of 10^{-10} (i.e., after spatial adaption, particles with $\Gamma_i < 10^{-10}$ are deleted).

Several initial values of the inter-particle spacing h were chosen, and a run was performed for each value. Then for each run, the pointwise errors were measured at each time step, using the maximum norm. The largest value of the error in the time-marching calculation was used for each run to produce the plot in Figure 1. As shown by the functional fit added in the plot, the results of this numerical convergence study are consistent with the spectral convergence of radial basis function interpolation using Gaussians. Note that the measure of error chosen for the plot is the maximum in both space and time, for each run. The curve slopes off around the level 10^{-10}, which is probably a result of our population control scheme (mentioned above).

A similar result to the one presented in Figure 1 was included in [6], but for stationary tests (no time stepping). There, it was simply a case of proving that the discretization of the vorticity field using vortex particles could exhibit spectral convergence. Here, we have incorporated time marching, with periodic spatial adaption using radial basis functions, in a viscous calculation.

The numerical convergence result of Figure 1 likely shows the best accuracy that can be obtained with the vortex method. This is because the Lamb–Oseen vortex flow is a benign test, due to its infinite smoothness and more importantly to its immunity

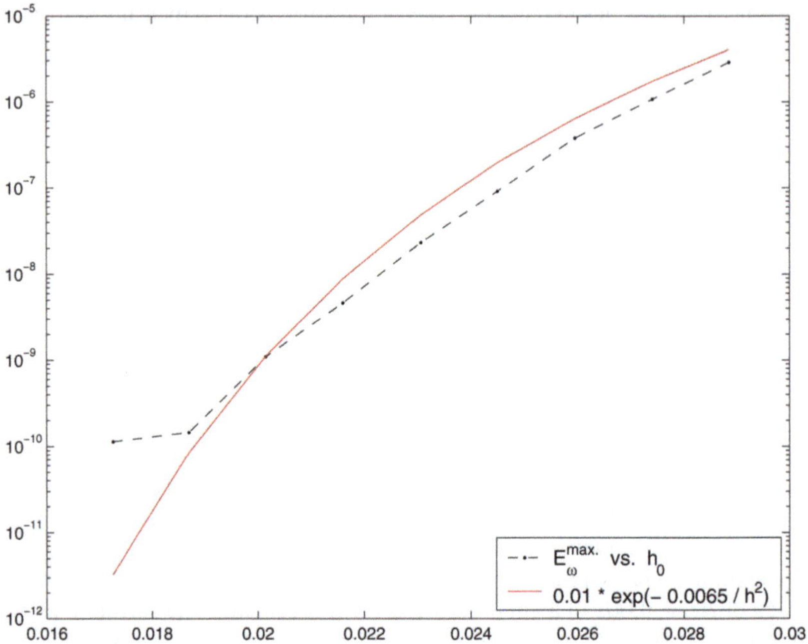

Fig. 1. Plot of the maximum vorticity errors during 200 time steps vs. the inter-particle spacing h and an exponential fit to the error behavior.

to convection error. Nonetheless, it is an important result because it demonstrates that the RBF-based spatial adaption scheme does not limit the accuracy of the vortex method. This is in contrast to the standard remeshing schemes, based on tensor products of 1D kernels (which furthermore require a Cartesian grid). It has been recognized that the remeshing schemes do introduce some error; these were clearly displayed in numerical experiments presented in [5].

Note again that the results above are produced by a *viscous* vortex method. The use of core spreading with adequate core size control, therefore, does not slow the rate of convergence of the spatial discretization. The core size control is provided here in the RBF interpolation step, where the core sizes can be reset to any desired value. This automatic form of core size control does not have an effect on the accuracy, unlike the vortex splitting technique of [15], which is numerically diffusive.

3.2 Non-Axisymmetric Test Flow

Next, a convergence study is presented using a non-axisymmetric flow, and thus subject to convection error. The initial vorticity consists of a quadrupole perturbation ω' on a Gaussian vortex ω_o, resulting in localized elliptical deformation of the core, and an evolution that develops filaments and exhibits a quasi-steady tripole for large amplitude of the perturbation. This flow was studied in [17], and used as proof-of-concept for the vortex method developed in [2]. Furthermore, detailed parameter studies for this flow problem have been performed recently [4]. The initial conditions are obtained from the superposition of an axisymmetric eddy and the non-axisymmetric perturbation, as $\omega = \omega_o + \omega'$:

$$\omega_o(\mathbf{x}) = \frac{1}{4\pi}\exp\left(\frac{-|\mathbf{x}|^2}{4}\right), \quad \omega'(\mathbf{x}) = \frac{\delta}{4\pi}\,|\mathbf{x}|^2 \exp\left(\frac{-|\mathbf{x}|^2}{4}\right)\cos 2\theta. \tag{11}$$

For the present calculations, the Reynolds number is 1000, defined as $Re = \Gamma/\nu$, and the amplitude of the perturbation is $\delta = 0.25$.

The use of a non-axisymmetric test problem to perform convergence studies is usually hampered by the lack of an analytical solution. But, as explained in [13], it is possible to extract the *observed order of convergence* from a grid-convergence study, using three grid solutions. This verification technique was of course developed for grid-based methods, but there is not obstacle to applying the same concepts in a meshless method, if one uses an equivalent measure of "grid-refinement".

The technique is as follows: if one obtains three numerical solutions of a given problem with the same code, using three different spatial discretizations given by h_1, h_2, and h_3 with a fixed grid-refinement ratio $r = h_3/h_2 = h_2/h_1$ (the subscript 1 refering to the finest resolution), then one can obtain an *empirical* convergence order p by the following relation:

$$p = \ln\left(\frac{u_{h_3} - u_{h_2}}{u_{h_2} - u_{h_1}}\right) \Big/ \ln(r). \tag{12}$$

In the case of the vortex particle method, we take as the resolution parameter the initial inter-particle spacing, at the time of discretization. In reality, the vortex method has two resolution parameters: the inter-particle spacing, and the overlap ratio (defined as the inter-particle spacing divided by the particle size). We chose to leave the value of the overlap ratio unchanged for this study, and thus base the refinement study on the inter-particle spacing. Moreover, as the discretization used is based on placing the vortex particles on a triangular lattice, we use a measure h which corresponds to the equivalent square lattice, providing the same cell area as in the triangular lattice used.

Three solutions were computed of the flow with initial vorticity as in (11), with a grid-refinement ratio of $r = 1.4$. The runs were carried for 740 time steps, with $\Delta t = 0.05$. The vorticity was sampled on the same mesh for all runs, corresponding to the finer value of h. Spatial adaption was performed every 10 steps with the same procedure described before, using radial basis function interpolation. Table 2, below, shows the problem sizes for this convergence study, in terms of the total number of vortex particles needed to cover the vorticity support down to a level of particle circulation of 10^{-10}.

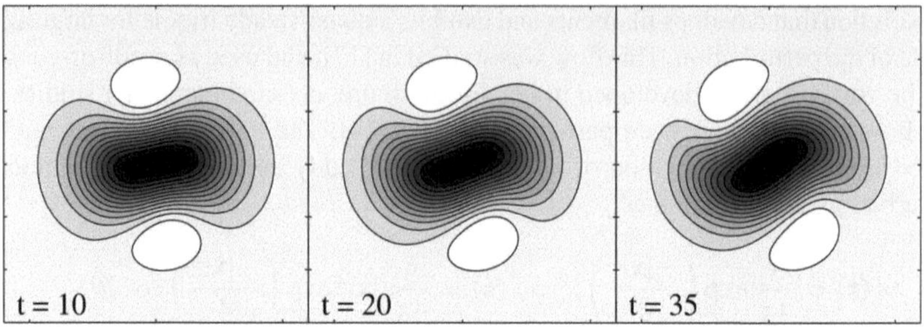

Fig. 2. Vorticity contours for the non-axisymmetric test flow, at the times used for the convergence study. Note that the white filled contour corresponds to negative inclusions in the vorticity field. This flow is undergoing a markedly non-steady re-organization of vorticity.

Table 1. Problem sizes for convergence study runs.

particle spacing	$N(t=0)$	$N(t=37)$
$h_1 = 0.09$	20389	21978
$h_2 = 0.126$	10361	11418
$h_3 = 0.1764$	5309	5957

Table 2. Observed order of convergence for time-marching runs.

time-slice \ norm used:	$E_\omega^{L^2}$	$E_\omega^{\text{rel,max}}$
$t = 10$	2.0282	2.0271
$t = 20$	2.026	2.027
$t = 35$	2.0175	2.0191

By means of a comparison of the point-by-point value of vorticity at each three time-slices, $t = 10, 20, 35$, and taking both a discrete L^2-norm and a maximum norm (normalized by the maximum vorticity), measurements of the observed order of convergence were performed by applying (12). The results are presented in Table 2, where the table values correspond to the computed value of p.

The results above are perhaps surprisingly consistent, giving an observed second-order of convergence. Since the principal difference between this test problem, and the Lamb–Oseen vortex used previously is the lack of axisymmetry in the present case, we conclude that the observed order of convergence is fundamentally due to convection error.

In other words, the vortex method exhibits spectral order of convergence in the spatial discretization, as demonstrated in the experiments of Section 3.1, but in a time-marching algorithm it is limited by the second order of the convection error.

This result is specific for the second order Gaussian cutoff used in the particle discretization of *these* simulations, however. There have been several proposals for using higher-order cutoff functions, but as shown in practice by the numerical experiments in [5], higher-order cutoffs are more sensitive to overlap, and thus require more frequent spatial adaption. This results in a considerable penalty, to the point that some authors opt for spatially adapting on *every* time step. Another means of achieving higher-order with the vortex method including convection error is the use of deformable basis functions; the recent work of Rossi [16] demonstrates fourth-order accuracy with elliptical Gaussian bases.

As a final note, we add that the theory of convergence of vortex methods (see [7] for details and further references) establishes second order convergence when using a Gaussian cutoff. However, the analysis assumes a standard initialization based on the local vorticity value multiplied by h^2 on a Cartesian grid, as an estimate of the particle circulation strengths. Thus, initialization of the method is already limited to second order. Using instead RBF interpolation to obtain the initial particle circulations allows for spectral accuracy in space in the absence of convection.

4 Conclusion

Extensive numerical experiments performed previously [2], both with standard remeshing and with the radial basis function interpolation, demonstrate that the remeshing schemes impose an accuracy limitation on the vortex method. By using instead radial basis function interpolation, the accuracy limitation of the spatial adaption process is removed, and convection error is left to dominate on the rate of convergence. We have demonstrated using numerical experiments that in the absence of convection error, spectral accuracy is possible is space. With convection error being present in a non-axisymmetric unsteady flow, we have shown by means of a refinement study an observed second order of convergence of the method. This result is a manifestation of the convection error produced when using simple Gaussians as basis functions. Higher order vortex methods are thus possible with techniques to reduce convection error, such as using deformable basis functions. Two important implications of the present study are the fact that there is no accuracy limitation imposed by the method used for spatial adaption, as radial basis function interpolation is spectral order, and that the viscous scheme of core spreading is also able to preserve the high accuracy and convergence.

Acknowledgements

Computations were carried out in the Beowulf cluster of the Laboratory for Advanced Computation in the Mathematical Sciences, University of Bristol (http://lacms.maths.bris.ac.uk/). Thanks to Tony Leonard for many helpful discussions and thanks to the PETSc team for prompt and helpful technical support. The author's travel was made possible by a grant from the Nuffield Foundation.

References

1. S. Balay, K. Buschelman, W. D. Gropp, D. Kaushik, M. Knepley, L. Curfman-McInnes, B. F. Smith, and H. Zhang. PETSc User's Manual. Technical Report ANL-95/11 – Revision 2.1.5, Argonne National Laboratory, 2002.
2. L. A. Barba. *Vortex method for computing high-Reynolds number flows: Increased accuracy with a fully mesh-less formulation.* PhD Thesis, California Institute of Technology, 2004.
3. L. A. Barba. Discussion: Three-dimensional vortex method for gas-particle two-phase compound round jet (Uchiyama, T., and Fukase, A. 2005 ASME J. Fluids Eng., 127, pp. 32–40). *J. Fluids Eng.*, 128:643–645, May 2006.
4. L. A. Barba and A. Leonard. Emergence and evolution of tripole vortices from non-shielded initial conditions. Submitted, 2006.
5. L. A. Barba, A. Leonard, and C. B. Allen. Numerical investigations on the accuracy of the vortex method with and without remeshing. AIAA #2003-3426, 16th CFD Conference, Orlando FL, June 2003.
6. L. A. Barba, A. Leonard, and C. B. Allen. Advances in viscous vortex methods – Meshless spatial adaption based on radial basis function interpolation. *Int. J. Num. Meth. Fluids*, 47(5):387–421, 2005.
7. G.-H. Cottet and P. Koumoutsakos. *Vortex Methods. Theory and Practice.* Cambridge University Press, 2000.
8. P. Degond and S. Mas-Gallic. The weighted particle method for convection-diffusion equations. Part 1. The case of an isotropic viscosity. *Math. Comp.*, 53:485–507, 1989.
9. C. Greengard. The core spreading vortex method approximates the wrong equation. *J. Comp. Phys.*, 61:345–348, 1985.
10. L. Greengard and V. Rokhlin. A fast algorithm for particle simulations. *J. Comp. Phys.*, 73:325–348, 1987.
11. P. Koumoutsakos. Inviscid axisymmetrization of an elliptical vortex. *J. Comp. Phys.*, 138:821–857, 1997.
12. A. Leonard. Vortex methods for flow simulation. *J. Comp. Phys.*, 37:289–335, 1980.
13. P. J. Roache. Quantification of uncertainty in computational fluid dynamics. *Ann. Rev. Fluid Mech.*, 29:123–160, 1997.
14. L. Rosenhead. The formation of vortices from a surface of discontinuity. *Proc. R. Soc. Lond. A*, 134:170–192, 1931.
15. L. F. Rossi. Resurrecting core spreading vortex methods: A new scheme that is both deterministic and convergent. *SIAM J. Sci. Comput.*, 17:370–397, 1996.
16. L. F. Rossi. Achieving high-order convergence rates with deforming basis functions. *SIAM J. Sci. Comput.*, 26(3):885–906, 2005.
17. L. F. Rossi, J. F. Lingevitch, and A. J. Bernoff. Quasi-steady monopole and tripole attractors for relaxing vortices. *Phys. Fluids*, 9:2329–2338, 1997.

18. T. Sarpkaya. Computational methods with vortices. *J. Fluids Eng.*, 11:5–52, 1989.
19. R. Schaback. Improved error bounds for scattered data interpolation by radial basis functions. *Math. Comp.*, 68(225):201–216, 1999.

A Hybrid Meshless/Spectral-Element Method for the Shallow Water Equations on the Sphere

Christopher D. Blakely

Center for Scientific Computation and Mathmematical Modeling, University of Maryland, College Park, MD 20742-3289, U.S.A.; E-mail: cblakely@cscamm.umd.edu, www.cscamm.umd.edu/~cblakely

Abstract. A hybrid approximation scheme for the shallow-water equations on the sphere is proposed which utilizes spectral-element approximation coupled with regional meshless collocation. The issue of satisfying continuity conditions across spectral-element to meshless collocation interfaces for this domain decomposition method is discussed and gives an example of a meshless collocation framework which can be successfully coupled with spectral-element approximation. We conclude the paper with numerical examples using the proposed hybrid scheme on two well-known standardized test problems for the rotational shallow-water equations on the sphere.

Key words: Meshless methods, spectral-element approximation, shallow-water equations.

1 Introduction and Motivations

The purpose of this paper will focus on constructing an innovative hybrid approximation method for geophysical fluid dynamics. To accomplish such a task, we will focus on the shallow-water equations which provide a useful model to global climate modeling because their solutions include nonlinear effects and wave structures similar to those of the full primitive equations of the atmosphere. The main backbone of this hybrid meshless/spectral-element shallow-water model will be focusing on incorporating a unique regional scale approximation method. This regional scale method will be accomplished by using a robust meshless approximation scheme called the empirical Backus-Gilbert reproducing kernel developed by Blakely in [5]. The advantage of such a hybrid approximation is two-fold: (1) high-order approximation results can be obtained in complex shaped geometries without the need of a mesh. Thus, no remeshing of a local region into smaller rectangles is needed, ultimately speeding up the computation time; (2) the Backus–Gilbert reproducing kernel method has been shown to be endowed with the unique power of ignoring oscillatory effects

V.M.A. Leitão et al. (eds), Advances in Meshfree Techniques, 199–219.

in scattered data. This will be useful when the spectral approximation forms high oscillations due to discontinuities in the data.

This article is organized as follows. Section 2 begins with a brief review of the shallow-water equations defined on the cubed-sphere and its discretization in time using a semi-implicit time stepping scheme. We follow this discussion in Section 3 with a brief review of the spectral-element and Backus-Gilbert reproducing kernel discretization methods of the time-discrete shallow-water equations. Next, we present the three-field variational formulation which effectively couples the two types of approximations in a weak sense by introducing two additional approximation spaces on the interfaces between the approximation, akin to domain decompositon and the mortar element method. Implementation of the coupling is then given and then finally, in order to verify the mathematical correctness of the algorithms presented in this paper and to validate the performance hybrid model, we conclude the paper with some standardized test cases which were proposed by Williamson et al. in [22].

2 The Shallow-Water Equations on the Cubed-Sphere

Being the simplest form of motion equations that can approximate the horizontal structure of the atmosphere or the dynamics of oceans, the shallow-water equations have been used as a robust testing model in atmospheric and oceanic sciences. The solutions can represent certain types of motion including Rossby waves and inertia-gravity waves while describing an incompressible fluid subject to gravitational and rotating acceleration. The governing equations for the inviscid flow of a thin layer of fluid in 2-D are the horizontal momentum and continuity equations for the velocity $\mathbf{u} = (u_1, u_2)$ and the geopotential height η .

While there are many different ways of defining the shallow-water equations, we focus in this model on cubed-sphere geometry originally proposed by Sadourny in [17] and used in other global models in recent years such as [20] and [21]. We begin by a brief review of the cubed-sphere while adopting notational conventions from [20]. Consider a cube inscribed inside a sphere where each corner of the cube is a pointin the sphere and where each face of the cube is subdivided into N_E subregions. The goal is to project each face of the cube onto the sphere and in effect, obtain a quasi-uniform spherical grid of $6 \times N_E$ subregions which can be further subdivided into many spectral element and meshless collocation subregions. In the mapping of the cube to sphere, each face of the cube is constructed with a local coordinate system and employs metric terms for transforming between the cube and the sphere which will now be defined.

Let (α, β) be equal angular coordinates such that $-\pi/4 \leq \alpha, \beta \leq \pi/4$. Then any x_1 and x_2 on a face P_i of the cube is related through $x_1 = \tan\alpha, x_2 = \tan\beta$. We denote $\mathbf{r}$ the corresponding position vector on the sphere with longitude λ and latitude θ. For such an equiangular projection, we define basis vectors $\mathbf{a}_1 = \mathbf{r}_\alpha$ and $\mathbf{a}_2 = \mathbf{r}_\beta$ which may be written as

$$\mathbf{r}_\alpha = \frac{1}{\cos^2\alpha}\mathbf{r}_{x_1}, \qquad \mathbf{r}_\beta = \frac{1}{\cos^2\beta}\mathbf{r}_{x_2}, \tag{1}$$

where $\mathbf{r}_{x_1}$ and $\mathbf{r}_{x_2}$ are defined as $\mathbf{r}_{x_1} = (\cos\theta\,\lambda_{x_1}, \theta_{x_1})$ and $\mathbf{r}_{x_2} = (\cos\theta\,\lambda_{x_2}, \theta_{x_2})$ The metric tensor $g_{ij},\ i, j \in [1, 2]$ can be derived as

$$g_{ij} = \mathbf{a_i}\cdot\mathbf{a_j} = \frac{1}{r^4\cos^2\alpha\cos^2\beta}\begin{bmatrix} 1+\tan^2\alpha & -\tan\alpha\tan\beta \\ -\tan\alpha\tan\beta & 1+\tan^2\beta \end{bmatrix},$$

where $r^2 = 1+\tan^2\alpha+\tan^2\beta$ and the Jacobian of the transformation and the matrix $\tilde{\mathbf{A}}$ are, respectively,

$$\sqrt{g} = [\det(g_{ij})]^{1/2} = \frac{1}{r^3\cos^2\alpha\cos^2\beta}, \qquad \tilde{\mathbf{A}} = \begin{bmatrix} \cos\theta\ \lambda_\alpha & \cos\theta\ \lambda_\alpha \\ \theta_\alpha & \theta_\beta \end{bmatrix}.$$

While using the definition of g_{ij} given in (2), we can write transformations between covariant and contravariant components of a vector $\mathbf{v}$ as

$$\begin{bmatrix} u_2 \\ u_2 \end{bmatrix} = \begin{bmatrix} g_{11}\ g_{12} \\ g_{21}\ g_{22} \end{bmatrix}\begin{bmatrix} u^1 \\ u^2 \end{bmatrix}, \qquad \begin{bmatrix} u^1 \\ u^2 \end{bmatrix} = \begin{bmatrix} g^{11}\ g^{12} \\ g^{21}\ g^{22} \end{bmatrix}\begin{bmatrix} u_1 \\ u_2 \end{bmatrix}. \tag{2}$$

With the metric terms defined, we can now write the shallow water equations in in the curvilinear coordinates system to be integrated on the cubed-sphere. In such a coordinate system, the shallow-water equations can be written as follows

$$\frac{\partial u^i}{\partial t} = -g^{ij}\left[\epsilon_{jk}u^k g(f+\zeta) + \frac{\partial}{\partial x^j}\left(\frac{1}{2}u_k u^k\right) + \frac{\partial\eta}{\partial x^j}\right],$$
$$\frac{\partial\eta'}{\partial t} = -u^j\frac{\partial\eta}{\partial x^j} - \frac{\eta}{g}\frac{\partial}{\partial x^j}(g\,u^j).$$

Here, we define $\eta = \eta' + \eta_0$, f is the Coriolis force and ζ is the relative vorticity. Covariant and contravariant vectors are defined through the short-hand metric tensor notation $u^i = g^{ij}u_j$, $g^{ij} = (g_{ij})^{-1}$. Furthermore, using ϵ_{ij} as the two-dimensional permutation matrix.the divergence and relative vorticity can be calculated as

$$g\,\nabla\cdot\mathbf{v} = \frac{\partial}{\partial x^j}(g\,u^j), \qquad g\zeta = \epsilon_{ij}\frac{\partial u_j}{\partial x^i}. \tag{3}$$

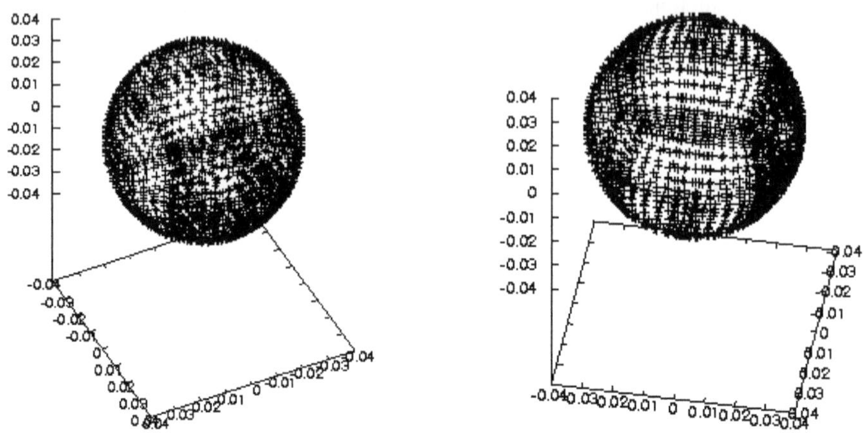

Fig. 1. Hybrid cube.

We note that the metric terms can be precalculated and stored once the issue of discretizing the cube has been resolved. To this end, we discuss the discretization of the cubed-sphere using the spectral element method in Section 3.1 and the meshless collocation method in Section 3.2. An example of the resulting discretized cube is shown in Figure 1.

2.1 Semi-Implicit Time Discretization

As an integral part of the hybrid meshless/spectral-element model, the semi-implicit time stepping scheme which we discuss in this subsection has many computational advantages. Semi-implicit time-stepping schemes were first used in atmospheric models in order to aleviate the problem of stability constraints ultimately due to the fast moving gravity waves in the discrete shallow water equations [20]. They have been successfully applied for allowing an increase in the time step without affecting the atmospherically important Rossby waves. Such a semi-implicit method is described in this subsection and was originally proposed in the spectral element model developed in [20].

In the hybrid meshless/spectral-element method, the semi-implicit time stepping is composed of an explicit leapfrog scheme for the advection terms combined with a Crank–Nicholson scheme for the gradient and divergence terms. Adopting the difference notation $\delta u^i = u^{i(n+1)} - u^{i(n-1)}$ and $\delta \eta^i = \eta^{(n+1)} - \eta^{(n-1)}$, the time discretized shallow water equations in curvilinear coordinates can be written as

$$\delta u^i + \Delta t g^{ij} \frac{\partial}{\partial x^j}(\delta\, \eta) = 2\Delta t\Big[- g^{ij} \frac{\partial}{\partial x^j}(\eta)^{n-1} + f_u^{i(n)} \Big],$$

$$\delta \eta^i + \Delta t \frac{\eta_0}{g} \frac{\partial}{\partial x^j}(g \delta u^j) = 2\Delta t\Big[\frac{\eta_0}{g} \frac{\partial}{\partial x^j}(g u^j)^{n-1} + f_\eta^{i(n)} \Big],$$

where the tendencies f_u and f_η contain nonlinear terms along with the Coriolis term, namely

$$f_u = -g^{ij}\Big[\epsilon_{jk} u^{k(n)} g(f + \zeta^n) + \frac{\partial}{\partial x^j}\Big(\frac{1}{2} u_k u^k\Big)^n \Big],$$

and

$$f_\eta = -u^j \frac{\partial \eta}{\partial x^j}.$$

Lastly, bringing the implicit terms to the left hand side of the equation and the explicit terms to the right, we end up with the time discrete evolution form of the shallow water equations

$$u^{i(n+1)} + \Delta t g^{ij} \frac{\partial}{\partial x^j}(\eta)^{n+1} = u^{i(n-1)} - \Delta t g^{ij} \frac{\partial}{\partial x^j}(\eta)^{n-1} + 2\Delta t f_u^{i(n)} \tag{4}$$

$$\eta^{n+1} + \Delta t \frac{\eta_0}{g} \frac{\partial}{\partial x^j}(g u^j)^{n+1} = \eta^{n-1} - \Delta t \frac{\eta_0}{g} \frac{\partial}{\partial x^j}(g u^j)^{n+1} + 2\Delta t f_\eta^n \tag{5}$$

which is now in the form needed for spatial discretization using the hybrid meshless/spectral-element. Because of the Crank–Nicholson terms, the storage of two previous time steps is needed. The overall performance of this semi-implicit method is reduced to the performance of a robust solver that can ultimately be parallelizable in an efficient manner for obtaining the solution at the $n + 1$ time step. The preconditioned conjugate gradient method using a block-jacobi type preconditioner offers such an approach but at the cost of inter-elemental communication at every iteration step of the iterative solver. The method is of course highly dependent on the spatial approximation scheme and will be discussed in the next two subsections. To this end, we first briefly review the construction of the nodal spectral element spproximation on the cubed-sphere.

3 Hybrid Meshless/Spectral-Element Discretization

3.1 Spectral-Element Discretization

The spectral element formulation for the cubed sphere begins by decomposing each face on the unit cube denoted by P_i, $i = 1, \dots, 6$ into N_e nonoverlapping subdomain elements Ω^e of equal area which are each referenced to a standard element $\Omega_{st} = [-1, 1]^2$ by a mapping $\mathbf{x}^e(r, s) \in \Omega^e$ for $(r, s) \in \Omega_{st}$. The mapping also has a well defined inverse $(r, s)^e(\mathbf{x}) \in \Omega_{st}$, $\mathbf{x} \in \Omega^e$. Since two fields are being approximated in the discrete shallow water equations, namely the velocity and the geopotential, two different approximation spaces are needed. As in most shallow water models, we adapt the so-called staggered grid approach to discretization. We begin by defining the approximation space for the velocity as $V_N := \mathbb{P}_{N,N_e} \cap H^1(P_i)$, where $\mathbb{P}_{N,N_e}$ is the space of piecewise continuous functions that map to polynomials of degree less than or equal N to the reference element on face i. Namely,

$$\mathbb{P}_{N,N_E}(P_i) := v(\mathbf{x}^e(r,s))|_{\Omega^e} \in \mathbb{P}_N(r) \otimes \mathbb{P}_N(s), \quad e = 1, \dots, N_e,$$

where $\mathbb{P}_N(r)$ is the space of all polynomials of degree less than or equal to N. In order to facilitate inter-element continuity on Ω^e for all $e \in [1, N_e]$ and more globally, inter-face continuity on P_i, $i = 1, \dots, 6$, nodal Lagrangian interpolants are used to construct the basis within each element. In this paper, we use Lagrangian interpolants constructed from orthogonal Legendre functions of degree p.

While many spectral-element and spectral collocation models of fluid dynamics such as the ones found in [19] have utilized a staggered grid where the pressure/geopotential field on each element is discretized on an $N-2$ Gauss–Legendre distribution which does not include the boundaries of the element, this hybrid meshless/spectral-element model relies on boundary information of the geopotential field as explained in the next section on the three-field formulation for the shallow water equations. We thus build a staggered grid for the geopotential field which includes the boundaries of each element Ω^e by considering the space $M_N := \mathbb{P}_{(N-2,N_e)} \cap L^2(P_i)$ and distributing $(N-2)^2$ Gauss–Lobotto–Legendre points (ξ_i^1, ξ_j^2).

Using the space M_N, the geopotential is expanded in a similar manner to the velocity components by using the $(N-2)$-th degree Lagrangian interpolants $\tilde{\pi}_i$ as

$$\eta(r,s)|_{\Omega^e} = \sum_{i=0}^{N-2} \sum_{j=0}^{N-2} \eta_{ij}^e \tilde{\pi}_i(r) \tilde{\pi}_j(s),$$

with η_{ij}^e being the nodal values of the geopotential at (ξ_i^1, ξ_j^2). This expansion will require evaluation not only on the geopotential grid, but on the velocity grid as well. As will be shown later, the geopotential field provides the means for coupling the spectral element and meshless collocation approximations.

With the definitions of the Lagrangian nodal expansion and quadrature rule in place for each element, we can now apply weak formulation to the semi-implicit time discretized shallow water equations. Find $(u^i, \eta) \in V_N \times M_N$ such that for all $(v_i, q) \in V_N \times M_N$

$$\langle u^{i(n+1)}, v_i \rangle - \Delta t g^{ij} \Big\langle \eta^{n+1}, \frac{\partial}{\partial x^j} v_i \Big\rangle$$
$$= \langle u^{i(n-1)}, v_i \rangle + \Delta t g^{ij} \Big\langle \eta^{n-1}, \frac{\partial}{\partial x^j} v_i \Big\rangle + 2\Delta t \langle f_u^{i(n)}, v_i \rangle,$$
$$\langle \eta^{n+1}, q \rangle + \Delta t \frac{\eta_0}{g} \Big\langle q, \frac{\partial}{\partial x^j} (g u^j)^{n+1} \Big\rangle$$
$$= \langle \eta^{n-1}, q \rangle - \Delta t \frac{\eta_0}{g} \Big\langle q, \frac{\partial}{\partial x^j} (g u^j)^{n+1} \Big\rangle + 2\Delta t \langle f_\eta^n, q \rangle. \tag{6}$$

With the matrices, cubed-sphere metrics, and Coriolis forcing constructed on each element, the semi-implicit scheme can now be formulated into N_e local Helmholtz problems where the geopotential is solved at every timestep from a discrete Helmholtz problem and then 'communicated' to the velocity field. Writing the assembled discrete shallow water system from the previous subsection in matrix-vector form, we get

$$\begin{bmatrix} \mathbf{B}^t & -\mathbf{D}^t \\ \mathbf{D}^t & \tilde{B}^t \end{bmatrix} \begin{bmatrix} \mathbf{u}^{n+1} \\ \eta^{n+1} \end{bmatrix} = \begin{bmatrix} \mathbf{R}_u^t \\ R_\eta^t \end{bmatrix}, \tag{7}$$

where

$$\mathbf{B}^t = \frac{\mathbf{B}}{\Delta t}, \quad \mathbf{R}_u^t = \frac{\mathbf{R}_u}{\Delta t}, \quad \tilde{B}^t = \frac{\tilde{B}}{\Delta t \eta_0}, \quad R_\eta^t = \frac{R_\eta}{\Delta t \eta_0}. \tag{8}$$

The Helmholtz problem for the geopotential perturbation at each timestep is obtained by solving for the velocity $\mathbf{u}^{n+1}$ in the above block system to arrive at

$$\mathbf{u}^{n+1} = \mathbf{B}^{-1}(\mathbf{R}_u^t + \Delta t g^{ij} \mathbf{D}^T \eta^{n+1}), \tag{9}$$

and then applying back-substitution to obtain an equation for the geopotential

$$g \tilde{B} \eta^{n+1} + \Delta t^2 \eta_0 \mathbf{D} g \mathbf{B}^{-1} g^{ij} \mathbf{D}^T \eta^{n+1} = R_\eta', \tag{10}$$

where

$$R_\eta' \equiv g R_\eta - \Delta t \eta_0 \mathbf{D} g \mathbf{B}^{-1} \mathbf{R}_u. \tag{11}$$

Once the geopotential η^{n+1} is computed, the velocity components u^1, u^2 are computed from (9) where thereafter, shared local nodal values on element boundaries of the velocity components are then averaged.

Furthermore, due to the fact that η_0, g, $\tilde{B}$ and $\mathbf{B}^{-1}$ are diagonal, and g^{ij} can be shown to be symmetric, it is easy to see that the matrix $\mathbf{H}_{SE}$ is symmetric and positive definite. In effect, an efficient preconditioned conjugate gradient method can be constructed by using local element direct solvers for the Helmholtz problem with zero Neumann pressure gradient boundary conditions. The inverse of each local Helmholtz operator matrix restricted to an element $\mathbf{H}|_{\Omega^e}$ is computed once and stored for use as a block-Jacobi preconditioner. This preconditioning technique enjoys a computational structure ideal for parallel processors due to the fact that the preconditioner is strictly local to an element and requires no global communication.

3.2 The Empirical Backus-Gilbert Reproducing Kernel Discretization

Coupled with the global spectral-element method for use in regional approximation of the shallow-water model, the empirical Backus–Gilbert reproducing kernel discretization method, originally introduced in Blakely [5] has been demonstrated to produce highly accurate solutions to time-dependent nonlinear PDEs while being endowed with great freedom in choosing the approximation space for building the reproducing kernel. Furthermore, as the name of the method suggests, the EBGRK is completely empirical with respect to the distribution of meshless nodes in the domain of interest. For complete details of the method the reader is referred to [5].

The EBGRK method considers a quasi-interpolant of the form

$$Pu(\mathbf{x}) = \sum_{i=1}^{N} u(\mathbf{x}_i)\Psi_i(\mathbf{x}), \tag{12}$$

where $\mathbf{u} = [u(\mathbf{x}_1), \ldots, u(\mathbf{x}_N)]^T$ represents the given data on a set of N distinct evaluation nodes $\mathcal{X} = \{\mathbf{x}_1, \ldots, \mathbf{x}_N\}$ on a bounded domain $\Omega \subset \mathbb{R}^2$. The finite set of nodes $\mathcal{X}$ is endowed with a *separation distance* defined as

$$q_{\mathcal{X}} := \frac{1}{2} \min_{\mathbf{x}_j \neq \mathbf{x}_i} \|\mathbf{x}_i - \mathbf{x}_j\|_2.$$

The quasi-interpolant $\Psi_i(\mathbf{x})$, or discrete reproducing kernel in some literature, is constructed to be minimized in a discrete quadratic expression subject to some approximation space reproduction constraints. Details on constructing the empirical reproducing kernel is out of the scope of this paper. We refer the reader to Blakely [5] for the construction of the kernel and efficient numerical implementation.

To initiate regional meshless approximation on the cubed-sphere, consider the domain $\Omega^M = \cup_{i=1}^{M} \Omega^{e_i}$ constructed of M contiguous elements on the discretized cubed-sphere. For simplicity, we will assume Ω^M lies on only one face of the the cube. Building a meshless approximation space $\mathcal{V}_{\Psi,\mathcal{X}}$ begins by randomly distributing two sets of N_M distinct collocation nodes in Ω^M and on its boundary $\partial\Omega^M$ giving two sets $\mathcal{X}_M^V$, $\mathcal{X}_M^\eta$ such that $\mathcal{X}_M^V = \mathcal{X}_M^\eta$. Using the EBGRK method, the kernels $\Psi_i(\cdot)$ and $\Phi_i(\cdot)$ are constructed to form the discrete spaces $\mathcal{V}_{\Psi,\mathcal{X}} = \text{span}\{\Psi_i(\cdot),\ i = 1, \ldots, N_M\}$ and $\mathcal{M}_{\Phi,\mathcal{X}} = \text{span}\{\Phi_i(\cdot),\ i = 1, \ldots, N_M\}$ and with respect to the sets $\mathcal{X}_M^V$ and $\mathcal{X}_M^\eta$, respectively.

For writing the cubed-sphere shallow water equations in strong form, we utilize the matrix-vector form of the equations originally given in [21]. As in any other collocation method, we construct the set of Dirac delta test functionals $\Pi_{\mathcal{X}_M^V} = \{\delta_{\mathbf{x}_1}, \ldots, \delta_{\mathbf{x}_{N_M}}\} \subset (H^1)'(\Omega^M)$ and multiply each of the velocity components and geopotential by each Dirac delta test functional $\delta_{\mathbf{x}_i} \in \Pi_{\mathcal{X}_M^V}$ evaluated which gives

$$\langle\delta_{\mathbf{x}_j}, u^{i(n+1)}\rangle + \Delta t\Big\langle\delta_{\mathbf{x}_j}, g^{ij}\frac{\partial}{\partial x^j}(\eta)^{n+1}\Big\rangle$$
$$= \langle\delta_{\mathbf{x}_j}, u^{i(n-1)}\rangle - \Delta t g^{ij}\Big\langle\delta_{\mathbf{x}_j}, \frac{\partial}{\partial x^j}(\eta)^{n-1}\Big\rangle + 2\Delta t\langle f_u^{i(n)}, \delta_{\mathbf{x}_j}\rangle,$$
$$\langle\delta_{\mathbf{x}_j}, \eta^{n+1}\rangle + \Delta t\Big\langle\delta_{\mathbf{x}_j}, \frac{\eta_0}{g}\frac{\partial}{\partial x^j}(gu^j)^{n+1}\Big\rangle$$
$$= \langle\delta_{\mathbf{x}_j}, \eta^{n-1}\rangle - \Delta t\Big\langle\delta_{\mathbf{x}_j}, \frac{\eta_0}{g}\frac{\partial}{\partial x^j}(gu^j)^{n+1}\Big\rangle + 2\Delta t\langle\delta_{\mathbf{x}_j}, f_\eta^n\rangle \tag{13}$$

with

$$\langle\delta_{\mathbf{x}_j}, f_u^{i(n)}\rangle = -\epsilon_{jk}\langle\delta_{\mathbf{x}_j}, g^{ij}u^{k(n)}gf\rangle - \Big\langle\delta_{\mathbf{x}_j}, g^{ij}u_k^n\frac{\partial}{\partial x^j}u^{k(n)}\Big\rangle,$$

and

$$\langle\delta_{\mathbf{x}_j}, f_\eta^n\rangle = -\Big\langle\delta_{\mathbf{x}_j}, u^j\frac{\partial\eta^n}{\partial x^j}\Big\rangle.$$

In order to approximate these equations spatially with the EBGRK method, we look for a solution $\mathbf{u} \in (\mathcal{V}_{\Psi,\mathcal{X}})^2 \subset (H^1(\Omega^M))^2$ and $\eta \in \mathcal{V}_{\Psi,\mathcal{X}}$ by taking for all $n \geq 0$

$$u_k^n(\mathbf{x}_j) = \sum_{i=1}^{N_M}\tilde{u}_k^n(\mathbf{x}_i)\Psi_i(\mathbf{x}_j), \quad \eta^n(\mathbf{x}_j) = \sum_{i=1}^{N_M}\tilde{\eta}^n(\mathbf{x}_i)\Phi_i(\mathbf{x}_j) \text{ for } \mathbf{x}_k \in \mathcal{X}_M^V,$$

where $\tilde{u}_k^n(\mathbf{x}_i)$ and $\tilde{\eta}^n(\mathbf{x}_i)$ are the approximated values at the collocation nodes $\mathbf{x}_i$ at time step n. Substituting these into (13) and applying the Dirac delta functionals, we get for all $\delta_{\mathbf{x}_j} \in \Pi_{\mathcal{X}_M^V}$

$$\Big\langle\delta_{\mathbf{x}_j}, \sum_{i=1}^{N_M}\tilde{u}^{k(n+1)}(\mathbf{x}_i)\Psi_i(\mathbf{x})\Big\rangle + \Big\langle\Delta t g^{ij}\frac{\partial}{\partial x^j}\delta_{\mathbf{x}_j}, \sum_{i=1}^{N_M}\tilde{\eta}^{n+1}(\mathbf{x}_i)\Phi_i(\mathbf{x})\Big\rangle$$
$$= \Big\langle\delta_{\mathbf{x}_j}, \sum_{i=1}^{N_M}\tilde{u}^{k(n-1)}(\mathbf{x}_i)\Psi_i(\mathbf{x})\Big\rangle + \Big\langle\Delta t g^{ij}\frac{\partial}{\partial x^j}\delta_{\mathbf{x}_j}, \sum_{i=1}^{N_M}\tilde{\eta}^{n-1}(\mathbf{x}_i)\Phi_i(\mathbf{x})\Big\rangle$$
$$+ \langle 2\Delta t\delta_{\mathbf{x}_j}, f_u^{i(n)}\rangle,$$
$$\Big\langle\delta_{\mathbf{x}_j}, \sum_{i=1}^{N_M}\tilde{\eta}^{n+1}\Phi_i(\mathbf{x})\Big\rangle + \Big\langle\Delta t\frac{\eta_0}{g}\frac{\partial}{\partial x^j}\delta_{\mathbf{x}_j}, g\sum_{i=1}^{N_M}\tilde{u}^{j(n+1)}(\mathbf{x}_i)\Psi_i(\mathbf{x})\Big\rangle$$
$$= \Big\langle\delta_{\mathbf{x}_j}, \sum_{i=1}^{N_M}\tilde{\eta}^{n-1}(\mathbf{x})\Phi_i(\mathbf{x})\Big\rangle - \Big\langle\Delta t\frac{\eta_0}{g}\frac{\partial}{\partial x^j}\delta_{\mathbf{x}_j}, g\sum_{i=1}^{N_M}\tilde{u}^{j(n-1)}(\mathbf{x}_i)\Psi_i(\mathbf{x})\Big\rangle$$
$$+ 2\langle\Delta t\delta_{\mathbf{x}_j}, f_\eta^n\rangle. \tag{14}$$

The calculation of

$$g_{\mathbf{x}_j}\frac{\partial \Psi_i(\mathbf{x}_j)}{\partial x^k}, \quad \mathbf{x}_j \in \mathcal{X}_M^V,$$

which is used in the divergence terms of the strong form shallow water system is made prior to time stepping and stored as matrices in the form

$$D_k = \begin{pmatrix} g_{\mathbf{x}_1}\Upsilon_1(\mathbf{x}_1) & g_{\mathbf{x}_1}\Upsilon_2(\mathbf{x}_1) & \cdots & g_{\mathbf{x}_1}\Upsilon_{N_M}(\mathbf{x}_1) \\ g_{\mathbf{x}_2}\Upsilon_1(\mathbf{x}_2) & g_{\mathbf{x}_2}\Upsilon_2(\mathbf{x}_2) & \cdots & g_{\mathbf{x}_2}\Upsilon_{N_M}(\mathbf{x}_2) \\ \vdots & & & \\ g_{\mathbf{x}_{N_M}}\Upsilon_1(\mathbf{x}_{N_M}) & g_{\mathbf{x}_{N_M}}\Upsilon_2(\mathbf{x}_{N_M}) & \cdots & g_{\mathbf{x}_{N_M}}\Upsilon_N(\mathbf{x}_{N_M}) \end{pmatrix}, \tag{15}$$

where $\Upsilon_j(\cdot)$ denotes the differential operator $\frac{\partial}{\partial x^k}$ acting on the kernel $\Psi_j(\cdot)$, which was shown how to be constructed in Section 3.2. A similar matrix $g_{ij}D_k^T$ used in calculating the gradient of the geopotential in strong form is also computed and stored prior to time stepping. These matrices are akin to the two-dimensional derivative matrix $\mathbf{D}_i$ of size $N^2 \times N^2$ in the spectral element formulation.

Using notation borrowed form the spectral element formulation, we write the discretized equations in matrix form with $\mathbf{D} = (D_1, D_2)$ as the derivative matrices with respect to the collocation nodes and $\mathbf{B}$, $\tilde{B}$ as the collocation matrices for the velocity and geopotential, respectively. This leads to the system

$$\begin{bmatrix} \mathbf{B}^t & -\mathbf{D}^t \\ \mathbf{D}^t & \tilde{B}^t \end{bmatrix} \begin{bmatrix} \tilde{\mathbf{u}}^{n+1} \\ \tilde{\eta}^{n+1} \end{bmatrix} = \begin{bmatrix} \mathbf{R}_u^t \\ R_\eta^t \end{bmatrix},$$

where

$$\mathbf{B}^t = \frac{\mathbf{B}}{\Delta t}, \quad \mathbf{R}_u^t = \frac{\mathbf{R}_u^t}{\Delta t}, \tag{16}$$

$$\tilde{B}^t = \frac{\tilde{B}}{\Delta t \eta_0}, \quad R_\eta^t = \frac{R_\eta}{\Delta t \eta_0}. \tag{17}$$

Performing the Uzawa velocity-geopotential decoupling algorithm where we solve for the velocity $\mathbf{u}^{n+1}$ in the above block system to arrive at

$$\tilde{\mathbf{u}}^{n+1} = \mathbf{B}^{-1}(\mathbf{R}_u^t + \Delta t g^{ij}\mathbf{D}^T\tilde{\eta}^{n+1}), \tag{18}$$

and then applying back-substitution to obtain an equation for the geopotential at the time step $n+1$.

$$g\tilde{B}\tilde{\eta}^{n+1} + \Delta t^2\eta_0\mathbf{D}g\mathbf{B}^{-1}g^{ij}\mathbf{D}^T\tilde{\eta}^{n+1} = R'_\eta, \tag{19}$$

where

$$R'_\eta \equiv gR_\eta - \Delta t\eta_0\mathbf{D}g\mathbf{B}^{-1}\mathbf{R}_u.$$

As with the spectral element case, we are now concerned with the solution to the Helmholtz problem for the geopotential. Although the meshless collocation method yields a strong form of the discrete shallow water model, the resulting discrete Helmholtz equation is similar to the spectral element Helmholtz equation in that the matrix operator

$$H_{MM} = g\tilde{B} + \Delta t^2 \eta_0 \mathbf{D} g \mathbf{B}^{-1} g^{ij} \mathbf{D}^T \tag{20}$$

is symmetric and positive definite and of size $N_M \times N_M$. It is important to notice that the term

$$\mathbf{D} g \mathbf{B}^{-1} g^{ij} \mathbf{D}^T$$

is a discrete pseudo-Laplacian operator on Ω^M, a local domain on the sphere S^2. Because of the fact that the domain is locally defined on the sphere, essential boundary conditions on $\partial\Omega^M$ are needed in order to show direct equivalence to a local Helmholtz elliptic problem on Ω^M. This was not the case with the spectral element discretization on the sphere since no boundary conditions are needed for the global shallow water model. As a result, the boundary information on $\partial\Omega^M$ must come from the spectral element approximation in order to produce a unique solution to the discrete Helmholtz problem (19). As we propose in the next section, if $\tilde{\eta}^{n+1}$ is the unique solution to this local Helmholtz problem at time $n+1$ with respect to the boundary information given by a global spectral element discretization at time $n+1$, then $\tilde{\eta}^{n+1}$ is an approximation to the geopotential field restricted to Ω^M at time step $n+1$. To accomplish this task, we adapt the three-field domain decomposition algorithm developed in [7] for coupling the two Helmholtz discretizations, which is discussed in the next section.

4 Coupling the Meshless and SE Approximations

Because the meshless approximation is done locally utilizing the strong formulation of the shallow-water equations on a local domain Ω^M, certain transition conditions are needed on the boundary of the subdomain connecting the meshless and spectral-element approximations in order to satisfy continuity and flux conditions of the solution along with the artificial fluxes of the field variables. In 1994, Brezzi and Marini (see [7]) developed a method termed the *three-field formulation* for *hybrid* finite-element formulations where the goal was to give the possibility of coupling different finite-element approximations using different meshes and basis functions from one subdomain to another.

In this paper, we extend the idea of the three-field technique to couple spectral-element and meshless collocation methods. As shown in the previous sections, the manner in which we couple the two approximation schemes is done implicity. Namely, after deriving the semi-implicit method, a symmetric positive definite discrete Helmholtz type equation was left to be solved at each time step for the geopotential. With the solution of the geopotential at hand, it could then be used to approximate the velocity field at the same time step. So the question that remains is how to solve Helmholtz equations for the coupled meshless/spectral-element approximation. In this section we consider solving the elliptic problem

$$Hu = \Delta u(\mathbf{x}) + g(\mathbf{x})u(\mathbf{x}) = f(\mathbf{x}), \quad \in S^2, \tag{21}$$

where S^2 is the unit sphere which is discretized via the cubed-sphere method discussed in Section 2. The fact that spatial discretization is not performed with spherical harmonics but rather on a cubed-sphere mesh allows for an adaptive localized approximation using meshless collocation via domain decomposition. Thus the heart of the hybrid shallow water model lies in the efficient handling of the Helmholtz equation on the sphere using the meshless/spectral-element formulation.

For proper stability analysis of this new three-field formulation for coupling spectral-elements and meshless collocation including Babuška–Brezzi inf-sup type conditions, the reader is referred to the paper by Blakely [6]. In order to introduce the method, we must first discuss the necessary approximation spaces that will be used in the formulation and discretization and their relevant physical meaning.

4.1 The Continuous Three-Field Formulation

Using the notation from the previous subsections, suppose we have are given a subdomain of M unioned elements $\Omega^M = \cup_{i=1}^M \Omega^{e_i}$. For simplicity of exposition of the three-field method, we assume Ω^M is on one face of the cube. Let $\Gamma_{SE/M}$ denote the boundary of Ω^M which we will call the *interface* of the hybrid method between the spectral element and meshless collocation approximations. Finally, we denote $\Omega^{SE} = \Omega - \Omega^M$, namely the collection of spectral elements not in Ω^M and then set $\Omega_1 := \Omega^{SE}$, $\Omega_2 := \Omega^M$ and $\Gamma_i := \partial\Omega_i$, for $i = [1, 2]$, which are the boundaries of these domains sharing the interface $\Gamma_{SE/M}$.

In addition to the Sobolev space $H_0^1(\Omega_i)$ on each domain Ω_i, utilizing the interface $\Gamma_{SE/M}$ leads to two additional types of spaces that will be needed for domain decomposition. We define a trace space and two dual spaces on $\Gamma_{SE/M}$ by considering $H^{1/2}(\Gamma_{SE/M})$ with corresponding norm $\|\cdot\|_\Gamma := \|\cdot\|_{H^{1/2}(\Gamma_{SE/M})}$ and denote the dual of this space as $H^{-1/2}(\Gamma)$. Furthermore we introduce two spaces of Lagrangian multipliers which provide the role of enforcing necessary boundary continuity over the interface $\Gamma_{SE/M}$ and are defined as $\Lambda_i := H^{-1/2}(\Gamma_i)$ for $i = [1, 2]$ which can be regarded as the dual of the trace spaces associated with the two Hilbert spaces $H^1(\Omega_1)$ and $H^1(\Omega_2)$. The Lagrangian multiplier space is endowed with the standard scalar inner product $L^2(\Gamma)$, $\langle \Lambda^i, H^{1/2}\rangle_\Gamma = \int_\Gamma \lambda_i u^i ds$ for $u^i \in H^1(\Omega)$. The third function space which acts as the global continuity space for the hybrid approximation is defined on the interface Γ as restrictions of functions on the sphere S^2 to the interface. Namely

$$\Phi := \{v \in L^2(\Gamma) : \exists u \in H^1(S^2), u = v \text{ on } \Gamma\}. \tag{22}$$

Global norms for the spaces $V_i := H_0^1(\Omega_i)$ and Λ_i can be given as broken norms over Ω_i

$$\|\mathbf{u}\|_V := \left(\sum_{i=1}^{2} \|u^i\|^2_{1,\Omega_i}\right), \quad \|\lambda\|_\Gamma := \left(\sum_{i=1}^{2} \|\lambda^i\|^2_{H^{-\frac{1}{2}}}(\Gamma_i)\right),$$

and can easily be shown to be Hilbert spaces with these induced norms. Furthermore, with the use of extension operators, the interface continuity space is endowed with the norm

$$\|\varphi\|_\Phi := \inf_{u|_\Gamma=\varphi} \|u\|_{1,\Omega}.$$

With the three approximation spaces at hand, the three-field formulation of the Helmholtz problem can be written for the two subdomains utilizing the additional two interface spaces Λ_i and the global continuity space Φ. Using the dual product notation $\langle\cdot,\cdot\rangle_i = \langle H^{-1/2}(\Gamma_i), H^{1/2}(\Gamma_i)\rangle$ the following variational form is called the three-field formulation. Find $\mathbf{u} \in \mathbf{V}$, $\lambda \in \Lambda$, and $\varphi \in \Phi$ such that

$$\begin{cases} \text{(i)} \ \sum_{i=1}^{2} \left(a_{\Omega_i}(u^i, v^i) - \langle\lambda^i, v^i\rangle_{\Gamma_i}\right) = \sum_{i=1}^{2} (f, v^i)_{\Omega_i}, \quad \forall \mathbf{v} \in \mathbf{V}, \\ \text{(ii)} \ \sum_{i=1}^{2} \langle\mu^i, u^i - \varphi\rangle_{\Gamma_i} = \mathbf{0}, \quad \forall \mu \in \Lambda \\ \text{(iii)} \ \sum_{i=1}^{2} \langle\lambda^i, \psi\rangle_{\Gamma_i} = \mathbf{0}, \quad \forall \psi \in \Phi \end{cases} \tag{23}$$

The bilinear operator a_{Ω_i} stems from the weak formulation of the Helmholtz equation and is defined as

$$a_{\Omega_i}(u^i, v^i) = \int_{\Omega_i} \nabla u^i \nabla v^i + g u^i v^i d\Omega_i.$$

Furthermore, the inner products of the form

$$\langle H^{-1/2}(\Gamma_i), H^{1/2}(\Gamma_i)\rangle_{\Gamma_i}$$

signify the artificial boundary or interface matching conditions. To be more specific, the second equation enforces weak continuity along the interface Γ_i with the solution u^i on Ω_i with respect to the interface continuity variable φ. The third equation serves two purposes: (1) it further constraines the space of Lagrangian multipliers Λ by adding orthogonality conditions with the interface space Φ, and (2) it renders the discrete formulation of the above system as a symmetric positive definite system which can then be solved for the global solution $(\mathbf{u}, \lambda, \varphi)$ using a preconditioned conjugate gradient method as will be shown in the next subsection.

We first note that a key observation in the three-field formulation comes from the first two equations of (23). For a given φ on the skeleton Γ, the first two equations are local Dirichlet problems where the boundary conditions on Γ_i are imposed in the weak sense. Because of this, one can show that the local problems are well-posed for a given sufficient φ. For a complete analysis of the three field method for coupling meshless and spectral-element approximations for elliptic equations, the reader is referred to the paper by Blakely [6]. The thesis by Rabin [16] and relevent references therein also give much insight to the three-field variational formulation in the finite-element context.

4.2 Discrete Version of the Three-Field Formulation

The difficulty in passing to the discrete formulation the variational problem (23) is in choosing the appropriate discrete subspaces of $\mathbf{V}$, Λ, and Φ. Arbitrarily choosing the subspaces can lead to unstable solutions of the discrete variational problem primarily due to not satisfying the discrete versions of the inf-sup conditions, so careful consideration of the spaces is necessary. In past approaches to the method, usually the discretization of the space $\mathbf{V}$ is chosen first and then Λ and Φ are chosen thereafter to satisfy the inf-sup requirements. In this section, we propose a discrete approximation to the three-fields formulation by considering the spectral-element and meshless collocation methods as the discretization tools which will then lead to the hybrid meshless/spectral-element method for the shallow water equations on the sphere.

With Ω_1 defining the domain for the spectral element approximation and the regional domain Ω_2 being allocated for meshless collocation, we define the space $V_N^1 := \mathbb{P}_{N,N_e} \cap H^1(\Omega_1)$, where $\mathbb{P}_{N,N_e}$ is the space of piecewise continuous functions that map to polynomials of degree less than or equal N to the reference element Ω^e. Namely,

$$\mathbb{P}_{N,E}(\Omega_1) := \left\{ v(\mathbf{x}^e(\mathbf{r}))|_{\Omega^e} \in \mathbb{P}_N(r) \otimes \mathbb{P}_N(s),\ e = 1, \ldots, N_e \text{ such that } \Omega^e \in \Omega_1 \right\},$$

where $\mathbb{P}_N(r)$ is the space of all polynomials of degree less than or equal to N. To restrict this space to Ω_1, we include all Ω^e such that $\Omega_1 \cap \Omega^e \neq 0$. This approximation space will provide each component of the velocity field on the spectral element partition Ω_1. As described in Section 3.1, the discrete geopotential space is obtained by utilizing the staggered grid approach and setting $M_N^1 := \mathbb{P}_{N-2,N_e} \cap H^1(\Omega_1)$. Since the bounderies on each element Ω^e are essential to the three-field method, the $N-2$ Gauss–Lobatto–Legendre distribution of nodes is used for the geopotential grid, which is contrary to many spectral element staggered grids which use Gauss–Legendre nodes for the geopotential/pressure field.

The regional domain Ω_2 allocates a collocation approximation by considering a random (or uniform) distribution of N_M distinct collocation nodes Ω^M and on its boundary $\partial\Omega^M$ giving two sets $\mathcal{X}_M^V$, $\mathcal{X}_M^\eta$ such that $\mathcal{X}_M^V = \mathcal{X}_M^\eta$. We then construct the approximation space $M_{N_M}^2 = V_{N_M}^2 := \text{span}\{\Psi_1(\mathbf{x}), \ldots, \Psi_{N_M}(\mathbf{x})\}$ as defined in Section 3.2.

With the spaces defined for the velocity and geopotential fields on each subdomain Ω_i, the Lagrangian multiplier spaces Λ^i for the interface boundaries Γ_i can now be constructed by using the spaces M_N^1 and $M_{N_M}^2$. Since M_N^1 defines a spectral approximation of order $N-2$, we define the Lagrangian multiplier space for Γ_1 as the space of Lagrangian interpolants of order less that or equal to N and restricted to Γ_1. This is given by

$$\begin{aligned}\Lambda_N^1 &= \mathbb{P}_{N,E}(\Gamma_1) \\ &:= \left\{\lambda(\mathbf{x}^e(r))|_{\Omega^e} \in \mathbb{P}_{N-4}(r)|_{\Gamma_1},\ e = 1, \ldots, N_e \text{ such that } \Omega^e \cap \Gamma_1 \neq 0\right\}.\end{aligned} \tag{24}$$

Using such a space for $H^{-\frac{1}{2}}(\Gamma_1)$, it can be shown that the discrete inf-sup condition for the interface inner product on Γ_1 is satisfied. Namely, for some constant $C_{1,N}$ dependent on the degree N of the spectral elements, we have

$$\inf_{\lambda_N^1 \in \Lambda_N^1/\{0\}} \sup_{\eta_N^1 \in M_N^1/\{0\}} \frac{\langle B\lambda_N^1, \eta_N^1\rangle_{\Gamma_1}}{\|\eta_N^1\|_{M_{N_M}^1} \ \|\lambda_N^1\|_{\Lambda_\mathbf{N}^1}} = \frac{\langle \lambda_N^1, \eta_N^1\rangle_1}{\|\eta_N^1\|_{M_{N_M}^1} \ \|\lambda_N^1\|_{\Lambda_\mathbf{N}^1}} > C_{1,N}$$

is satisfied. This result is proved in the paper on the Mortar Spectral Element method by Ben Belgacem et al. [3] in a similar interface inner product using Lagrangian multipliers.

In order to complete the space Λ we need the additional interface space on Ω_2. On the boundary Γ_2, a second meshless collocation space for Λ_N^2 is constructed using a random distribution of N_{M_Γ} nodes restricted to the interface Γ_2 producing the set X_{Γ_2}. Using the EBGRK method presented in Section 3.2, the Lagrangian multiplier space for Γ_2 is taken to be $\Lambda_{N_M}^2 = \text{span}\{\Psi_1^\lambda(\mathbf{x}), \ldots, \Psi_{N_M}^\lambda(\mathbf{x}) : \mathbf{x} \in \Gamma_2\} \subset H^{-1/2}(\Gamma_2)$ where $\Psi_i^\lambda(\cdot)$ denotes the i-th discrete reproducing kernel function on X_{Γ_2}.

Lastly, in order to connect the two pairs of approximation spaces (M_N^1, Λ_N^1) and $(M_{N_M}^2, \Lambda_{N_M}^2)$ on Ω_1 and Ω_2, respectively, we build a suitable discrete subspace of Φ by taking the Lagrangian interpolants constructed from Legendre polynomials of degree $N-2$ restricted to Γ. Namely,

$$\Phi_N := \{\varphi(\mathbf{x}^e(r))|_{\Omega^e} \in \mathbb{P}_{N-2}(r)|_\Gamma,\ e = 1, \ldots, N_e, \quad \text{such that } \ \Omega^e \cap \partial\Omega \neq 0\}.$$

This will ensure that the discrete inf-sup condition for Φ and Λ_N^1 on Γ_1 is satisfied.

The last issue we need to resolve in this three field formulation is complying with the strong form of the shallow water equations on Ω_2. To this end, since Ω_2 utilizes a meshless collocation technique, we define the set of test distributions on Ω_2 as $M_{\delta,N_M}^2 = \{\delta_{\mathbf{x}_i} : \mathbf{x}_i \in X_\eta^{N_M}\}$ where $\delta_{\mathbf{x}_i}$ is the Dirac delta function at node $\mathbf{x}_i$.

The original variational formulation in (23) can now be modified as follows. Find $(\eta_N^1, \lambda_N^1, \eta_N^2, \lambda_N^2, \varphi) \in M_N^1 \otimes \Lambda_N^1 \otimes M_{N_M}^2 \otimes \Lambda_{N_M}^2 \otimes \Phi$ such that

$$\begin{cases} \text{(i)} \quad a_{\Omega_1}(\eta_N^1, \chi_N^1) - \langle \lambda_N^1, \chi_N^1 \rangle_{\Gamma_1} = (f, \chi_N^1)_{\Omega_1}, \quad \forall \chi_N^1 \in M_N^1, \\ \text{(ii)} \quad \langle \mu_N^1, \eta_N^1 - \varphi_N \rangle_{\Gamma_1} = \mathbf{0}, \quad \forall \mu_N^1 \in \Lambda_N^1, \\ \text{(iii)} \quad \langle \lambda_N^1, \psi_N \rangle_{\Gamma_1} = \mathbf{0}, \quad \forall \psi_N \in \Phi, \end{cases} \tag{25}$$

and

$$\begin{cases} \text{(i)} \quad a_{\Omega_2}(\eta_N^2, \chi_N^2) - \langle \lambda_N^2, \chi_N^2 \rangle_{\Gamma_2} = (f, \chi_N^2)_{\Omega_2}, \quad \forall \chi_N^2 \in M_{\delta,N_M}^2, \\ \text{(ii)} \quad \langle \mu_N^2, \eta_N^2 - \varphi_N \rangle_{\Gamma_2} = \mathbf{0}, \quad \forall \mu_N^2 \in \Lambda_{N_M}^2, \\ \text{(iii)} \quad \langle \lambda_N^2, \psi_N \rangle_{\Gamma_1} = \mathbf{0} \quad \forall \psi_N \in \Phi. \end{cases} \tag{26}$$

Once the discrete approximation spaces have been chosen and numerical integration has been done, an efficient manner in solving this is to construct the Schur compliment system and then apply a conjugate gradient method. To do this, we first write (25) and (26) in algebraic form as:

$$\begin{aligned} A_i \eta_i - B_i^T \lambda_i &= \mathbf{f}_i, \\ -B_i \eta_i + C_i^T \varphi &= \mathbf{0}, \\ C_i \lambda_i &= \mathbf{0}, \end{aligned}$$

for $i = 1, 2$. Now applying block Gaussian elimination to the linear system, we obtain a linear system for φ as

$$\mathcal{S} \varphi = \mathbf{g}, \tag{27}$$

where $\mathcal{S} = \mathcal{S}_1 + \mathcal{S}_2$, $\mathbf{g} = \mathbf{g}_1 + \mathbf{g}_2$ and

$$\mathcal{S}_i := C_i D_i^{-1} C_i^T, \ \mathbf{g}_i := C_i D_i^{-1} B_i A_i^{-1} \mathbf{f}_i, \ D_i := B_i A_i^{-1} B_i^T, \quad i = 1, 2. \tag{28}$$

The $\mathcal{S}$ matrix can be considered as the Schur compliment matrix with respect to $\mathbf{u}$ and λ of the entire system defined above. Furthermore, it was shown by Brezzi in [7] that the Schur compliment $\mathcal{S}$ is symmetric and positive definite if the matrices B_i^T and C_i have full rank. One can then apply a conjugate gradient method to the system (27) to obtain the solution of the elliptic problem on the global domain. It can be remarked that by definition of $\mathbf{g}_i$, the calculation of a conjugate gradient iteration requires the solution to the local Helmholtz equation in each subdomain Ω_i, $i = 1, 2$. Block-Jacobi preconditioning is used to solve each of these local Helmholtz problems by considering zero Neumann conditions for each local problem. This way, each local

Helmholtz problem has a unique solution and in effect, the matrix A_i has an inverse which can be calculated before time-stepping.

The last issue of the discrete three-field formulation is related to the efficient construction of the matrices C and B. As they include the integration of the basis functions for the Lagrangian multiplier spaces and the interface space and are independent of the data, they can be calculated and stored prior to time stepping as well. The matrices have the form

$$\mathbf{C}_i(j,k) = \langle \mu_{i,j}, \phi_k \rangle_\Gamma, \qquad \mu_{i,j} \in \Lambda_N^i, \quad \phi_{i,k} \in \Phi^N, \tag{29}$$

$$\mathbf{B}_i(j,k) = \langle \eta_{i,j}, \mu_{i,k} \rangle_{\Gamma_i}, \quad \mu_{i,j} \in \Lambda_N^i, \quad \eta_{i,k} \in M_N^i. \tag{30}$$

For $i = 2$, the above calculations involve integration on a spectral grid using meshless reproducing kernels. The choice of the β parameter and N_M for a given radial basis that constructs the reproducing kernel determines the stability of the entire hybrid model. A study on these parameters is given in full detail in Blakely [6].

5 Numerical Experiments

As is it well established that the spherical rotational shallow-water equations represent a simplified model of the dynamica of the atmosphere, Williamson et al. [22] have proposed a series of eight test cases for the equations in spherical geometry. It is proposed by the authors that in order to have any type of success with a new numerical scheme for an a climate model, successful integrations of the numerical scheme with these test cases are imperative. The purpose of the tests are to examen the sensitivites of a numerical scheme with many computational challenges faced in atmospheric modeling such as stabilization of the scheme for large time steps over a long period of time, the pole problem, simulating flows which have discontinuous first-derivatives in the potential vorticity, and simulating flows over mountain topographies.

5.1 Test Case 2: Global Steady State Nonlinear Zonal Geostrophic Flow

As the second test case, a steady state flow to the full non-linear shallow water equations is prescribed and the challenge for a numerical scheme is simply to test its numerical stability with respect to l_1, l_inf errors over time. Since the flow is steady, the numerical scheme should be able to integrate the model for many steps without the addition any filtering. The velocity field is given as

$$u = u_0(\cos\theta\ \cos\alpha + \cos\lambda\ \sin\theta\ \sin\alpha),$$

$$v = -u_0 \sin\lambda\ \sin\alpha,$$

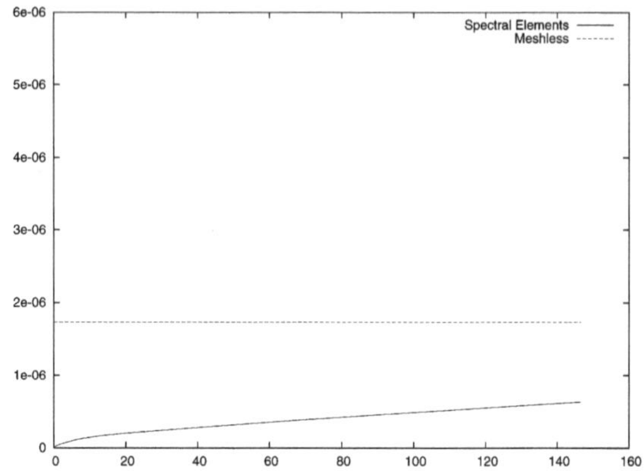

Fig. 2. Plot of l_1 errors of the spectral element and meshless geopotential solutions on face 2 of the cubed sphere using 256 (64 per element) nodes for the evaluation of the solutions.

which is non-divergent. The analytic geopotential field is give by

$$\eta = gh_0 - \left(a\Omega u_0 + \frac{u_0^2}{2}\right)(-\cos\lambda\cos\theta\ \sin\alpha + \sin\theta\ \cos\alpha)^2,$$

with constants $u_0 = 2\pi a/(12 \text{ days})$ and $gh_0 = 2.94 \times 10^4\ \text{m}^2/\text{s}^2$.

For this numerical experiment, we began with 24 total spectral elements (4 on each face) and ran the model for 121 days without any additional filtering. A second integration was performed using on face number 2 of the cube the meshless collocation approximation built from compactly supported radial functions (see [18]). Figure 2 shows the l_1 errors over time of the geopotential solution on face 2 for both the spectral element and meshless collocation approximations.

Notice how errors in the meshless approximation do not grow nearly as fast as in the spectral element approximation, despite not being as accurate. This is due to the collocation properties of the radial basis used. For the geopotential grid, a total 256 Gaussian–Lobatto–Legendre nodes (64 per element) were used at each time step. Furthermore, to obtain an accurate error comparison between the methods, the collocation approximation was evaluated at the spectral element nodes. Similar results for the l_{inf} error were also obtained.

5.2 Test Case 6: Rossby-Haurwitz Waves

The most interesting of the test cases features an initial condition for the velocity which is actually the analytical solution to the non divergent nonlinear Barotropic equation on the sphere, given as a vorticity equation. We refer the reader to [22] for the initial conditions of the velocity components and geopotential field.

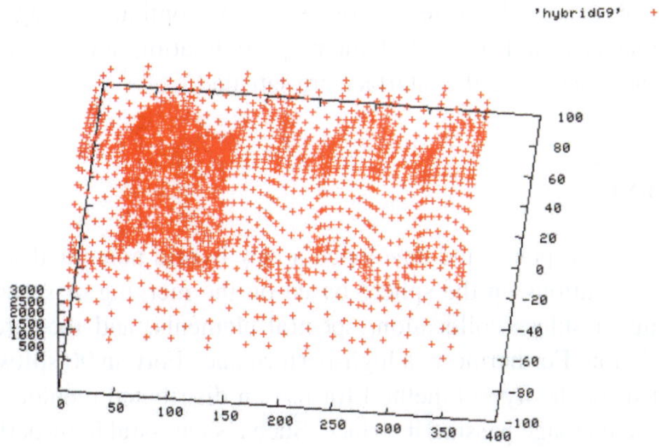

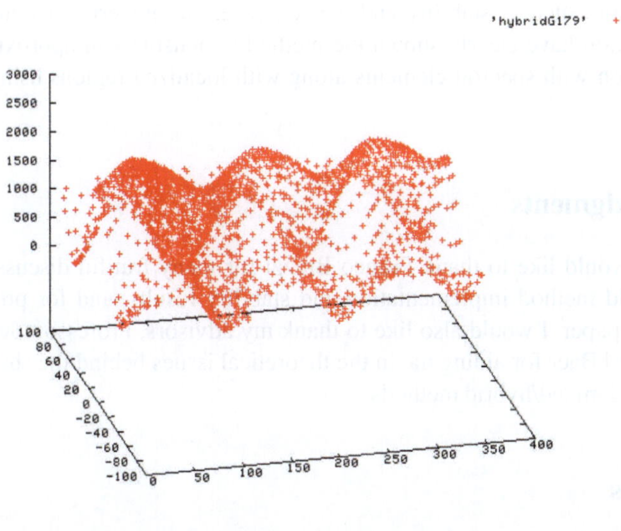

Fig. 3. Plots of geopotential approximation using 20 spectral elements and 4 elements allocated to meshless collocation. Plot after 10 and 60 days.

As originally proposed, these waves were expected to evolve nearly steadily with wavenumber 4. However, Thuburn and Li showed that this case is actually weakly unstable in that it will eventually break down once perturbed. This usually occurs after about 40 days depending on the model and parameters used. Figure 3 shows the geopotential layed out on rectangular coordinates for easier viewing. The figure to the left shows the field after 10 days at an angle where the hybrid mesh structure on the cubed-sphere can easily be seen. Notice that the continuity along the interfaces between the meshless and spectral-element approximations are preserved, meaning the inf-sup conditions along the interface are satisfied.

6 Conclusion

In this article, we proposed and developed a new hybrid numerical scheme for the shallow water equations on the sphere based on the merging of several numerical tools including meshless collocation, spectral elements, and the three-field variational formulation. Furthermore, a high-performance Fortran 90 software suite has been developed for the hybrid method for use on distributed memory parallel processors with the message passing interface. Such a successful high-performance implementation ultimately required the use of other Fortran 90 numerical packages for almost half of the computational tasks in the model, such as the domain decomposition. Although much work still remains with theoretical issues of the hybrid approximation scheme such as stability and convergence, the numerical examples in the previous section have clearly shown the method's robustness in approximating the global solution with spectral elements along with localized regions using meshless collocation.

Acknowledgments

The author would like to thank Franco Brezzi for many frutiful discussions about the three-field method implementation and stability results, and for proof-reading much of the paper. I would also like to thank my advisors, Professors John Osborn and Ferdinand Baer for aiding me in the theoretical issues behind the shallow-water equations and mixed/hybrid methods.

References

1. F. Baer, H. Wang, J.J. Tribbia and A. Fournier, *Climate Modeling with Spectral Elements*, in progress, 2005.
2. I. Babuška, The finite element method with Lagrangian multipliers. *Numerische Mathematik*, 20:179–192, 1973.
3. F. Ben Belgacem, C. Bernardi, N. Chorfi and Y. Maday, Inf-Sup conditions for the mortar spectral element discretization of the Stokes problem. *Numerische Mathematik*, 85(2):257–281, 2000.
4. S. Bertoluzza, Analysis of a stabilized three-fields domain decomposition method. *Numerische Mathematik*, 93:611–634, 2003.
5. C. Blakely, A new Backus–Gilbert meshless method for initial-boundary value problems. *International Journal on Computational Methods*, to appear, 2006.

6. C. Blakely, A hybrid meshless/spectral-element approximation method for elliptic partial differential equations. *International Journal on Computer Mathematics*, submitted, 2006.
7. F. Brezzi and L.D. Marini, A three-field domain decomposition method. In: A. Quarteroni et al. (Eds), Decomposition Methods in Science and Engineering, 1994, pp. 27-34.
8. F. Brezzi and L.D. Marini, The three-field formulation for elasticity problems. *GAMM Mitteilungen*, 38:124–153, 2005.
9. M.D. Buhmann, *Radial Basis Functions: Theory and Implementations*. Cambridge University Press, Cambridge, 2003.
10. G.E. Fasshauer, Matrix-free multilevel moving least-squares methods. In: *Approximation Theory*, X St. Louis, MO, 2002, pp. 271–278.
11. A. Fournier, M.A. Taylor and J.J. Tribbia, The spectral-element atmosphere model: High-resolution parallel computation and localized resolution of regional dynamics. *Monthly Weather Review*, 123:726–748, 2004.
12. J. Galewsky, L.M. Polvani and R.K. Scott, An initial-value problem to test numerical models of the shallow water equations. *Monthly Weather Review*, 2003.
13. A. Gelb and E. Tadmor, Adaptive edge detectors for piecewise smooth data based on the minmod limiter, CSCAMM Report CS-05-06, 2005.
14. E. Kalnay, *Atmopheric Modeling, Data Assimilation and Predictibility*. Cambridge University Press, 2003.
15. G.E. Karniadakis and S.J. Sherwin, *Spectral/hp Element Methods for Computational Fluid Dynamics*. Oxford University Press, 1999, 404 pp.
16. G. Rapin, The three-field formulation for elliptic equations: stabilization and decoupling strategies. PhD Thesis, Universität Göttingen, 2003.
17. R. Sadourny, Conservative finite-difference approximations of the primitive equations on quasi-uniform spherical grids. *Monthly Weather Review*, 100:136–144, 1972.
18. R. Schaback, Convergence of unsymmetric kernel based meshless collocation methods. Preprint from http://www.num.math.uni-goettingen.de/schaback/research/group.html, 2005.
19. M. Taylor, J. Tribbia and M. Iskandarani, Performance of a spectral-element atmospheric model on the HP Exemplar SPP2000. NCAR Tech. Rep TN-439 + EDD, 16 pp.
20. S.J. Thomas and R.D. Loft, Semi-implicit spectral element atmospheric model. *Journal of Scientific Computing*, 17:339–350, 2002.
21. M. Taylor, J. Tribbia and M. Iskandarani, The spectral element method for the shallow water equations on the sphere. *Journal of Computational Physics*, 130:92–108, 1997.
22. D.L. Williamson, J.B. Drake, J.J. Hack, R. Jakob and P.N. Swarztrauber, A standard test set for numerical approximations to the shallow water equations in spherical geometry. *Journal of Computational Physics*, 102:211–224, 1992.

Iterated Approximate Moving Least Squares Approximation

Gregory E. Fasshauer and Jack G. Zhang

Department of Applied Mathematics, Illinois Institute of Technology, Chicago, IL 60616, U.S.A.; E-mail: fasshauer@iit.edu

Abstract. The radial basis function interpolant is known to be the best approximation to a set of scattered data when the error is measured in the native space norm. The approximate moving least squares method, on the other hand, was recently proposed as an efficient approximation method that avoids the solution of the system of linear equations associated with the radial basis function interpolant. In this paper we propose and analyze an algorithm that iterates on the residuals of an approximate moving least squares approximation. We show that this algorithm yields the radial basis interpolant in the limit. Supporting numerical experiments are also included.

Key words: RBP interpolation, MLS approximation, approximate approximation, residual iteration.

1 Introduction

In this paper we will be interested in solving the following approximation problem. For a given set of data $\{(x_i, f(x_i), i = 1, 2, \ldots, N, \; x_i \in \Omega \subseteq \mathbb{R}^s, \; f(x_i) \in \mathbb{R}\}$, we seek a continuous function $\mathcal{P}_f : \mathbb{R}^s \to \mathbb{R}$ that either interpolates the data, i.e.,

$$\mathcal{P}_f(x_i) = f(x_i), \qquad i = 1, 2, \ldots, N, \tag{1}$$

or such that $\mathcal{P}_f$ provides a close approximation to f measured in some appropriate norm.

We will be using *radial basis functions* (RBFs) to solve this problem, and one of the main features of RBFs is the fact that they can be applied without any restriction on the location of the data sites. *Approximate moving least squares* (AMLS) approximation on the other hand is in its current form mainly applicable to uniformly spaced data. This is due to the fact that it is known that the formulation of an approximate

V.M.A. Leitão et al. (eds), *Advances in Meshfree Techniques*, 221–239.

approximation method for scattered data is significantly more complicated than in the case of uniform data (see, e.g., [8, 9]).

We are interested in the use and comparison of these two multivariate approximation methods. Radial basis function interpolation, on the one hand, is known to yield the best approximation to given (scattered) data with respect to the native space norm of the basic function used. The benefits of this optimality property are somewhat reduced by the need to solve a (generally) large system of linear equations which can also be ill-conditioned. To avoid the solution of such a system of linear equations we recently proposed an alternative meshfree method which we refer to as the approximate moving least squares method (see, e.g., [2–5]). Using the AMLS method the solution is obtained via a simple sum based directly on the given data. Thus, the AMLS method is a quasi-interpolation approach. The drawback associated with the simplicity of the AMLS method is its lesser degree of accuracy.

We will see later that an algorithm which iterates on AMLS residuals converges to the RBF interpolant, and therefore a few iterations can be considered as an efficient and numerically stable alternative to the RBF interpolation approach. While the initial iterate of the algorithm will be an AMLS approximation designed for uniformly spaced data, we will see that the algorithm can generate an equivalently nice solution even when the data sites are irregularly distributed.

The remainder of the paper is organized as follows. In Section 2 we set our notation and present some of the salient facts for the two approximation methods we are interested in. The iterative algorithm is described in Section 3 which also contains an analysis of its convergence. Numerical experiments that demonstrate the performance of the algorithm are presented in Section 4. The paper is concluded with some remarks and an outlook on future work in Section 5.

2 The Two Approximation Methods

2.1 RBF Interpolation

The standard RBF interpolation approach for our data fitting problem is to assume that the interpolant $\mathcal{P}_f$ is a linear combination of radial basis functions Φ_j, i.e.,

$$\mathcal{P}_f(x) = \sum_{j=1}^{N} c_j \Phi_j(x) . \tag{2}$$

where the $\Phi_j : \mathbb{R}^s \to \mathbb{R}$ are defined by shifting a single basic function to the data sites, i.e.,

$$\Phi_j(x) = \varphi\left(\left\|\frac{x - x_j}{h}\right\|\right) \tag{3}$$

for some univariate function $\varphi : [0, \infty) \to \mathbb{R}$. Note that we include a scale factor h in the definition of the basic function. This scale factor is given by the fill distance

$h = \sup_{x\in\Omega} \min_{x_j,\ j=1,\ldots,N} \|x - x_j\|_2$ of the data sites. Note that our definition of the radial basis functions is reminiscent of the *stationary* approximation paradigm, i.e., the basis functions are scaled proportional to the fill distance. On the one hand it is known that most RBFs do not yield a convergent stationary approximation scheme (except when we use such functions as polyharmonic splines). However, it is exactly the stationary setting that is studied in the *approximate approximation* context, and there one can observe convergence subject to a saturation error whose size can be controlled by an initial scaling of the basic function (see, e.g., [5, 9]).

Now, equation (2) can be rewritten as

$$\mathcal{P}_f(x) = \sum_{j=1}^{N} c_j \varphi\left(\left\|\frac{x - x_j}{h}\right\|\right), \tag{4}$$

and with condition (1) this leads to finding the coefficients c_j from the linear system

$$\mathbf{A}\mathbf{c} = \mathbf{f}. \tag{5}$$

Here $\mathbf{A}$ is an $N \times N$ interpolation matrix given by

$$\mathbf{A} = \left\{\varphi\left(\left\|\frac{x_i - x_j}{h}\right\|\right)\right\}_{i,j=1}^{N} \quad \text{or} \quad \mathbf{A} = \left\{\Phi_j\left(x_i\right)\right\}_{i,j=1}^{N},$$

$\mathbf{c} = [c_1, c_2, \ldots, c_N]^T$, and $\mathbf{f} = [f(x_1), f(x_2), \ldots, f(x_N)]^T$. The interpolation matrix $\mathbf{A}$ is guaranteed to be non-singular if φ is a strictly positive definite radial function on $\mathbb{R}^s$.

Examples of basic functions we could consider for the purpose of interpolation are the strictly positive definite Gaussians $\varphi(r) = e^{-\varepsilon^2 r^2}$ or the inverse multiquadrics $\varphi(r) = \frac{1}{\sqrt{1+\varepsilon^2 r^2}}$. Note that the shape parameter ε will play the role of the initial scaling of the basic function just mentioned in the context of saturated approximate approximation. However, throughout this paper we will use the following set of strictly positive definite functions defined as

$$\varphi(r) := \frac{1}{\sqrt{\pi^s}}\, e^{-r^2} L_n^{s/2}(r^2), \tag{6}$$

with

$$L_n^{s/2}(t) := \frac{e^t t^{-s/2}}{n!} \frac{d^n}{dt^n}\left(e^{-t} t^{n+s/2}\right), \quad n = 0, 1, 2, \ldots,$$

being the generalized Laguerre polynomials. We will refer to φ as a *Laguerre–Gaussian*. Note that we did not include the shape parameter ε here in order to keep the formulas transparent. Later we will consider the functions $\varphi_\varepsilon = \varphi(\varepsilon\cdot)$.

The Laguerre–Gaussian functions are *oscillatory*. They can be explicitly written as

$$\varphi(r) = \sum_{k=0}^{n} \frac{(-1)^k (n+s/2)!}{n!(k+s/2)!} \binom{n}{k} r^{2k} e^{-r^2}.$$

It is known that a (radial) function φ is strictly positive definite and radial on $\mathbb{R}^s$ if its (radial) Fourier transform is non-negative and not identically equal to zero. For the Laguerre–Gaussians one can show

$$\hat{\varphi}(t) = \frac{e^{-\frac{t^2}{4}}}{\sqrt{2^s}} \sum_{k=0}^{n} \frac{t^{2k}}{k!4^k} \geq 0,$$

and it is obvious that equality holds only when $t = 0$.

Note that the definition of φ depends on the space dimension s, and therefore φ is strictly positive definite and radial only on $\mathbb{R}^s$ for certain values of s. However, in the special case $n = 0$ the basic function φ becomes the regular Gaussian whose definition is independent of the space dimension s and thus it is strictly positive definite on $\mathbb{R}^s$ for all s.

The primary motivation for us to investigate Laguerre–Gaussians lies in the fact that they satisfy certain *continuous moment conditions*. These moment conditions come up when one generates basis functions for approximate moving least squares approximation. Details are given in next section.

2.2 Approximate Moving Least Squares Approximation

Roughly speaking, approximate moving least squares approximation is an approximate version of the standard moving least squares method which does not require the solution of any linear systems. The concept of approximate approximations was first suggested by Maz'ya in the early 1990s. A key ingredient in this approach are the *continuous moment conditions* for the basic function φ. A radial version of this requirement may be described as

$$\int_{\mathbf{R}^s} \|x\|^k \varphi(\|x\|) dx = \delta_{\alpha,0} \qquad \text{for } 0 \leq k \leq d. \tag{7}$$

According to the theory (see, e.g., [9]), a basic function φ that satisfies these conditions provides the following results.

For uniformly spaced $x_j \in \mathbb{R}^s$ and $\varepsilon > 0$, the quasi-interpolant

$$\mathcal{Q}_f(x) := \varepsilon^s \sum_{j=1}^{N} f(x_j) \varphi\left(\varepsilon \left\| \frac{x - x_j}{h} \right\|\right) \tag{8}$$

approximately solves the data fitting problem (1) with a guaranteed convergence

$$\|f - \mathcal{Q}_f\|_\infty = \mathcal{O}(h^{d+1}) + \epsilon(\varphi, \varepsilon).$$

As before, h is the fill distance of the given data points. The quantity $\epsilon(\varphi, \varepsilon)$ is referred to as a *saturation error*, and it depends only on the basic function φ and the initial scale factor ε. By choosing an appropriate scaling parameter ε, this saturation error may be pushed to the order of machine accuracy on any given computer.

One major advantage of the AMLS method is that the continuous moment conditions (7) provide a possibility to explicitly derive such a basic function φ. For example, any normalized integrable φ will satisfy (7) for $d = 0$. In fact, as mentioned earlier, the Laguerre–Gaussian functions satisfy (7) with $d = 2n + 1$ for each corresponding s-dimensional space. Thus, Laguerre–Gaussians are admissible for *both* RBF interpolation and AMLS approximation.

Unlike the standard RBF interpolation or moving least squares method, AMLS approximation is a completely matrix free method and hence significantly improves computational efficiency and successfully avoids the difficulties associated with ill-conditioned system matrices. However, if we are interested in an approximation that exactly interpolates the data given in (8), then the quasi-interpolant $\mathcal{Q}_f$ requires that its generating functions are cardinal functions. In other words, the approximant $\mathcal{Q}_f$ will naturally involve an error at the data sites in addition to the saturation error $\epsilon(\varphi, \varepsilon)$. On the other hand, as we pointed out earlier, the interpolant $\mathcal{P}_f$ is the "best" solution to problem (1) in the Hilbert function space defined by the chosen basis functions. Therefore, we are motivated to seek a method that comes close to both ideals, i.e., to find a solution that, on the one hand, does not require solving a linear system but, on the other hand, is closer to the RBF interpolant than the plain AMLS approximant.

Next, we will formulate a residual iteration algorithm that can achieve this goal with an acceptable amount of additional computations.

3 Interpolation via Iterated AMLS Approximation

As shown in the previous section, it is possible that a standard RBF interpolant $\mathcal{P}_f$ and an AMLS quasi-interpolant $\mathcal{Q}_f$ share the same set of basis functions, e.g., basis functions generated by a basic Laguerre–Gaussian function. In this section we will explain how an initial approximant $\mathcal{Q}_f$ can be pushed closer towards the corresponding interpolant $\mathcal{P}_f$ by a residual iteration process. We will also provide some theoretical analysis of the convergence of such an iteration.

3.1 The Iterative Algorithm

We will now combine all constant parameters h, ε, and ε^s into the definition of the basic function φ. That is, we redefine (4) and (8) in the simple form

$$\mathscr{P}_f(x) := \sum_{j=1}^{N} c_j \varphi\left(\|x - x_j\|\right), \tag{9}$$

$$\mathscr{Q}_f(x) := \sum_{j=1}^{N} f(x_j) \varphi\left(\|x - x_j\|\right). \tag{10}$$

Note that now each of $\mathscr{P}_f$ and $\mathscr{Q}_f$ is a linear combination with the *same* basic function φ. For example, we can use the scaled s-dimensional Gaussian,

$$\varphi(r) = \frac{\varepsilon^s}{\sqrt{\pi^s}} e^{-\varepsilon^2 r^2/h^2}.$$

Clearly, the interpolation matrix based on (9) becomes

$$\mathbf{A} = \left\{\varphi\left(\|x_i - x_j\|\right)\right\}_{i,j=1}^{N}, \tag{11}$$

and the linear system is still in the form $\mathbf{Ac} = \mathbf{f}$, where the vectors $\mathbf{c}$ and $\mathbf{f}$ are as defined earlier.

To construct an algorithm that iterates on residuals, we start with an initial AMLS quasi-interpolant

$$\mathscr{Q}_f^{(0)}(x) = \sum_{j=1}^{N} f(x_j) \varphi\left(\|x - x_j\|\right). \tag{12}$$

We then iteratively define

$$\mathscr{Q}_f^{(n)}(x) = \mathscr{Q}_f^{(n-1)}(x) + \sum_{j=1}^{N} \left[f(x_j) - \mathscr{Q}_f^{(n-1)}(x_j)\right] \varphi\left(\|x - x_j\|\right). \tag{13}$$

That is, the current approximant is successively updated by a residual function which is also constructed by AMLS approximation on the same set of data points.

Certainly, so far we have no evidence that the sequence of these approximating functions $\{\mathscr{Q}_f^{(n)}\}$ converges in any form as $n \to \infty$, or what its limit will be if it is convergent. Next, we will show that under some appropriate assumptions $\mathscr{Q}_f$ does converge to the interpolant $\mathscr{P}_f$ as $n \to \infty$.

3.2 A Necessary and Sufficient Condition

Theorem 1. *The sequence of functions* $\{\mathscr{Q}_f^{(n)}\}$ *defined by (12) and (13) converges to the interpolant* $\mathscr{P}_f$ *defined in (9) if and only if the chosen basic function* φ *generates an interpolation matrix* $\mathbf{A}$ *that satisfies* $\|\mathbf{I} - \mathbf{A}\|_2 < 1$ *for a given set of distinct data points* $\{x_j\} \subseteq \Omega \subseteq \mathbb{R}^s$.

Proof. First, we convert our notation to matrix-vector form. Define a column vector of functions

$$\Psi(x) := [\varphi(\|x - x_1\|), \varphi(\|x - x_2\|), \ldots, \varphi(\|x - x_N\|)]^T.$$

Clearly, $\Psi(x)$ is related to the interpolation matrix $\mathbf{A}$, i.e., due to the symmetry,

$$\mathbf{A}^T = [\Psi(x_1) \mid \Psi(x_2) \mid \cdots \mid \Psi(x_N)] \ (= \mathbf{A}). \tag{14}$$

The interpolant $\mathcal{P}_f$ defined in (9) can also be expressed in matrix-vector form using the vector $\Psi(x)$, i.e.,

$$\mathcal{P}_f(x) = \Psi(x)^T \mathbf{c}. \tag{15}$$

Next, we will inductively prove that the functions $\mathcal{Q}_f^{(n)}$ defined in (13) can also be explicitly expressed in matrix-vector notation

$$\mathcal{Q}_f^{(n)}(x) = \Psi(x)^T \left[\sum_{k=0}^{n} (\mathbf{I} - \mathbf{A})^k \right] \mathbf{f}, \quad \text{for all } n = 0, 1, 2, \ldots, \tag{16}$$

where $\mathbf{f} = [f(x_1), f(x_2), \ldots, f(x_N)]^T$ as before.

The initial case $n = 0$ is clear. Suppose (16) holds up to an index n. We need to show that

$$\mathcal{Q}_f^{(n+1)}(x) = \Psi(x)^T \left[\sum_{k=0}^{n+1} (\mathbf{I} - \mathbf{A})^k \right] \mathbf{f}.$$

Using the induction hypothesis, the definition of $\mathbf{f}$ and the relation (14) between the interpolation matrix $\mathbf{A}$ and the vector function Ψ we have

$$\begin{aligned}
\mathcal{Q}_f^{(n+1)}(x) &= \mathcal{Q}_f^{(n)}(x) + \sum_{j=1}^{N} \left[f(x_j) - \mathcal{Q}_f^{(n)}(x_j) \right] \varphi\left(\|x - x_j\|\right) \\
&= \Psi(x)^T \left[\sum_{k=0}^{n} (\mathbf{I} - \mathbf{A})^k \right] \mathbf{f} \\
&\quad + \sum_{j=1}^{N} \left[f(x_j) - \Psi(x_j)^T \left[\sum_{k=0}^{n} (\mathbf{I} - \mathbf{A})^k \right] \mathbf{f} \right] \varphi\left(\|x - x_j\|\right) \\
&= \Psi(x)^T \left[\sum_{k=0}^{n} (\mathbf{I} - \mathbf{A})^k \right] \mathbf{f} \\
&\quad + \Psi(x)^T \left[\mathbf{f} - \mathbf{A}^T \left[\sum_{k=0}^{n} (\mathbf{I} - \mathbf{A})^k \right] \mathbf{f} \right].
\end{aligned}$$

Now, straightforward algebra yields

$$\begin{aligned}\mathcal{Q}_f^{(n+1)}(x) &= \Psi(x)^T \left[\mathbf{I} + \sum_{k=0}^{n} (\mathbf{I}-\mathbf{A})^{k+1}\right]\mathbf{f} \\ &= \Psi(x)^T \left[\sum_{k=0}^{n+1} (\mathbf{I}-\mathbf{A})^{k}\right]\mathbf{f}.\end{aligned}$$

As (16) shows, it is clear that each updated approximant $\mathcal{Q}_f^{(n)}$ is still a linear combination of the same basis functions but with an updated coefficient vector of the form

$$\left[\sum_{k=0}^{n} (\mathbf{I}-\mathbf{A})^{k}\right]\mathbf{f}.$$

Thus, the fact that $\mathcal{Q}_f^{(n)} \to \mathcal{P}_f$ as $n \to \infty$ is equivalent to

$$\left[\sum_{k=0}^{n} (\mathbf{I}-\mathbf{A})^{k}\right]\mathbf{f} \to \mathbf{c} \quad \text{as } n \to \infty.$$

Since $\mathbf{c}$ is determined by the linear system (5) defined by the interpolation problem, i.e., $\mathbf{c} = \mathbf{A}^{-1}\mathbf{f}$, the convergence is therefore equivalent to

$$\left[\sum_{k=0}^{n} (\mathbf{I}-\mathbf{A})^{k}\right]\mathbf{f} \to \mathbf{A}^{-1}\mathbf{f} \quad \text{as } n \to \infty,$$

or,

$$\sum_{k=0}^{n} (\mathbf{I}-\mathbf{A})^{k} \to \mathbf{A}^{-1} \quad \text{as } n \to \infty.$$

The proof is completed by noting that a Neumann series satisfies

$$\sum_{k=0}^{\infty} (\mathbf{I}-\mathbf{A})^{k} = \mathbf{A}^{-1}$$

if and only if $\|\mathbf{I}-\mathbf{A}\|_2 < 1$. □

If the assumption of Theorem 1 holds, then a discrete ℓ_2 error of this iterated approximation with respect to the given data can be computed.

Corollary 1. *Let*

$$\mathbf{q}^{(n)} = [\mathcal{Q}_f^{(n)}(x_1), \mathcal{Q}_f^{(n)}(x_2), \ldots, \mathcal{Q}_f^{(n)}(x_N)]^T.$$

If $\|\mathbf{I}-\mathbf{A}\|_2 < 1$ *then* $\|\mathbf{f}-\mathbf{q}^{(n)}\|_2 \to 0$ *for* $n \to \infty$.

Proof.

$$\|\mathbf{f}-\mathbf{q}^{(n)}\|_2 = \left\| \mathbf{A}\mathbf{A}^{-1}\mathbf{f} - \left[\mathbf{A}\sum_{k=0}^{n}(\mathbf{I}-\mathbf{A})^k\right]\mathbf{f}\right\|_2$$

$$\leq \|\mathbf{A}\|_2 \left\|\mathbf{A}^{-1} - \sum_{k=0}^{n}(\mathbf{I}-\mathbf{A})^k\right\|_2 \|\mathbf{f}\|_2.$$

Since the matrix $\mathbf{A}^{-1}$ is invertible, using the formula for the sum of a finite geometric series, we have

$$\|\mathbf{f}-\mathbf{q}^{(n)}\|_2 \leq \|\mathbf{A}\|_2 \left\|\mathbf{A}^{-1}(\mathbf{I}-\mathbf{A})^{n+1}\right\|_2 \|\mathbf{f}\|_2$$

$$\leq \|\mathbf{A}\|_2 \|\mathbf{A}^{-1}\|_2 \|\mathbf{I}-\mathbf{A}\|_2^{n+1} \|\mathbf{f}\|_2$$

$$= \text{cond}(\mathbf{A}) \|\mathbf{I}-\mathbf{A}\|_2^{n+1} \|\mathbf{f}\|_2 \to 0,$$

where cond($\mathbf{A}$) is the ℓ_2-condition number of $\mathbf{A}$. □

If the assumption of Theorem 1 holds, then we can also estimate the norm of the residual functions.

Corollary 2. *If* $\|\mathbf{I}-\mathbf{A}\|_2 < 1$ *then*

$$\|R^{(n)}\|_2 = \left\| \sum_{j=1}^{N}\left[f(x_j) - \mathcal{Q}_f^{(n-1)}(x_j)\right]\varphi\left(\|\cdot - x_j\|\right)\right\|_2 \to 0$$

for $n \to \infty$.

Proof. From (13) we have

$$R^{(n)} = \sum_{j=1}^{N}\left[f(x_j) - \mathcal{Q}_f^{(n-1)}(x_j)\right]\varphi\left(\|\cdot - x_j\|\right)$$

$$= \mathcal{Q}_f^{(n)} - \mathcal{Q}_f^{(n-1)}.$$

But now, using matrix-vector notation,

$$\left\|\mathcal{Q}_f^{(n)} - \mathcal{Q}_f^{(n-1)}\right\|_2 = \left\|\Psi(\cdot)^T\left[\sum_{k=0}^{n}(\mathbf{I}-\mathbf{A})^k\right]\mathbf{f} - \Psi(\cdot)^T\left[\sum_{k=0}^{n-1}(\mathbf{I}-\mathbf{A})^k\right]\mathbf{f}\right\|_2$$

$$\leq \|\Psi(\cdot)\|_2 \|\mathbf{I}-\mathbf{A}\|_2^n \|\mathbf{f}\|_2 \to 0,$$

and thus the statement follows. □

Clearly, the speed of convergence is governed by the entries in the interpolation matrix **A**, and thus by the choice of φ (see (11)). A basic function that is strictly positive definite and satisfies the continuous moment conditions does not automatically guarantee such a matrix **A**. This may easily be demonstrated by counterexamples.

3.3 A Sufficient Condition

Theorem 2. *Using the same notation as in Theorem 1, if a basic function φ is strictly positive definite and generates an interpolation matrix* **A** *so that*

$$\max_{i=1,2,\ldots,N} \left\{ \sum_{j=1}^{N} \mid \mathbf{A}_{i,j} \mid \right\} < 2, \tag{17}$$

then $\mathcal{Q}_f^{(n)}$ converges to $\mathcal{P}_f$ as $n \to \infty$.

Proof. Let λ_k for $k = 1, 2, \ldots, N$ be the eigenvalues of **A**. That is, $1 - \lambda_k$ are the eigenvalues of the matrix $\mathbf{I} - \mathbf{A}$ for $k = 1, 2, \ldots, N$.

Since φ is a strictly positive definite function, the matrix **A** is positive definite, i.e.,

$$\lambda_k > 0 \quad \text{for } k = 1, 2, \ldots, N.$$

Thus,

$$1 - \lambda_k < 1 \quad \text{for } k = 1, 2, \ldots, N. \tag{18}$$

Recall that

$$\mathbf{A}_{i,j} = \varphi \|x_i - x_j\|, \quad \text{for } i, j = 1, 2, \ldots, N.$$

This shows that all diagonal entries in **A** are identical and equal to $\varphi(0)$. Hence, Gerschgorin's Theorem implies that

$$|\lambda_k - \varphi(0)| \leq \max_{i=1,2,\ldots,N} \left\{ \sum_{j=1}^{N} \mid \mathbf{A}_{i,j} \mid \right\} - \varphi(0), \quad \text{for } k = 1, 2, \ldots, N.$$

Using the assumption (17) we get

$$\mid \lambda_k - \varphi(0) \mid < 2 - \varphi(0),$$

which is equivalent to

$$\varphi(0) - 2 < \lambda_k - \varphi(0) < 2 - \varphi(0).$$

Rearranging these inequalities yields

$$-1 < 1 - \lambda_k < 3 - 2\varphi(0), \quad \text{for } k = 1, 2, \ldots, N. \tag{19}$$

Combining (18) and (19) we have

$$-1 < 1 - \lambda_k < 1, \quad \text{for } k = 1, 2, \ldots, N,$$

or

$$\max_{k=1,2,\ldots,N} \{|1 - \lambda_k|\} < 1.$$

Since the matrix $\mathbf{I} - \mathbf{A}$ is also symmetric, standard results from linear algebra tell us that the 2-norm of the matrix $\mathbf{I} - \mathbf{A}$ is given by

$$\|\mathbf{I} - \mathbf{A}\|_2 = \max_{k=1,2,\ldots,N} \{|1 - \lambda_k|\}.$$

So

$$\|\mathbf{I} - \mathbf{A}\|_2 < 1,$$

and the convergence of $\mathcal{Q}_f^{(n)}$ to $\mathcal{P}_f$ follows from Theorem 1. □

A simple example that illustrates Theorem 2 is obtained if we use Shepard's partition of unity functions. We define the set of basis functions as

$$w_j(x) = \varphi(\|x - x_j\|) \Big/ \sum_{\ell=1}^{N} \varphi(\|x - x_\ell\|)$$

with a strictly positive definite and positive basic function φ. If an interpolant is constructed with these basic functions, i.e.,

$$\mathcal{P}_f(x) = \sum_{j=1}^{N} c_j w_j(x),$$

then one can easily show that the interpolation matrix $\mathbf{W} = \{w_j(x_i)\}_{i,j=1}^{N}$ is invertible. Moreover, summation of the entries in any row of $\mathbf{W}$ always results in a row sum equal to one due to the partition of unity property. Therefore, if the residual iteration algorithm is started with a quasi-interpolant using the w_j as generating functions, then it will converge to $\mathcal{P}_f$.

In order to ensure convergence of the residual iteration algorithm via Theorem 2 in the first place, and to increase its speed of convergence as much as possible, Theorem 2 and its proof tell us that we want to have an interpolation matrix $\mathbf{A}$ whose entries are not too large in magnitude. From the definition of the basic functions φ we know that the entries of the matrix $\mathbf{A}$ carry a multiplicative factor ε^s. This simply implies that at least the diagonal entries in $\mathbf{A}$ could become arbitrarily large for $\varepsilon > 0$ being large (recall that $\mathbf{A}_{j,j}$ are indeed positive since φ is strictly positive definite). Therefore, ε must be chosen to be small. On the other hand, however, it is known that for many commonly used strictly positive definite functions (e.g., the Gaussian) a small shape parameter ε implies that the smallest eigenvalue of $\mathbf{A}$, $\lambda_{\min}$ (which again is always positive), will likely be very close to zero (see, e.g., Chapter 12 in [10]). Therefore, again according to the proof of Theorem 2, $\|\mathbf{I}-\mathbf{A}\|_2$ will be at least $1-\lambda_{\min}$ which is very close to 1. Thus, for such a choice of ε the convergence of the iteration will be extremely slow. We will use a set of experiments to demonstrate this trade-off phenomenon caused by the scaling parameter ε.

To conclude this section, we would like to come back to the data point distribution issue as mentioned earlier. Once $\mathcal{Q}_f^{(n)}$ is guaranteed to converge to $\mathcal{P}_f$ it is not so crucial how the initial approximant was constructed since $\mathcal{P}_f$ is still the "best" solution to the problem even when data points are scattered. That is to say, although $\mathcal{Q}_f^{(0)}$ is an AMLS quasi-interpolant formulated with a scaling designed for the uniform data problem, the iterated approximant is still good when the data points are scattered. From the experimental results presented in the next section, we will see that this iterative method does work equivalently well for both regular and irregular grid points.

4 Numerical Experiments

In this sections, we use a set of numerical experiments to illustrate some of the advantages and features of the iterated AMLS approximation method described in previous sections. We will study the behavior of the method with respect to the data size N and the shape parameter ε used in the definition of the basic function φ. Throughout the rest of this section the test function f we use for our experiments is a mollified linear combination of exponentials similar to the famous Franke function. More precisely, in the case $s=1$ we use

$$f(x) = 15e^{-\frac{1}{1-(2x-1)^2}}\left(\frac{3}{4}e^{-\frac{(9x-2)^2}{4}} + \frac{3}{4}e^{-\frac{(9x+1)^2}{49}} + \frac{1}{2}e^{-\frac{(9x-7)^2}{4}} - \frac{1}{5}e^{-(9x-4)^2}\right) \tag{20}$$

on the interval $[0, 1]$, and for $s=2$ we let

$$\begin{aligned} g(x,y) &= \frac{3}{4}e^{-1/4((9x-2)^2+(9y-2)^2)} + \frac{3}{4}e^{-(1/49)(9x+1)^2-(1/10)(9y+1)^2} \\ &\quad + \frac{1}{2}e^{-1/4((9x-7)^2+(9y-3)^2)} - \frac{1}{5}e^{-(9x-4)^2-(9y-7)^2} \\ f(x,y) &= 15g(x,y)e^{-\frac{1}{1-(2x-1)^2}}e^{-\frac{1}{1-(2y-1)^2}} \end{aligned} \tag{21}$$

which we will sample in the unit cube $[0, 1]^2$. The basic function used to generate the basis functions for all experiments presented here is the scaled s-dimensional

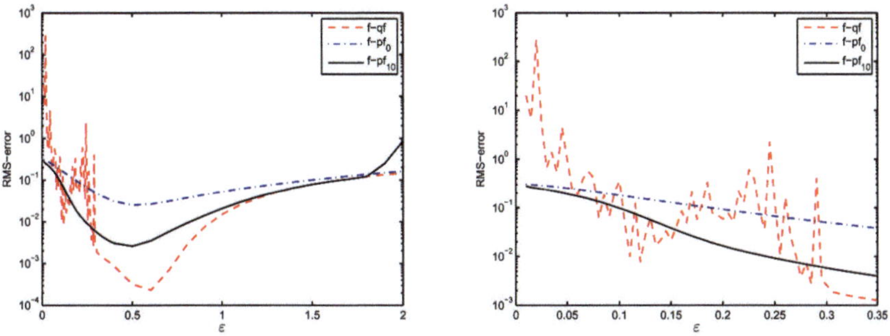

Fig. 1. Comparison of accuracy and stability of the RBF interpolant, AMLS approximant, and iterated AMLS approximant for 1089 Halton data points in 2D.

Gaussian

$$\varphi(r) = \frac{\varepsilon^s}{\sqrt{\pi^s}} e^{-\varepsilon^2 r^2/h^2}, \tag{22}$$

where h is the fill-distance corresponding to a *uniform* distribution of points in the domain. This manner of scaling the basic function proportional to the fill-distance is known as the stationary approximation paradigm. We have performed the same set of experiments also with first-order Laguerre–Gaussians (see (6)). Due to the similarity of these results to those for Gaussians we focus only on Gaussians here.

Although it is known that RBF interpolation theoretically yields the native space best approximation for scattered data fitting problems, the results obtained in practice are sometimes unreliable due to poor numerical stability of the solution of the associated linear system (5). For a given set of data the accuracy of an interpolant may strongly depend on the scale parameter ε that is often part of the definition of the basic function, e.g., the Gaussian function (22).

Generally speaking, when ε is too large, the basic function is very peaked, and thus the resulting interpolant will be very "spiky" so that it can not be considered as a good approximant to the true function. On the other hand, a small ε makes a flat basic function which will generate an ill-conditioned interpolation matrix $\mathbf{A}$ giving rise to computational difficulty (in the extreme case, when $\varepsilon \to 0$, the matrix $\mathbf{A}$ becomes a constant matrix which is singular). The interpolant resulting from such an ill-conditioned calculation is obviously to be trusted less. Figure 1 illustrates both of these observations.

In both plots of Figure 1 we display the root-mean-squared error of the interpolant (`f-qf`, dashed line), the AMLS approximant (`f-pf`$_0$, dash-dotted line), and the result of ten iterations of the iterated AMLS method (`f-pf`$_{10}$, solid line) versus the scale parameter ε as used in (22). The data were obtained by sampling the two-dimensional modified Franke function (21) at 1089 non-uniform Halton points (see,

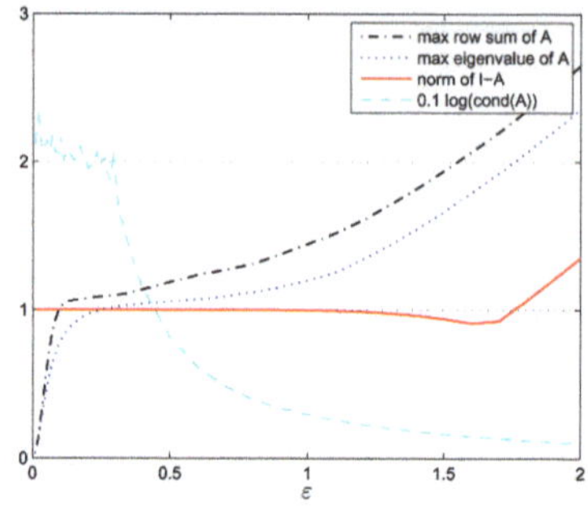

Fig. 2. Maximum row sum, maximum eigenvalue, $\|\mathbf{I} - \mathbf{A}\|$, and cond($\mathbf{A}$), for the matrix $\mathbf{A}$ used in Figure 1.

e.g., [11] for more information on Halton points). The error was computed on a grid of 40×40 equally spaced evaluation points.

In this example ε should be characterized as being "large" if its value is greater than approximately 1.2. For these values the error of the interpolant is not significantly different from that of the AMLS approximation, and a plot of the interpolant would be a very "spiky" surface as mentioned earlier. Within this range for the scale parameter we can distinguish two different behaviors of the iterative algorithm. If ε is so large that convergence of the iterative algorithm is no longer ensured then the iterated approximant blows up. Figure 2 shows that the maximum row sum of the matrix $\mathbf{A}$ is greater than two if $\varepsilon > 1.5$. In fact, the iterated algorithm does not blow up until about $\varepsilon = 1.8$ (when the maximum eigenvalue of $\mathbf{A} > 1$). This also indicates that the maximum row sum criterion is a relatively easy to check and safe criterion to ensure convergence of the iterative algorithm. Clearly, one wants to avoid the use of these "large" ε values.

Moreover, within the "large" ε range we can usually find those values for which the iterative algorithm converges rapidly to the interpolant. For the example shown in Figure 1 this corresponds to about $1.2 < \varepsilon < 1.8$. Since the interpolant is still rather "spiky" for these ε-values neither the interpolant nor the iterated approximant are desirable in this case. However, the (iterated) AMLS apprimant is usually smoother for this range of ε-values and there may be certain circumstances in which this may be more desirable than the interpolant.

A third range of values of the scale parameter corresponds to a "good" interpolant – usually accompanied by slow convergence of the iterative algorithm. For the example shown in Figure 1 this corresponds to about $0.4 < \varepsilon < 1.2$. For most problems of small to modest size (for which we were able to compute the interpolant in Matlab) the smallest achievable RMS-error for the interpolant falls into this range. However, for larger problems the optimal interpolant may be associated with an ε-value that leads to an ill-conditioned system matrix. This brings us to the last ε-range.

If ε is in the range that causes instability (roughly $\varepsilon < 0.4$ in Figure 1), then the iterated AMLS method can successfully overcome the computational difficulty associated with the solution of the ill-conditioned interpolation system. To illustrate this more clearly we have displayed a "zoomed-in" view for this range of small ε in the right part of Figure 1. Even pure AMLS quasi-interpolation may work better than interpolation for this range of ε. In addition, the error between the test function and the iterated AMLS approximant is significantly improved after only 10 iterations. It is clear that if the number of iterations is much smaller than the number of data points, then the iterated AMLS method requires far less computational work compared to solving for the interpolant. In the experiments reported here we always perform 10 iterations. That is, the computational complexity is of order $\mathcal{O}(10N^2)$ while direct computation of the interpolant usually requires a computational complexity on the order of $\mathcal{O}(N^3)$. Of course, fast summation techniques such as the fast (non-uniform) Fourier transform (see, e.g., [6, 7] or [5]) can be used to improve the efficiency of both approaches.

In order to study the connection between the convergence behavior of the interpolant and that of the iterated AMLS approximant we will now focus on an ε that falls into the "reasonable" range for both the interpolant and the iterative algorithm, say, $0.4 \leq \varepsilon \leq 0.8$ for the examples we present.

It is well-known (see, e.g., [1]) that stationary interpolation with Gaussians is saturated. However, to our knowledge until now no one has provided an explicit estimate for the saturation error. In the quasi-interpolation setting, on the other hand, which is discussed in the literature on approximate approximations (see, e.g., [9]) the saturation error is well understood. This begs the question whether the saturation error of stationary RBF interpolation can be explained by the one of AMLS approximation via the residual iteration process. Since we have shown convergence of iterated approximate MLS approximation to the RBF interpolant we may expect the saturation error of the approximate approximation setting to propagate to the interpolant. This may be intuitively reasonable since the residual function is constructed with the same AMLS functions as the initial AMLS approximant. Further detailed analysis of this phenomenon is required.

The graphs in Figure 3 illustrate how the convergence behavior of the iterated AMLS approximation matches that of the interpolant. The graphs shows RMS-errors versus N, the number of data points, for a 1D approximation problem with data function (20). The graphs in the left column are for uniformly spaced data, and in the right column for Halton points. Each row corresponds to a different (fixed) value of ε.

We see that the behavior is very similar in both the uniform and the non-uniform (Halton) setting. The main difference is the larger error for the basic AMLS approximation (`f-pf`$_0$, dash-dotted line) in the Halton setting. This is easily explained by the

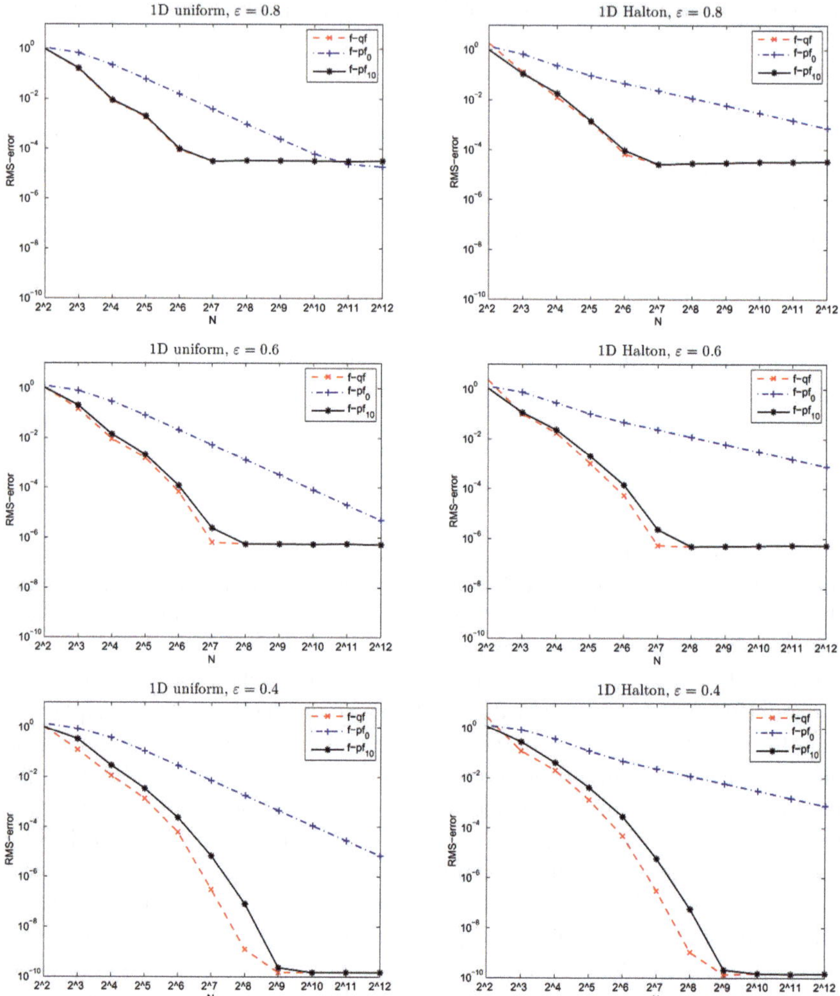

Fig. 3. Saturated convergence of stationary RBF interpolation and iterated AMLS approximation for $\varepsilon = 0.8, 0.6, 0.4$ from top down with uniform (left) and Halton data (right).

fact that we have not adapted the scaling of the basis functions for the non-uniform point distribution (see (22)). The similarity of the errors for the iterated AMLS approximant in the Halton and uniform settings clearly illustrates how the iterative algorithm automatically adapts the AMLS method to the non-uniform setting.

In Figure 4 we collect the error curves for the iterated AMLS approximants from the two columns of Figure 3 in one graph each. We can now clearly see that a larger

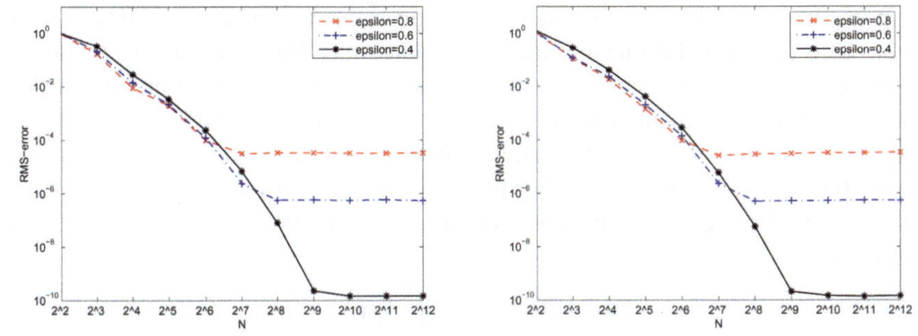

Fig. 4. Saturation errors of iterated AMLS approximation for uniform (left) and Halton (right) data in 1D.

ε makes the saturation error come in earlier, i.e., the drop in the error stops at a larger error value and at a smaller data size N.

5 Conclusions

In this report we have presented a residual iterative algorithm for solving the scattered data fitting problem. As shown by the analysis and numerical results in the previous sections, the method works well for both uniform data and Halton data (an example of irregular data). This is true in terms of accuracy, convergence speed, numerical stability and computational cost.

As illustrated above, the residual iterative algorithm provides a mechanism to transfer the saturation error associated with (stationary) AMLS approximation to RBF interpolation. We plan to investigate this connection more carefully in the future.

As is well known, RBF interpolation is the best approximation in the native space for the data fitting problem. However, in practice, this "best" solution might not be "good enough" – especially if the choice of the shape parameter ε is bad. Moreover, in some cases one may not really desire an interpolant as the "best" solution. This will be true, for example, when the data are obtained from some inaccurate noisy experiments. In this case the residual iterative method will definitely provide a more stable and more reliable solution for the problem.

Clearly, the success of our iterative method is determined by several components such as the scaling parameter ε and the basic function itself. As pointed out above, we want the basic function to be strictly positive definite to ensure uniqueness of the RBF interpolant. Moreover, we want the basic function to satisfy a set of continuous moment conditions defined in (7) required for AMLS approximation. As an example of a class of functions that carry both of these properties we presented the Laguerre–Gaussian functions. In order to obtain more insight and carry out a deeper analysis of the performance of this iterative method, it is desirable to test the method with some other basic functions besides the Laguerre–Gaussians. Presently we are not aware of any direct connection between strict positive definiteness of a function and the set of continuous moment conditions (7). Of course, it is a simple matter to verify that any integrable normalized strictly positive definite function satisfies condition (7) for (at least) $d = 0$. On the other hand, a function that satisfies condition (7) is not guaranteed to be strictly positive definite. For example, let $\varphi : [0, \infty) \to \mathbb{R}$ be defined as

$$\varphi(r) = \frac{1 + r^2}{1 + r^2 + r^4}.$$

Then its one-dimensional radial version

$$\varphi(\|x\|) = \frac{\sqrt{3}}{2\pi} \frac{1 + \|x\|^2}{1 + \|x\|^2 + \|x\|^4}, \qquad x \in \mathbb{R}$$

satisfies condition (7) for $d = 0$. But it can be verified that the Fourier transform of φ is negative at some points. Therefore φ is not strictly positive definite.

Since finding strictly positive definite functions that satisfy the continuous moment conditions seems to be a non-trivial task, we plan to search for additional function classes that carry both of these properties.

References

1. Buhmann, M. D., Multivariate interpolation using radial basis functions. Ph.D. Dissertation, University of Cambridge, 1989.
2. Fasshauer, G. E., Approximate moving least-squares approximation with compactly supported weights. In: *Meshfree Methods for Partial Differential Equations*, M. Griebel and M. A. Schweitzer (Eds.), Lecture Notes in Computer Science and Engineering, Vol. 26, Springer Verlag, Berlin, 2002, pp. 105–116.
3. Fasshauer, G. E., Approximate moving least-squares approximation: A fast and accurate multivariate approximation method. In: *Curve and Surface Fitting: Saint-Malo 2002*, A. Cohen, J.-L. Merrien, and L. L. Schumaker (Eds.), Nashboro Press, Nashville, 2003, pp. 139–148.
4. Fasshauer, G. E., Toward approximate moving least squares approximation with irregularly spaced centers. *Computer Methods in Applied Mechanics & Engineering*, 193:1231–1243, 2004.
5. Fasshauer, G. E. and Zhang, J. G., Recent results for moving least squares approximation. In: *Geometric Modeling and Computing: Seattle 2003*, M. L. Lucian and M. Neamtu (Eds.), Nashboro Press, Brentwood, TN, 2003, pp. 163–176.
6. Kunis, S. and Potts, D., NFFT, Softwarepackage (C-library). Universität Lübeck, http://www.math.uni-luebeck.de/potts/nfft/, 2002.
7. Kunis, S., Potts, D. and Steidl, G., Fast Fourier transforms at nonequispaced knots: A user's guide to a C-library. Universität Lübeck, http://www.math.uni-luebeck.de/potts/nfft/, 2002.

8. Lanzara, F., Maz'ya, V. and Schmidt, G., Approximate approximations from scattered data. Preprint, 2005.
9. Maz'ya, V. and Schmidt, G., On quasi-interpolation with non-uniformly distributed centers on domains and manifolds. *J. Approx. Theory*, 110:125–145, 2001.
10. Wendland, H., *Scattered Data Approximation*. Cambridge University Press, Cambridge, 2005.
11. Wong, T.-T., Luk, W.-S. and Heng, P.-A., Sampling with Hammersley and Halton points. *J. Graphics Tools*, 2:9–24, 1997.

A Kansa Type Method Using Fundamental Solutions Applied to Elliptic PDEs

Carlos J.S. Alves and Svilen S. Valtchev

CEMAT and Departamento de Matemática, Instituto Superior Técnico, TULisbon, 1049-001 Lisboa, Portugal; E-mail: {calves, ssv}@math.ist.utl.pt

Abstract. A Kansa type modification of the Method of Fundamental Solutions (MFS) is presented. This allows us to apply the MFS to a larger class of elliptic problems. In the case of inhomogeneous problems we reduce to a single linear system, contrary to previous methods where two linear systems are solved, one for the particular solution and one for the homogeneous solution of the problem. Here the solution is approximated using fundamental solutions of the Helmholtz equation. Several numerical tests in 2D will be presented in order to illustrate the convergence of the method. Mixed, Dirichlet–Neumann, boundary conditions will be considered.

Key words: Elliptic boundary value problems, Method of Fundamental Solutions, Helmholtz fundamental solution.

1 Introduction

Since its introduction, in 1964 by Kupradze and Aleksidze [9], the Method of Fundamental Solutions has been widely used by mathematicians and engineers for solving homogeneous problems (e.g. [4, 5, 10]). More recently the MFS has also been applied for inhomogeneous problems using radial basis functions (e.g. [10]), or using fundamental solutions from an associated eigenvalue equation, cf. [2]. The mathematical justification for the MFS approximation has been made using density results, cf. [1, 5].

Here we propose a modification of classical MFS methods, using Helmholtz fundamental solutions to simplify Laplacian terms, in a similar manner as RBFs are used in Kansa's method, e.g. [8]. This can be used for the numerical solution of a large class of homogeneous and inhomogeneous elliptic problems.[1] We consider a general

[1] In this work, a boundary value problem, for instance (1), is called homogeneous if the right hand side $f = 0$ and inhomogeneous if $f \neq 0$.

V.M.A. Leitão et al. (eds), Advances in Meshfree Techniques, 241–256.

elliptic PDE with mixed, Dirichlet–Neumann boundary conditions, given by:

$$\begin{cases} a\Delta u + bu = f \text{ in } \Omega \\ u = g_1 \qquad\qquad \text{on } \Gamma_1 \\ \dfrac{\partial u}{\partial \nu} = g_2 \qquad\quad \text{on } \Gamma_2, \end{cases} \tag{1}$$

where Ω is a bounded, simply connected domain with a smooth boundary $\Gamma = \partial\Omega = \Gamma_1 \bigcup \Gamma_2$ and a, b, f, g_1, g_2 are given functions.

2 Theoretical Aspects

We will not discuss in detail the well posedness of problem (1). From a simple classical approach (e.g. [7]) we know that for an elliptic PDE with $|a(x)| \geq \varepsilon > 0$ a.e. $x \in \Omega$, $a, b \in L^\infty(\Omega)$, $f \in L^2(\Omega)$ the Dirichlet problem has a weak solution $u \in H_0^1(\Omega)$, which is unique except for a countable number of eigenvalues for some differential operators (for instance, the Helmholtz operator with a, b positive constants). On the other hand for the Neumann problem, to ensure existence, a compatibility condition is sometimes required, for instance for $b = 0$, we must force $\int_\Gamma g_2 = \int_\Omega f/a$.

We will assume that the problem (1) is known to be well posed.

The application of the MFS for the solution of boundary value problems involving elliptic differential equations has been justified by density results on the boundary (eg. [5]) and in the whole domain cf. [2]. These density results are however restricted to some elliptic differential operators with constant coefficients.

Since here we consider general variable coefficients explicit fundamental solutions are not available, and we propose to use fundamental solutions of Helmholtz operators:

$$\Phi_k(x) = \begin{cases} \dfrac{i}{4} H_0^{(1)}(k|x|) \text{ in 2D} \\ \dfrac{e^{ik|x|}}{4\pi|x|} \qquad\quad \text{in 3D} \end{cases} \tag{2}$$

i.e. Φ_k satisfies $-(\Delta + k^2)\Phi_k = \delta$, where δ represents the Dirac delta and k is a positive frequency. Here $|x|$ represents the Euclidian norm of x and $H_0^{(1)} := J_0 + iY_0$ is the first Hänkel function, defined through the Bessel functions of first and second kind, J_0 and Y_0, respectively.

The main idea consists in using Helmholtz fundamental solutions since they reduce the application of the differential operator to algebraic manipulation. In fact, defining $P(1, x) = b + ax$, the differential operator is given by $P(1, \Delta)u = bu + a\Delta u$ and in particular for Φ_k we have

$$P(1,\Delta)\Phi_k = P(1,-k^2)\Phi_k$$

(except at the origin).

Moreover, considering an appropriate admissible domain $\hat{\Omega} \subset \bar{\Omega}^c$ with boundary $\hat{\Gamma} = \partial\hat{\Omega}$, for instance an open set $\hat{\Omega}$ such that $\hat{\Omega}^c \supset \Omega$ we have the density result, established in [2]:

$$S_\Omega = \text{span}\{\Phi_\kappa(\cdot - y)_{|\Omega} : y \in \hat{\Gamma}, \kappa \in I \subset \mathbf{R}^+\} \text{ is dense in } L^2(\Omega). \tag{3}$$

where I is an open interval.

This density result motivates the approximation of a solution u by a sequence of linear combinations of fundamental solutions. Assuming the $L^2(\Omega)$ sense in the differential equality, we may say that u is given by an infinite sum from S_Ω

$$u(x) = \sum_{\kappa,y}^{(\infty)} \beta_{\kappa,y}\Phi_\kappa(x-y), \qquad (x \in \Omega). \tag{4}$$

We first justify that the approximation of the differential equation, in the domain Ω is possible using Helmholtz fundamental solutions.

Theorem 1. (i) *If $f_b/b \in L^2(\Omega)$ there exists a sequence (v_n) with $v_n \in S_\Omega$ such that $bv_n \to f_b$ in $L^2(\Omega)$.*

(ii) *If $f_a/a \in L^2(\Omega)$ there exists a sequence (u_n) with $u_n \in S_\Omega$ such that $a\Delta u_n \to f_a$ in $L^2(\Omega)$.*

Thus if $f = f_a + f_b$ and $f_a/a, f_b/b \in L^2(\Omega)$ then $a\Delta u_n + bv_n \to f$ in $L^2(\Omega)$.

Proof. (i) is a trivial consequence of the density result (3). To prove (ii), since $f_a/a \in L^2(\Omega)$ we also consider (3) with

$$f_a(x)/a(x) = \sum_{\kappa,y}^{(\infty)} \alpha_{\kappa,y}\Phi_\kappa(x-y), \qquad (x \in \Omega)$$

then

$$u_n(x) = \sum_{\kappa,y}^{(n)} \frac{\alpha_{\kappa,y}}{-\kappa^2}\Phi_\kappa(x-y), \qquad (x \in \Omega)$$

and

$$\Delta u_n(x) = \sum_{\kappa,y}^{(n)} \alpha_{\kappa,y}\Phi_\kappa(x-y), \qquad (x \in \Omega)$$

converges in $L^2(\Omega)$ to f_a/a. Therefore $a\Delta u_n \to f_a$ in $L^2(\Omega)$. □

This does not prove that the solution exists, it only shows that it is possible to use sequences of fundamental solutions to approximate the right hand side of the equation. In simple situations, such as a, b constant, we can construct $u_n = v_n$ giving an explicit solution.

Theorem 2. *Consider a, b constants and S_Ω with $\sqrt{b/a} \notin I$. If $f \in L^2(\Omega)$ with*

$$f(x) = \sum_{\kappa,y}^{(\infty)} \alpha_{\kappa,y} \Phi_\kappa(x-y), \qquad (x \in \Omega)$$

then

$$u(x) = \sum_{\kappa,y}^{(\infty)} \frac{\alpha_{\kappa,y}}{b - a\kappa^2} \Phi_\kappa(x-y), \qquad (x \in \Omega)$$

is a particular solution for $a\Delta u + bu = f$ in $L^2(\Omega)$.

Proof. Applying the differential operator $a\Delta + bI$ to $\Phi_k(x-y)$ we obtain

$$(a\Delta + bI)\Phi_k(x-y) = (-a\kappa^2 + b)\Phi_k(x-y)$$

and applying it to the sum that defines u gives exactly f.

The choice of the set of frequencies I ensures that $\kappa^2 \neq b/a$. □

To obtain the solution of the full problem the boundary conditions must also be imposed on Γ using the expansion (4).

We will solve simultaneously the domain equation and the boundary conditions in a discrete linear system defined by a finite number of test frequencies and point sources.

3 The Numerical Method

We consider m source-points y_j, distributed on the admissible source set $\hat{\Gamma}$ (cf. [1]), and p positive test frequencies k_r, and approximate the solution by

$$\tilde{u}(x) = \sum_{r=1}^{p} \sum_{j=1}^{m} \alpha_{r,j} \Phi_{k_r}(x - y_j). \tag{5}$$

The unknown coefficients $\alpha_{r,j}$ will be fitted such that $\tilde{u}(x)$ satisfies the boundary conditions as well as the inhomogeneous elliptic equation, on some prescribed collocation points x_i. This ideia is similar to the Kansa technique, where instead of the RBFs we use fundamental solutions of the Helmholtz equation (e.g. [8]). This leads to a discrete problem that consists in solving a three block linear system of the form

$$\begin{bmatrix} \mathsf{A} \\ --- \\ \mathsf{B} \\ --- \\ \mathsf{C} \end{bmatrix}_{n\times(pm)} \begin{bmatrix} \alpha_{1,1} \\ \alpha_{1,2} \\ \cdots \\ \alpha_{p,m-1} \\ \alpha_{p,m} \end{bmatrix}_{(pm)\times 1} = \begin{bmatrix} f(x_i^d) \\ --- \\ g_1(x_i^*) \\ --- \\ g_2(x_i^{**}) \end{bmatrix}_{n\times 1} \tag{6}$$

where $x_i^d \in W \subseteq \Omega$, $x_i^* \in \Gamma_1$ and $x_i^{**} \in \Gamma_2$ are the n^d, n^* and n^{**} collocation points and $n := n^d + n^* + n^{**}$. The three submatrices are defined as follows:

- associated with the elliptic differential equation

$$\mathsf{A} = [(b(x_i^d) - k_r^2 a(x_i^d))\Phi_{k_r}(x_i^d - y_j)]_{n^d \times (pm)}$$

- associated with the Dirichlet boundary condition

$$\mathsf{B} = [\Phi_{k_r}(x_i^* - y_j)]_{n^* \times (pm)}$$

- associated with the Neumann boundary condition

$$\mathsf{C} = [\nu_{x_i^{**}} \cdot \nabla\Phi_{k_r}(x_i^{**} - y_j)]_{n^{**} \times (pm)}$$

Here $\nu_{x_i^{**}}$ represents the unitary normal vector at $x_i^{**} \in \Gamma_2$. The linear system (6) can be solved by direct collocation, taking $n = pm$, or in the least squares sense, taking $n > pm$. In this work we used the Gauss method with partial pivoting for the solution of the linear system.

4 Numerical Simulations

We focus on the 2D case of problem (1), the 3D case can be considered in a similar way. In the first three simulations, we will consider a circular domain $\Omega_1 = S^1 = \{x \in \mathbf{R}^2 : |x| \leq 1\}$ and smooth boundary data. Any other domain can be considered, noting that there should exist an uniquely defined normal vector ν at any point $x_i^{**} \in \Gamma_2$. In the other two examples we will test the performance of the method in more general settings. More precisely, we will consider irregular boundary and non-smooth boundary data.

As we will consider only real functions a, b, u, f, g_1, g_2, from an implementation point of view, we can take only the smooth, non-singular, J_0, part of the fundamental solution (2). Density results are still applicable, see [2]. Additionally, the source-points y_j can be chosen on the boundary $\partial\Omega$ (Boundary Knot-type Method, see [6]) or even inside the domain.

Note that the normal derivative of the non-singular part of the fundamental solution in 2D is given by

$$\frac{\partial}{\partial\nu}\left(\frac{J_0(k|x-y|)}{4}\right) = -k\frac{\nu_x \cdot (x-y)}{4|x-y|}J_1(k|x-y|).$$

4.1 Helmholtz-Type Equations

We start by taking $a(x) = b(x) = 1$. In this case equation (1) resumes to an unitary Helmholtz inhomogeneous problem with mixed boundary conditions. In [2] such

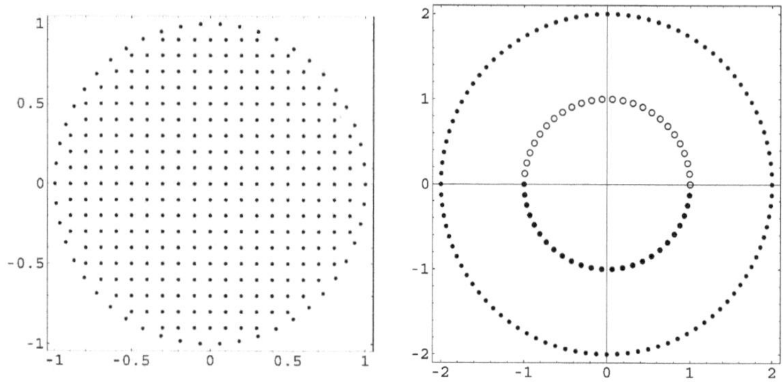

Fig. 1. The knots x_i^d, left and x_i^* (dots), x_i^{**} (circles) and $y_j \in \partial B(0, 2)$, right.

problem was studied, but only with Dirichlet boundary conditions. Let the exact solution be given by

$$u_1(x) = \sin(\pi x_1)x_2 + \cos(\pi x_2)x_1$$

and $f,\ g_1,\ g_2$, calculated accordingly.

From the numerical point of view, we consider $n^d = 1401$ collocation points $x_i^d \in \bar{\Omega}_1$ and $m = 100$ source-points, evenly distributed on $\partial B(0, 2) = \{x \in \mathbf{R}^2 : |x| = 2\}$, see Figure 1. We took $p = 10$ integer test frequencies $\{1, 2, \ldots, 10\}$. The knots x_i^* and x_i^{**} were also evenly distributed on Γ_1 and Γ_2, according to the type of boundary conditions (BC).

In Figure 2 we present the plots of the absolute error (the relative error is of the same order of magnitude) for three simulations. On the first plot, a pure Dirichlet BC problem ($n^* = 200,\ n^{**} = 0$), on the second plot a mixed BC problem ($n^* = n^{**} = 100$), where the Neumann condition is imposed on the upper half of $\partial\Omega_1$, and finally on the third plot a pure Neumann BC problem ($n^* = 0,\ n^{**} = 200$).

The method showed excellent numerical results, absolute (relative) errors of magnitude less than 10^{-6} were observed in the three cases. It is worth noting that our method performs worse when a Dirichlet BC problem is considered (Figure 2, first plot, error 10^{-6}) than in the pure Neumann case (Figure 2, third plot, error 10^{-8}). For mixed BC the numerical results are intermediate. This behavior is observed for any knot distribution and test frequencies as we illustrate in Table 1.

Here (see Table 1) we vary the number of boundary collocation points and the number of source-points, $y_j \in \partial B(0, 2)$. Two sets of $p = 10$ tests frequencies were also considered: $\mathcal{F}_1 = \{2, 4, \ldots, 20\}$ and $\mathcal{F}_2 = \{1, 2, \ldots, 10\}$. Note that the choice of the test frequencies is a delicate problem and the behavior of the exact solution should be studied carefully in order to choose an appropriate set. A highly oscillating exact solution will require higher test frequencies.

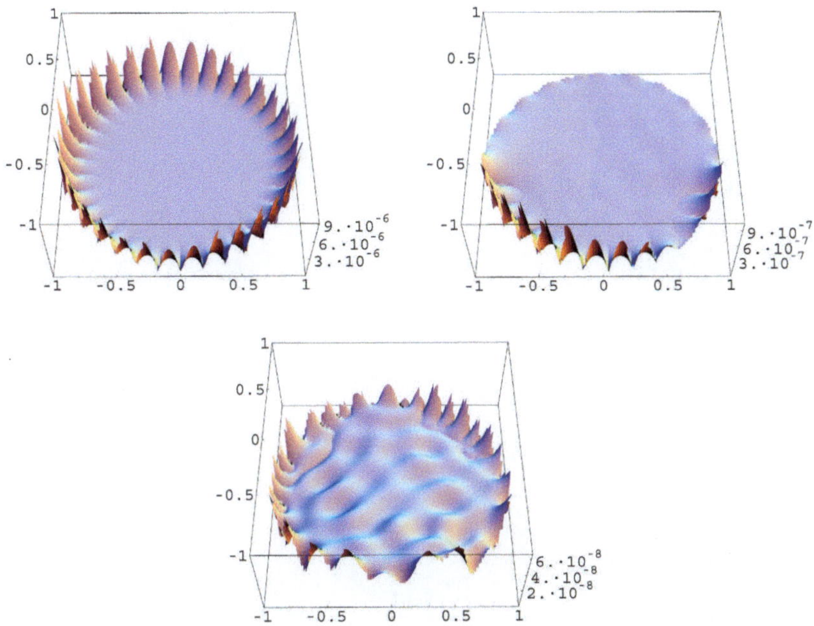

Fig. 2. Plots of the absolute error of $\tilde{u}_1$ for a Dirichlet (first plot), Dirichlet–Neumann (second plot) and Neumann (third plot) boundary conditions problem.

Table 1. The maximum absolute error of $\tilde{u}_1$ for several knot and frequency choices, measured over 2000 random points on the annulus $\{x \in \mathbf{R}^2 : 0.95 \leq |x| \leq 1.0\}$.

	$n^* + n^{**}$	m	Dirichlet BC	mixed BC	Neumann BC
$\mathcal{F}_1$	100	50	$1.005E-04$	$2.525E-05$	$1.933E-05$
	100	100	$5.075E-05$	$3.810E-04$	$1.970E-05$
	200	100	$5.515E-05$	$2.563E-05$	$2.256E-05$
$\mathcal{F}_2$	100	50	$3.310E-06$	$1.157E-06$	$4.823E-07$
	100	100	$2.189E-05$	$3.269E-06$	$2.655E-07$
	200	100	$8.895E-06$	$9.673E-07$	$6.574E-08$

The numerical results shown in Table 1 confirm the convergence of the method. The relative errors are lower than 0.04%.

Our method can also be applied to Poisson (take $a(x) = 1,\ b(x) = 0$) or modified Helmholtz (take $a(x) = 1,\ b(x) = -1$) problems. It is also not a problem to

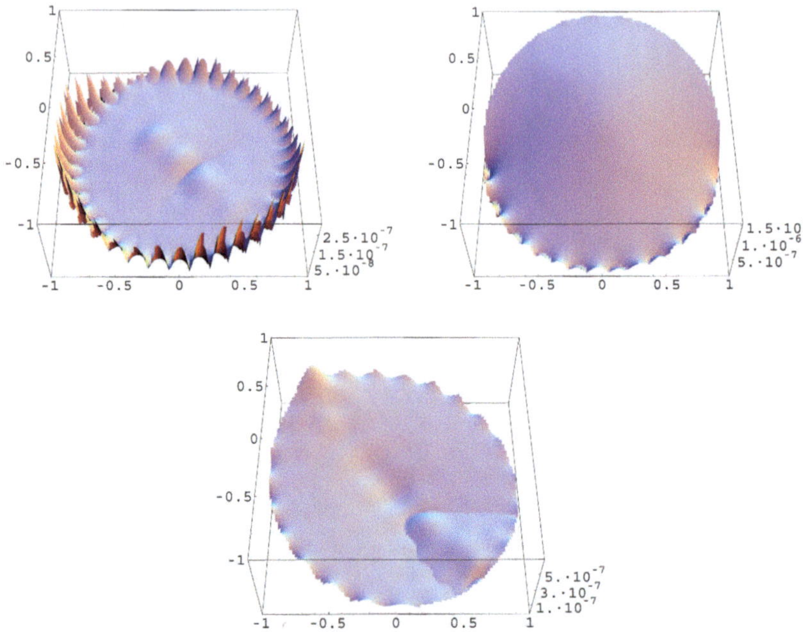

Fig. 3. Plots of the absolute error of $\tilde{u}_2$ for a Dirichlet (first plot), Dirichlet–Neumann (second plot) and Neumann (third plot) boundary conditions problem.

approximate the solution of a general homogeneous equation as we will see in the following subsection.

4.2 Homogeneous Elliptic Equations of Type (1)

Here we consider a homogeneous elliptic equation of type (1), i.e. $f = 0$, and take

$$a(x) = \sin(x_1 + x_2) \quad \text{and} \quad b(x) = \frac{4(1 - x_1^2 - x_2^2)\sin(x_1 + x_2)}{(1 + x_1^2 + x_2^2)^2}.$$

For this particular choice, the exact solution is given by

$$u_2(x) = \frac{1}{1 + x_1^2 + x_2^2}$$

and the boundary conditions are calculated accordingly.

In Figure 3 we show the numerical results for the three types of boundary conditions as in the previous example. We took $n^d = 1301$ and $n^* + n^{**} = 100$ collocation

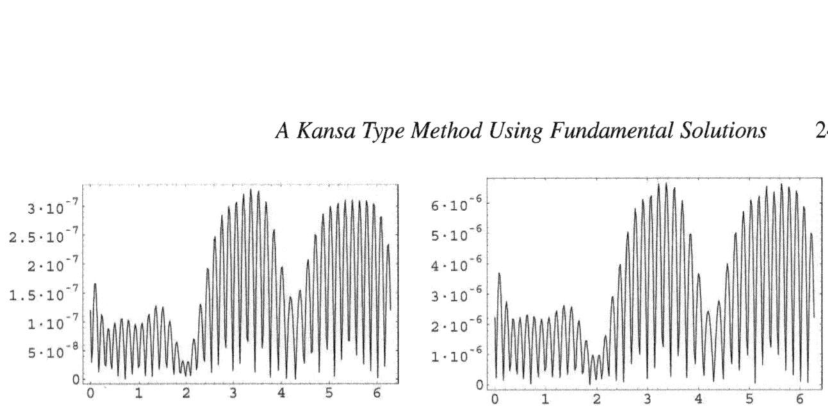

Fig. 4. Absolute error of $\tilde{u}_2$ (left) and $\frac{\partial \tilde{u}_2}{\partial \nu}$ (right), on the boundary $\partial\Omega_1$, for a Dirichlet boundary conditions problem.

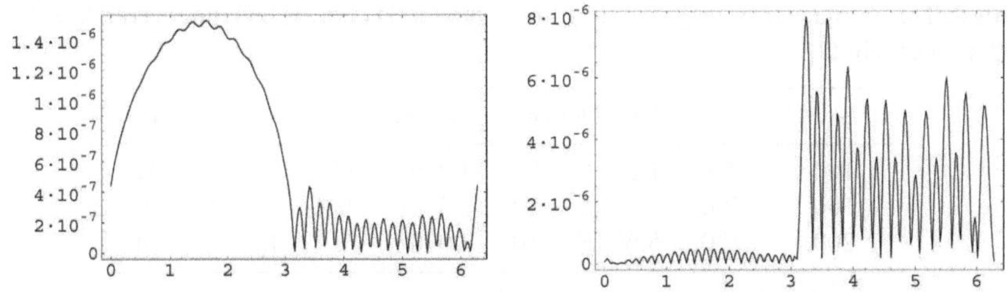

Fig. 5. Absolute error of $\tilde{u}_2$ (left) and $\frac{\partial \tilde{u}_2}{\partial \nu}$ (right), on the boundary $\partial\Omega_1$, for a Dirichlet–Neumann boundary conditions problem.

points and $p = 15$ integer test frequencies $\{1, 2, \ldots, 15\}$. For the source points y_j we considered $m = 100$ knots, uniformly distributed on $\partial B(0, 1.5)$.

For this example the numerical method shows absolute errors of magnitude lower than 10^{-6} in the three cases. The best results are observed in the Dirichlet BC case. On the other hand, the method shows higher numerical errors in the mixed BC case.

In Figures 4, 5 and 6 we illustrate the absolute error of the approximate solution $\tilde{u}_2$ (left plots) and its normal derivative $\partial\tilde{u}_2/\partial\nu$ (right plots) on the boundary $\partial\Omega_1$, for the three types of boundary conditions. In general, we may conclude that if we impose a Dirichlet type BC on a portion Γ_1 of the boundary then $\tilde{u}$ will show lower absolute error near Γ_1 than its normal derivative $\partial\tilde{u}/\partial\nu$.

As we mentioned in Section 2 the mathematical justification of the MFS is given in terms of density results on the $L^2(\Omega_1)$ norm. In Table 2 we present some numerical results for the absolute error of $\tilde{u}_2$, measured on a discrete l^2 norm. More precisely, we measured the quantity

$$||u - \tilde{u}||_{2,N} := \frac{1}{N}\left(\sum_{i=1}^{N} |u(t_i) - \tilde{u}(t_i)|^2\right)^{1/2},$$

where $\{t_i\}$ are N random error test points in $\bar{\Omega}_1$.

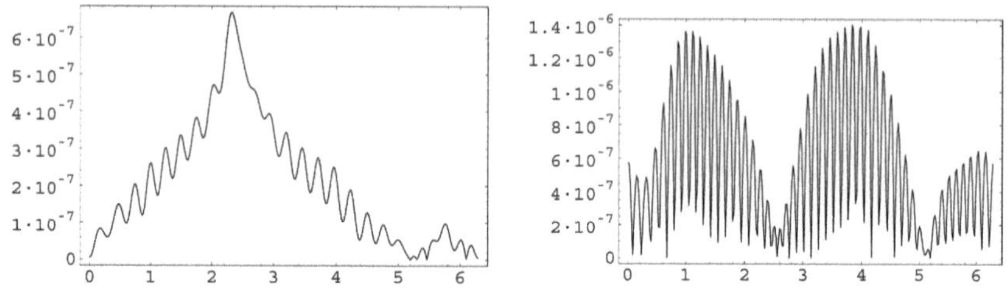

Fig. 6. Absolute error of $\tilde{u}_2$ (left) and $\frac{\partial \tilde{u}_2}{\partial \nu}$ (right), on the boundary $\partial\Omega_1$, for a Neumann boundary conditions problem.

Table 2. The absolute error $||u_2 - \tilde{u}_2||_{2,N}$ for several knot choices, measured over $N = 5000$ random points in $\bar{\Omega}_1$.

$n^* + n^{**}$	m	Dirichlet BC	mixed BC	Neumann BC
100	50	$1.657E-09$	$7.936E-09$	$1.123E-09$
100	100	$4.294E-10$	$1.544E-08$	$2.630E-09$
200	100	$5.650E-10$	$3.078E-10$	$4.344E-10$

The numerical results confirm the convergence of the method on the discrete norm. Note that as the exact and approximate solutions are smooth functions and N is finite the two norms $||\cdot||_\infty$ and $||\cdot||_{2,N}$ are equivalent.

4.3 Inhomogeneous Elliptic Equations of Type (1)

We now consider a more general, inhomogeneous, example of problem (1). Let $a(x) = e^{-(x_1^2+x_2^2)}$, $b(x) = \cos(x_1) + \cos(x_2)$ and for the exact solution we took

$$u_3(x) = \sin(x_1 x_2) x_1 x_2,$$

with f, g_1 and g_2 calculated accordingly, using equation (1).

For the mixed boundary condition case we will impose a Neumann condition on the first and third quadrants of $\mathbf{R}^2$ and Dirichlet BC on the second and fourth quadrants, see Figure 7.

In Figure 8 we present the numerical results for $n^d = 1401$ knots $x_i^d \in \bar{\Omega}_1$ and $m = 100$ source-points, evenly distributed on $\partial B(0, 2)$. We took $p = 10$ integer test frequencies $\{1, 2, \ldots, 10\}$ and $n^* + n^{**} = 200$ evenly distributed boundary knots.

The absolute error behavior for the Dirichlet BC problem (see Figure 8, first plot) is similar to the one observed for the previous two examples. On the other hand, the error plots for the mixed BC (second plot) and Neumann BC (third plot) problems differ significantly. For the current example we can no longer state that the maximum error occurs at the boundary as for the Helmholtz-type example. We also note that

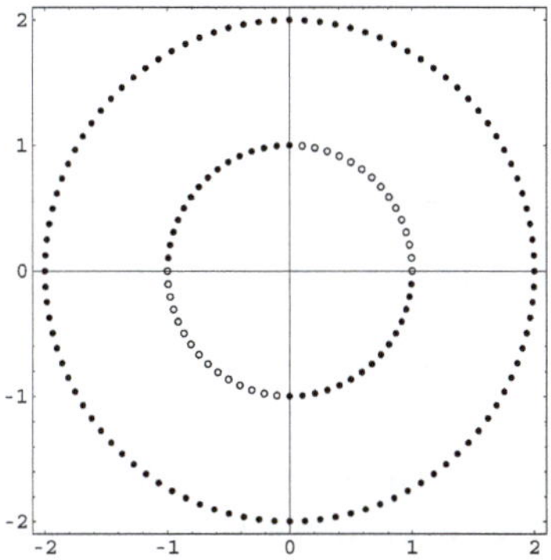

Fig. 7. The knots x_i^* (dots), x_i^{**} (circles) and $y_j \in \partial B(0, 2)$.

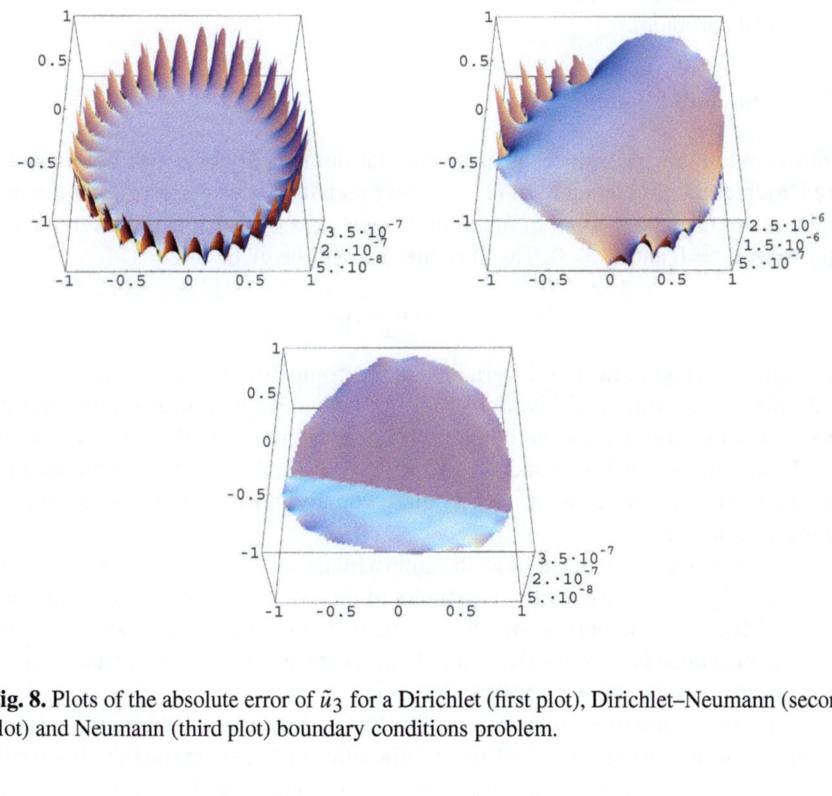

Fig. 8. Plots of the absolute error of $\tilde{u}_3$ for a Dirichlet (first plot), Dirichlet–Neumann (second plot) and Neumann (third plot) boundary conditions problem.

Table 3. The maximum absolute error of $\tilde{u}_3$ for several knot choices, measured over 10^4 random points on $\bar{\Omega}_1$.

$n^* + n^{**}$	m	Dirichlet BC	mixed BC	Neumann BC
100	50	$1.374E-06$	$4.226E-07$	$4.236E-07$
100	100	$3.723E-07$	$7.919E-07$	$1.895E-07$
200	100	$4.418E-07$	$2.533E-06$	$3.692E-07$

in the mixed BC problem the approximate solution is worse near Γ_2 than near Γ_1, contrary to what was observed in Section 4.1.

In Table 3 we present some numerical results varying the number of source and boundary collocation points. Note that, as $||u_3||_\infty \approx 0.24$, the relative error is of the same magnitude as the measured absolute error.

Numerical tests were also performed for source-points located inside the domain. For example we took $y_j \in \partial B(0, 0.99)$, mixed BC and the same collocation knots and test frequencies as in Figure 8, center. The observed absolute errors were slightly lower. The Boundary Knot-type method ($y_j \in \partial\Omega_1$) showed similar results for the Dirichlet BC problem.

4.4 Non-Smooth Boundary Data

In this simulation we consider the same circular domain, $\Omega_1 = S^1$, as before and a pure Dirichlet BC problem. Let $a(x) = x_1^2 + x_2^2$ and $b(x) = \sin(x_1 + x_2)$. For the forcing term we take a non-smooth function $f(x) = |x_1 x_2|$, which is not diferentiable at the axes $x_1 = 0$ and $x_2 = 0$. The Dirichlet BC will be given by

$$g_1(x) = |\sin(x_1)| + \cos(e^{x_2}),$$

which has two singularities of the gradient at the points $(0, 1)$ and $(0, -1)$.

As the exact solution of this problem is not known we will measure the error of the approximate solution on the boundary $\partial\Omega_1$. Here we took $n^* = 90$, $n^{**} = 0$, $m = 30$ with $y_j \in \partial B(0, 1.5)$ and $p = 15$ with $\mathcal{F} = \{1, 2, \ldots, 15\}$. For the domain collocation knots x_i^d we chose $n^d = 721$ quasi equally spaced points, see [3], shown in Figure 9, first plot.

In Figure 9, second plot, we show the approximate solution of this problem in Ω_1. In the same figure we present also a joint plot of the exact and approximate solutions on $\partial\Omega_1$. Clearly the numerical method fits the boundary data almost perfectly. The error is approximately $4.69E-02$ on the $||\cdot||_\infty$ norm and $3.28E-04$ on the $||\cdot||_{2,N}$ norm, when measured on $N = 500$ random points on $\partial\Omega_1$.

As the approximating functions, i.e. the Bessel functions of the first kind, are smooth functions we expect to find some difficulties in approximating the boundary data near the two singularities at $(0, 1)$ and $(0, -1)$. In Figure 10 (left) we show the

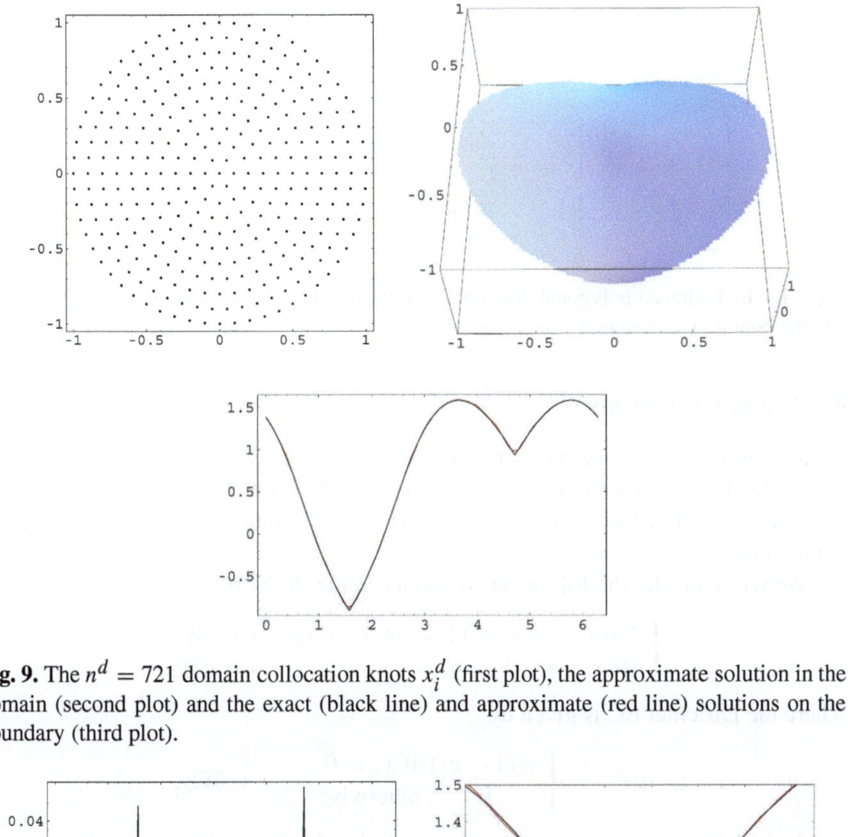

Fig. 9. The $n^d = 721$ domain collocation knots x_i^d (first plot), the approximate solution in the domain (second plot) and the exact (black line) and approximate (red line) solutions on the boundary (third plot).

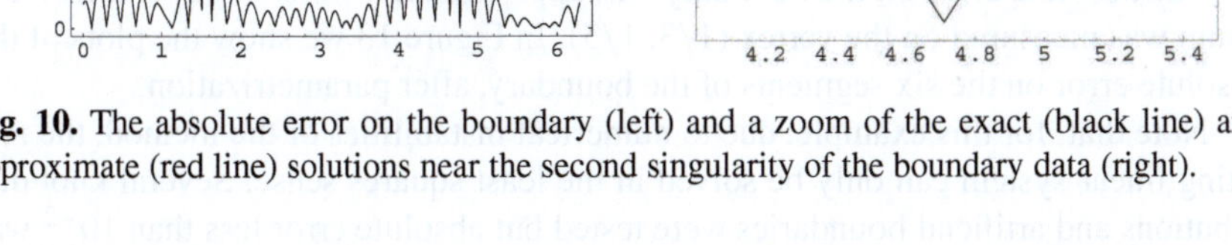

Fig. 10. The absolute error on the boundary (left) and a zoom of the exact (black line) and approximate (red line) solutions near the second singularity of the boundary data (right).

plot of the absolute error on the boundary. Two local maxima, corresponding to the values $\pi/2$ and $3\pi/2$ of the angular parameter of $\partial\Omega_1$ are clearly visible. In order to illustrate these local problems we also include a zoom of one of the referred regions (see Figure 10, right).

Note that this local problems do not affect the global convergence of the method on the discrete Euclidian norm.

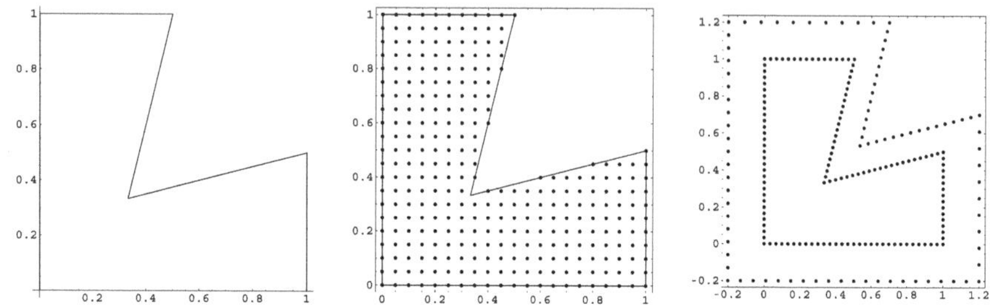

Fig. 11. The L-shaped polygonal domain Ω_2 (left), the domain collocation knots x_i^d (center) and the boundary and source points (right).

4.5 Polygonal Domains

In the last simulation we will apply our method for a Dirichlet BC problem on a polygonal, L-shaped domain (see Figure 11, left). The domain collocation points and the boundary collocation points, as well as the source points, are shown in Figure 11, center and right.

We will consider the following Boundary Value Problem

$$\begin{cases} \Delta u(x) + x_1 x_2 u(x) = \sin(x_1 + x_2), & x \in \Omega_2 \\ u(x) = g_1(x), & x \in \partial\Omega_2 \end{cases}$$

where the Dirichlet BC is given by

$$g_1(x) = \begin{cases} x_2(1-x_2) & \text{if } x_1 = 0 \\ 0 & \text{otherwise} \end{cases}, \quad x \in \partial\Omega_2.$$

As in the previous example, the exact solution for this problem is not known and we measure the absolute error only on the boundary $\partial\Omega_2$.

In Figure 12 we present the approximate solution for $n^* = 352$, $n^{**} = 0$, $n^d = 1134$, $m = 132$, $p = 5$ and $\mathcal{F} = \{1, 2, \ldots, 5\}$.

The absolute error on the boundary was approximately $3.5E-02$, and its maximum was measured on the vertex $(1/3, 1/3)$. In Figure 13 we show the plots of the absolute error on the six segments of the boundary, after parametrization.

Note that, for this example, due to numerical instabilities of the method, the resulting linear system can only be solved in the least squares sense. Several knot distributions and artificial boundaries were tested but absolute error less than 10^{-2} was not observed.

5 Concluding Remarks

We presented a Kansa type modification of the MFS that can be used for the numerical solution of a large class of elliptic problems. The method can be applied for ho-

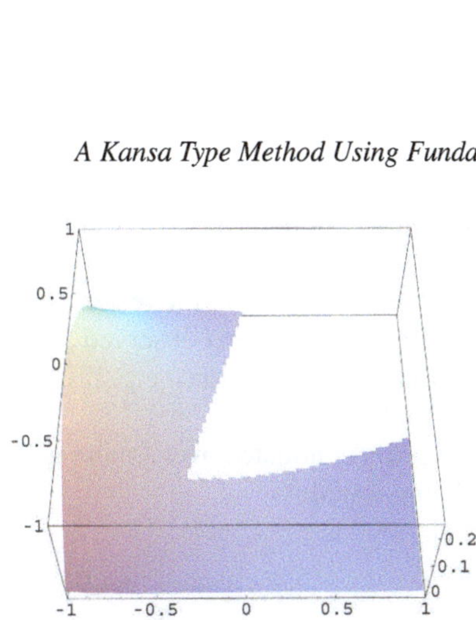

Fig. 12. The approximate solution in Ω_2.

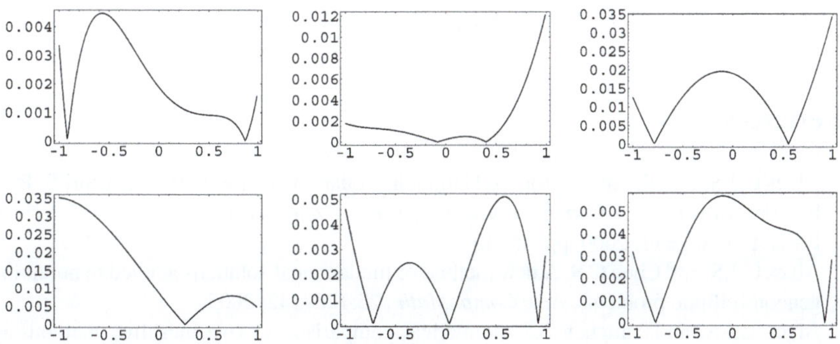

Fig. 13. The absolute error on $\partial\Omega_2$, starting from the vertex (0, 0) and proceeding in contra clockwise direction.

mogeneous or inhomogeneous problems and several boundary conditions (Dirichlet, Neumann or mixed). The method is based on the solution of a single linear system, contrary to previous two-stage methods, where a system is solved for the particular solution and another, for the homogeneous solution. The use of the fundamental solution of the Helmholtz equation avoids the calculation of the Laplacian operator. No numerical/analytical integration or derivation is necessary for the application of our method. The choice of the collocation and source points may be random, but we should avoid close knots in order to avoid linear dependence of rows or columns in the linear system.

In the first three simulations the method shows excellent numerical results and low approximation errors. We may conclude that our method is indicated for the solution of PDEs with smooth boundary conditions and in domains with regular boundaries. On the other hand, in the last two examples, the method failed to decrease the error. In this case, the lowest absolute (relative) error we measured was of order 10^{-2}, on the $\|\cdot\|_\infty$ norm.

One disadvantage to be mentioned is that the resulting linear systems are highly ill conditioned. The same problem was also observed for the classical Method of Fundamental Solutions and in its modifications, e.g. [2]. Regularization techniques may be a solution for this problem.

Current investigation is being conducted over a more general class of PDEs. Better knot distributions are being tested.

Acknowledgements

This work was partially supported by FCT through the projects POCTI-MAT/60863/2004 and POCTI/ECM/58940/2004, by NATO through Collaborative Linkage Grant PST.CLG 980398, and by a project CAPES-GRICES.

References

1. Alves C.J.S. Density results for the Helmholtz equation and the MFS. In: Atluri S, Brust F (Eds), *Advances in Computational Engeneering and Sciences*, Vol. I. Tech Science Press, Los Angeles, 2000, pp. 45–50.
2. Alves C.J.S. and Chen C.S. A new method of fundamental solutions applied to nonhomogeneous elliptic problems. *Adv. Comp. Math.*, 23:125–142, 2005.
3. Alves C.J.S. and Valtchev S.S. Numerical comparison of two meshfree methods for acoustic wave scattering. *Eng. Analysis Bound. Elements*, 29:371–382, 2005.
4. Arantes e Oliveira E.R. Plane stress analysis by a general integral method. *Proc. ASCE Eng. Mech. Div.*, 94:79–101, 1968.
5. Bogomolny A. Fundamental solutions method for elliptic boundary value problems. *SIAM J. Num. Anal.*, 22:644–669, 1985.
6. Chen W. and Tanaka M. Symmetry boundary knot method. *Eng. Analysis Bound. Elements*, 26:489–494, 2002.
7. Evans L. *Partial Differential Equations*. Graduate Studies in Mathematics, Vol. 19. AMS, Providence, RI, 1998.
8. Kansa E.J. Multiquadrics: A scattered data approximation scheme with applications to computational fluid-dynamics-II. Solutions to hyperbolic, parabolic and elliptic partial differential equations. *Comput. Math. Appl.*, 19:147–161, 1990.
9. Kupradze V.D. and Aleksidze M.A. The method of fundamental equations for the approximate solution of certain boundary value problems. *USSR Comp. Math. Math. Phys.*, 4:82–126, 1964.
10. Golberg M.A. and Chen C.S. The method of fundamental solutions for potential, Helmholtz and diffusion problems. In: Golberg M.A. (Ed.), *Boundary Integral Methods and Mathematical Aspects*. WIT Press, 1999, pp. 105–176.

From Global to Local Radial Basis Function Collocation Method for Transport Phenomena

Božidar Šarler

Laboratory for Multiphase Processes, University of Nova Gorica, Vipavska 13, SI-5000 Nova Gorica, Slovenia; E-mail: bozidar.sarler@p-ng.si

Abstract. This article introduces basic concepts of meshless methods for solving partial differential equations in their strong form by collocation or least squares approximation. Global and local formulations are defined. The current achievements, based on the local form and collocation with radial basis functions are explained in detail. Heat transfer and fluid flow problems are treated. These achievements represent a simple, and at the same time more efficient version of the classical meshless radial basis function collocation (Kansa) method. Instead of global, the collocation is made locally over a set of overlapping domains of influence and the time-stepping is performed in an explicit way. Only small systems of linear equations with the dimension of the number of nodes included in the domain of influence have to be solved for each node. The computational effort thus grows roughly linearly with the number of the nodes. The represented approach thus overcomes the principal large scale bottleneck of the original Kansa method and widely opens space for industrial applications of the method. The purpose of this article is to give a concentrated information on this new method, which has already been successfully applied in macroscopic and microscopic transport phenomena field, accompanied with research requirements for the future. It is devoted to practicing engineers and researchers.

Key words: Radial basis function collocation method, transport phenomena, strong formulation, multiquadrics.

Nomenclature

Latin Symbols

b	=	augmented vector of the system of linear equations
c	=	coefficient of the weight function
$\underline{\mathbf{D}}$	=	diffusion tensor
f	=	body force
K	=	number of global approximation functions
N	=	number of gridpoints

V.M.A. Leitão et al. (eds), *Advances in Meshfree Techniques,* 257–282.

$\mathbf{p}$	=	position vector
P	=	pressure
$\tilde{P}$	=	pressure correction
S	=	source
t	=	time
$\underline{\mathbf{T}}$	=	deviatoric part of the stress tensor
$\mathbf{v}$	=	velocity
$\tilde{\mathbf{v}}$	=	velocity correction
$\hat{\mathbf{v}}$	=	velocity estimate

Greek Symbols

α	=	coefficient of the global approximation function
Δt	=	time-step
Δt_P	=	pressure iteration time-step
ρ	=	density
ψ	=	global approximation function
$\underline{\boldsymbol{\psi}}$	=	system matrix
$\underline{\boldsymbol{\Psi}}$	=	system matrix (boundary condition information included)
Φ	=	general transport variable
θ	=	scalar inhomogeneous part in Poisson equation
$\boldsymbol{\Theta}$	=	vector inhomogeneous part in Poisson equation
A	=	transported-diffused variable relation function
Υ	=	boundary condition indicator function
Γ	=	boundary
ω	=	sub-domain
Ω	=	domain

Superscripts

l	=	sub-domain indicator
D	=	Dirichlet boundary indicator
N	=	Neumann boundary indicator
R	=	Robin boundary indicator

Subscripts

0	=	initial time
x	=	Cartesian coordinate
y	=	Cartesian coordinate
z	=	Cartesian coordinate
ς	=	coordinate indicator
ζ	=	coordinate indicator
ξ	=	coordinate indicator
τ	=	coordinate indicator
ι	=	time level indicator

Mathematical Symbols

∇	=	divergence operator

∇^2 = Laplace operator
$\sum$ = sum

1 Introduction

1.1 Motivation

The development of efficient as well as simple algorithms for the numerical solution of partial differential equations (PDEs) is of major interest in applied sciences and engineering. The most popular discrete approximate methods for PDEs are nowadays the finite difference (FDM) [22], the finite volume (FVM) [35], the finite element (FEM) [41], the spectral (SM) [3], and the boundary-domain integral (BDIM) [40] methods. Despite the powerful features of these methods, there are often substantial difficulties in applying them to realistic, geometrically complex three-dimensional transient problems. A common element of the mentioned methods is the need to create a polygonisation, either in the domain and/or on its boundary. This type of meshing is often the most time consuming part of the solution process and is far from being fully automated.

Figure 1 shows the most common types of space discretisation arrangements in numerical methods used for non-linear transport phenomena. The FDM discretisation shown in Figure 1(a), involves pointisation only. However, the points are restricted to coordinate directions and uniformity. The FVM discretisation shown in Figure 1(b) includes polygonisation with rectangles. The rectangles are restricted to coordinate directions. The FEM discretisation is shown in Figure 1(c) and includes polygonisation with triangles. The triangles might be of arbitrary dimension and orientation. The triangles can be exchanged with other types of polygons. The BDIM discretisation is shown in Figure 1(c) as well. It includes discretisation of the domain with polygons (cells) and discretisation of the boundary with straight lines (boundary elements). The DRBEM discretisation is shown in Figure 1(d). Instead of the discretisation of the domain with polygons (cells), the domain is discretised by pointisation and the boundary is discretised with straight lines (boundary elements). This method belongs to the so-called mesh reduction techniques, since the domain polygonisation is replaced by the domain pointisation. However, boundary polygonisation remains. A discretisation used in the Meshless or Mesh-free (MSM) methods is shown in Figure 1(e). It includes pointisation of the domain and boundary. The gridpoints can be arbitrarily spaced and non-uniform. The question mark at the right lower corner of Figure 1 stands for our inability of solving PDEs without gridpoints (Grid-less, Grid-free methods).

The numerical solution of coupled heat and mass transfer problems is becoming increasingly important as a result of the computational modelling needs in diverse modern technologies. A broad class of such heat, mass, momentum, and solute

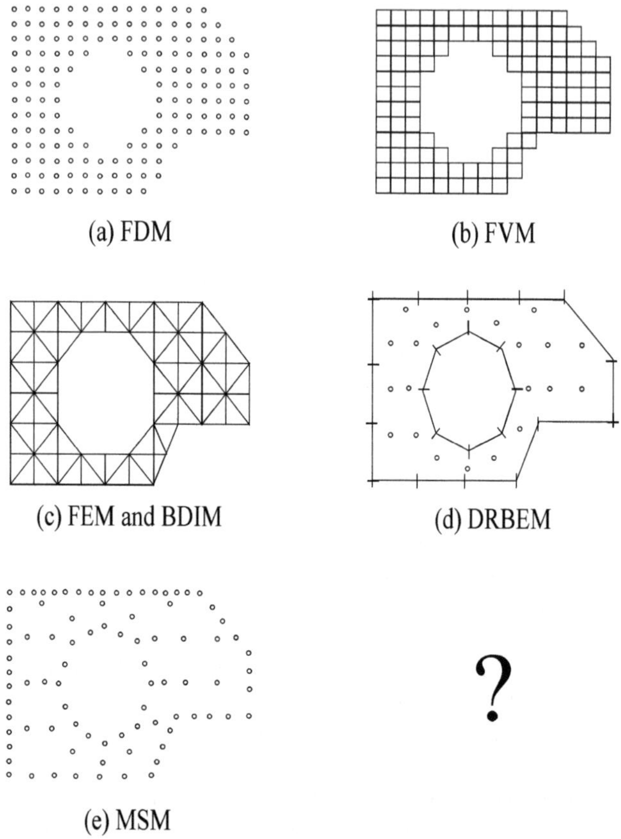

Fig. 1. Schematics of the domain discretisations used in numerical methods for the solution of partial differential equations. The question mark represents post-meshless methods.

transfer problems involves two or more phases, separated by free (steady-state) or moving (transient) interphase boundaries. Due to the existence of complex shaped interphase boundaries, most of the numerical simulations of engineering gas-liquid and liquid-solid two-phase flows conducted so far, have been based on averaged field equations with constitutive interphase relations solved on a fixed mesh. However, the diversity of the possibly involved length scales, inhomogeneities, and anisotropies, usually requires the adaptation of the mesh with respect to high field gradients and subsequent re-meshing. The most physically sound information can be directly perceived only from the numerical approaches that explicitly take into account the moving boundaries. The principal bottleneck in these types of numerical methods is the time consuming re-meshing of the evolving interphase boundaries and phase domains which limits such methods to problems with quite trivial phase patterns. The polygonisation problem is thus even more pronounced in such type of front-tracking approach.

1.2 Definition and Characteristics of Meshless Methods

The meshless, sometimes also called meshfree or mesh reduction methods establish a system of algebraic equations for the whole problem domain and boundary without polygonisation [1, 2, 15, 16]. Meshless methods use a set of nodes scattered within the problem domain as well as sets of nodes scattered on the boundaries of the domain to represent the problem domain and its boundaries. These sets of scattered nodes do not form a polygonisation (mesh), which means that no information on the geometrical connections between the nodes is required.

There exist a number of meshless methods such as the *Element free Galerkin methods, the Meshless local Petrov–Galerkin method, the Point interpolation method, the Point assembly method, the Finite point method, the Finite difference method with arbitrary irregular grids, Smoothed particle hydrodynamics, Reproducing kernel particle method, Kansa method*, etc.

The discussion in this article is limited to the very recent generation of meshless methods only. They are characterised by the following features:

- The governing equation is solved in its strong form.
- The formulation is almost independent on the problem dimension.
- The complicated geometry is easy to cope with.
- No polygonisation is needed.
- No integrations are needed.
- The method is very efficient.
- The method is very accurate.
- The methods is simple to learn.
- The method is simple to code.

2 Governing Equations

This part introduces the general transport equation, relevant types of boundary conditions and the reformulation of the governing transport equation into non-linear Poisson equation. It discusses the basic strategies of solving the non-linear Poisson equation in cases of weak and strong non-linearities.

2.1 General Transport Equation

Let us limit our discussion to solution of the general transport equation [32], defined on a fixed domain Ω with boundary Γ, standing for a reasonably broad spectra of mass, energy, momentum, and species transfer problems

$$\frac{\partial}{\partial t}[\rho \mathsf{A}(\Phi)] + \nabla \cdot [\rho \mathbf{v}\, \mathsf{A}(\Phi)] = \nabla \cdot (\underline{\mathbf{D}} \nabla \Phi) + S, \tag{1}$$

with ρ, Φ, t, $\mathbf{v}$, $\underline{\mathbf{D}}$, and S standing for density, transport variable, time, velocity, diffusion matrix

$$\underline{\mathbf{D}} \equiv \begin{bmatrix} D_{11} & D_{12} & D_{13} \\ D_{21} & D_{22} & D_{23} \\ D_{31} & D_{32} & D_{33} \end{bmatrix}, \tag{2}$$

and source, respectively. The scalar function A stands for possible more involved constitutive relations between the conserved $\mathsf{A}(\Phi)$ and diffused Φ quantities. The solution of the governing equation for the transport variable at the final time $t_0 + \Delta t$ is sought, where t_0 represents the initial time and Δt the positive time increment. The solution is constructed by the initial and boundary conditions that follow. The initial value of the transport variable $\Phi(\mathbf{p}, t)$ at a point with position vector $\mathbf{p}$ and time t_0 is defined through the known function Φ_0

$$\Phi(\mathbf{p}, t) = \phi_0(\mathbf{p}, t); \quad \mathbf{p} \in \Omega + \Gamma. \tag{3}$$

The boundary Γ is divided into not necessarily connected parts $\Gamma = \Gamma^D \cup \Gamma^N \cup \Gamma^R$ with Dirichlet, Neumann and Robin type boundary conditions, respectively. At the boundary point $\mathbf{p}$ with normal $\mathbf{n}_\Gamma$ and time $t_0 \leq t \leq t_0 + \Delta t$, these boundary conditions are defined through known functions Φ_Γ^D, Φ_Γ^N, Φ_Γ^R, $\Phi_{\Gamma\,\mathrm{ref}}^R$

$$\Phi = \Phi_\Gamma^D; \quad \mathbf{p} \in \Gamma^D, \tag{4}$$

$$\frac{\partial}{\partial n_\Gamma}\Phi = \Phi_\Gamma^N; \quad \mathbf{p} \in \Gamma^N, \tag{5}$$

$$\frac{\partial}{\partial n_\Gamma}\Phi = \Phi_\Gamma^R(\Phi - \Phi_{\Gamma\,\mathrm{ref}}^R); \quad \mathbf{p} \in \Gamma^R. \tag{6}$$

2.2 Poisson Reformulation of the General Transport Equation

The general transport equation (Equation 1) can be reformulated into Poisson form. This form sometimes permits an easier theoretical treatment. The inhomogeneous part of the Poisson equation can be split into a scalar part and a divergence of the vector part

$$\nabla^2 \Phi = \theta + \nabla \cdot \boldsymbol{\Theta}, \tag{7}$$

$$\theta = \frac{\left\{\frac{\partial}{\partial t}[\rho\, \mathsf{A}(\Phi)] - S\right\}}{D}, \tag{8}$$

$$\boldsymbol{\Theta} = \frac{[\rho\, \mathbf{v}\, \mathsf{A}(\Phi) - \underline{\mathbf{D}}' \nabla \Phi]}{D}. \tag{9}$$

The diffusion matrix is split into a constant isotropic part $D\underline{\mathbf{I}}$ and a remaining anisotropic part $\underline{\mathbf{D}}'$,

$$\underline{\mathbf{D}} + D\underline{\mathbf{I}} + \underline{\mathbf{D}}', \underline{\mathbf{D}}' \equiv \begin{bmatrix} D_{11} - D & D_{12} & D_{13} \\ D_{21} & D_{22} - D & D_{23} \\ D_{31} & D_{32} & D_{33} - D \end{bmatrix}. \tag{10}$$

The partial time derivative might be approximated by a two

$$\frac{\partial}{\partial t}[\rho \, \mathsf{A}(\Phi)] \approx \frac{1}{\Delta t}[\rho \, \mathsf{A}(\Phi) - \rho_0 \, \mathsf{A}(\Phi_0)], \tag{11}$$

or by a three-level finite difference scheme

$$\frac{\partial}{\partial t}[\rho \, \mathsf{A}(\Phi)] \approx \frac{1}{2\Delta t}[3\rho \, \mathsf{A}(\Phi) - 4\rho_0 \, \mathsf{A}(\Phi_0) + \rho_{-1} \, \mathsf{A}(\Phi_{-1})], \tag{12}$$

with the following notation

$$\mathsf{A}_{-1} \equiv \mathsf{A}(t_0 - \Delta t), \tag{13}$$

$$\mathsf{A}_0 \equiv \mathsf{A}(t_0), \tag{14}$$

$$\mathsf{A} \equiv \mathsf{A}(t_0 + \Delta t). \tag{15}$$

The unknown can be discretized in time by a two level scheme with a time level indicator $0 \leq \iota \leq 1$. This gives the fully implicit scheme for $\iota = 1$, the fully explicit scheme for $\iota = 0$, and the Crank–Nicolson scheme for $\iota = 1/2$.

$$\nabla^2 \Phi \approx \iota \nabla^2 \Phi + (\iota - 1)\nabla^2 \Phi_0, \tag{16}$$

$$\theta \approx \frac{\left\{\frac{\partial}{\partial t}[\rho \, \mathsf{A}(\Phi)] - \iota S - (\iota - 1)S_0\right\}}{D}, \tag{17}$$

$$\mathbf{\Theta} \approx \frac{[\iota \rho \mathbf{v} \, \mathsf{A}(\Phi) + (\iota - 1)\rho_0 \mathbf{v}_0 \, \mathsf{A}(\Phi_0) - \iota \underline{\mathbf{D}}' \nabla \Phi - (\iota - 1)\underline{\mathbf{D}}_0' \nabla \Phi_0]}{D}. \tag{18}$$

The involved parameters of the governing equation and boundary conditions are assumed to be dependent on the transport variable, space, and time. The solution of such type of non-linear equation requires iterations. Let us (for the sake of compact notation) assume further discussion in non-time discretized form. The inhomogeneous terms are, due to non-linearity, respectively expanded as

$$\theta = \theta + \frac{\partial \bar{\theta}}{\partial \Phi}(\Phi - \bar{\Phi}), \tag{19}$$

$$\mathbf{\Theta} = \bar{\mathbf{\Theta}} + \frac{\partial \bar{\mathbf{\Theta}}}{\partial \Phi}(\Phi - \bar{\Phi}), \tag{20}$$

with the over-bar denoting the value from previous iteration. The final form of the transformed equation, suitable for iterative solution then becomes

$$\nabla^2 \Phi = Q(\Phi), \tag{21}$$

$$Q(\Phi) = \bar{\theta} + \frac{\partial \bar{\theta}}{\partial \Phi}(\Phi - \bar{\Phi}) + \nabla \cdot \bar{\mathbf{\Theta}} + \nabla \cdot \frac{\partial \bar{\mathbf{\Theta}}}{\partial \Phi}(\Phi - \bar{\Phi}). \tag{22}$$

The source term can be treated by under-relaxation (recommended in case of weak non-linearity)

$$\nabla^2 \Phi = Q(\bar{\Phi}) + c_{\text{rel}} \left[Q(\bar{\Phi}) - Q(\bar{\bar{\Phi}}) \right], \tag{23}$$

with an under-relaxation coefficient $c_{\text{rel}} < 1$. The two-fold over-bar denotes the known value two iterations ago. Alternatively, the source term can be treated explicitly (recommended in case of strong non-linearity)

$$\left[\nabla^2 - \frac{\partial \bar{\theta}}{\partial \Phi} - \nabla \cdot \frac{\partial \bar{\mathbf{\Theta}}}{\partial \Phi} \right] \Phi = \bar{\theta} - \frac{\partial \bar{\theta}}{\partial \Phi}\bar{\Phi} + \nabla \cdot \bar{\mathbf{\Theta}} - \nabla \cdot \frac{\partial \bar{\mathbf{\Theta}}}{\partial \Phi}\bar{\Phi}. \tag{24}$$

The discretisation of Equations (23) and (24) and subsequent set-up of the algebraic equation systems is quite different. In case of Equation (23), the system matrix of the algebraic equation system need not to be recalculated each iteration and in the case of Equation (24), the system matrix of the algebraic equation system needs to be recalculated ateach iteration. Strategy used in Equation (24) usually requires less iterations than strategy used in Equation (23), however, each iteration is computationally more costly. Therefore, the strategy in Equation (23) is used for weak non-linear problems, and the strategy in Equation (24) is for the strong non-linear problems. For the strong non-linear problems the strategy in Equation (23) might require a very small under-relaxation and a huge number of iterations. The iterations over one time-step are completed when the criterion in Equation (25) is satisfied in all computational nodes $n = 1, 2, \ldots, N$

$$\max |\Phi_n - \bar{\Phi}_n| \leq \Phi_{\text{itr}}. \tag{25}$$

The steady-state is achieved when the criterion in Equation (26) is satisfied in all computational nodes $n = 1, 2, \ldots, N$

$$\max |\Phi_n - \Phi_0| \leq \Phi_{\text{ste}}. \tag{26}$$

The parameters Φ_{itr} and Φ_{ste} are denoted as iteration and steady-state limits. The iteration limit has to be an order of magnitude lower as the steady-state limit. If this is not true, a non-convergent virtual steady-state can be achieved. The derived formulas in this part are universally valid for all known spatial discretization types, including meshless.

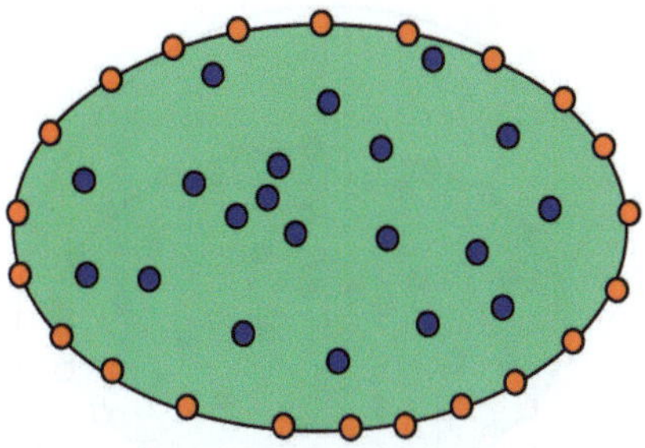

Fig. 2. Schematics of the polygon-free domain and boundary node arrangement.

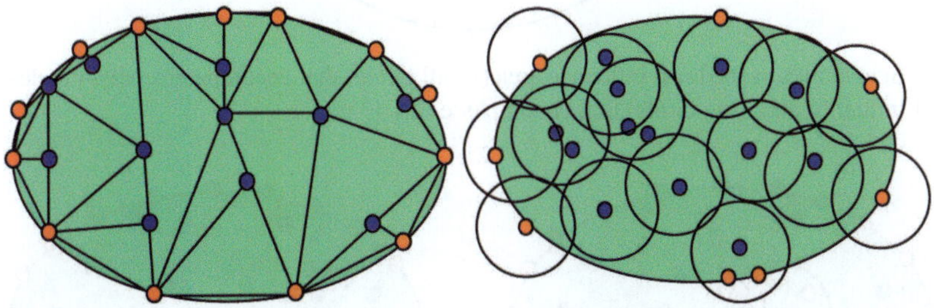

Fig. 3. Left: Schematics of the concept of contiguous (non-overlapping) sub-domains (typical for FEM). Right: Schematics of the concept of non-contiguous (overlapping) sub-domains (typical for MSM).

3 Spatial Discretisation in Meshless Methods

This part introduces the spatial discretisation (pointisation) in Meshless methods and the concept of overlapping sub-domain (Figures 2 to 6).

3.1 Pointisation

The value of the transport variable Φ_n is solved in a set of nodes $\mathbf{p}_n$; $n = 1, 2, \ldots, N$ of which N_Ω belong to the domain and N_Γ to the boundary, i.e. $N = N_\Gamma + N_\Omega$.

4 Representation of Function over a Set of Nodes

This part introduces the concept of the representation of the function with global shape functions and the concepts of the calculation of shape function coefficients by collocation (interpolation) and by the approximation.

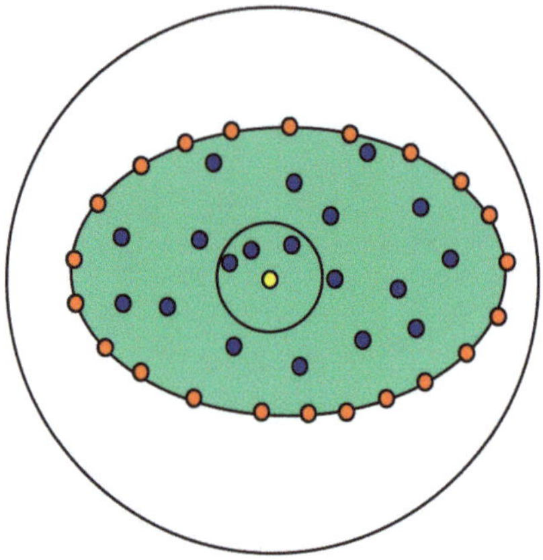

Fig. 4. Schematics of the local influence area (small circle) that encompasses only 4 nodes and global influence area (big circle) that encompasses all 43 nodes.

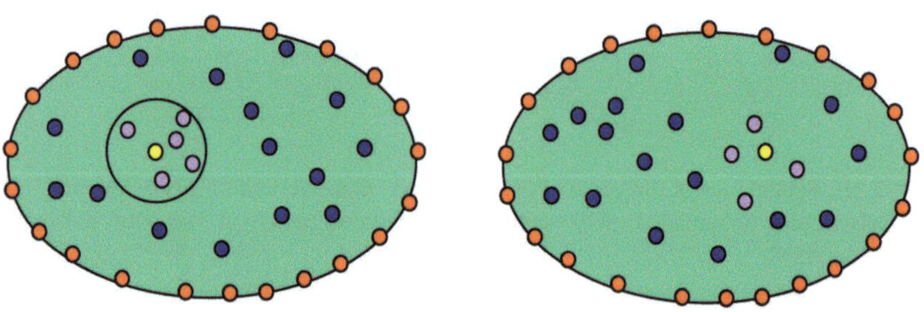

Fig. 5. Schematics of the sub-domains. In the left case the sub-domain is defined by the geometry of the circle. In the right case, the sub-domain is defined by the four nearest nodes to the central node.

4.1 Global Representation of Function

The representation of function over a set l of (in general) non-equally spaced ${}_lN$ nodes ${}_l\mathbf{p}_n$; $n = 1, 2, \ldots, {}_lN$ is made in the following way

$$\Phi(\mathbf{p}) \approx \sum_{k=1}^{{}_lK} {}_l\psi_k(\mathbf{p})\, {}_l\alpha_k \tag{27}$$

${}_l\psi_k$ stand for the shape functions, ${}_l\alpha_k$ for the coefficients of the shape functions, and ${}_lK$ represents the number of the shape functions. The left lower index on entries of Equation (27) represents the sub-domain ${}_l\omega$ on which the coefficients ${}_l\alpha_k$ are determined. The total number of sub-domains is denoted by L. The sub-domains ${}_l\omega$ can in general be contiguous (overlapping) or non-contiguous (non-overlapping).

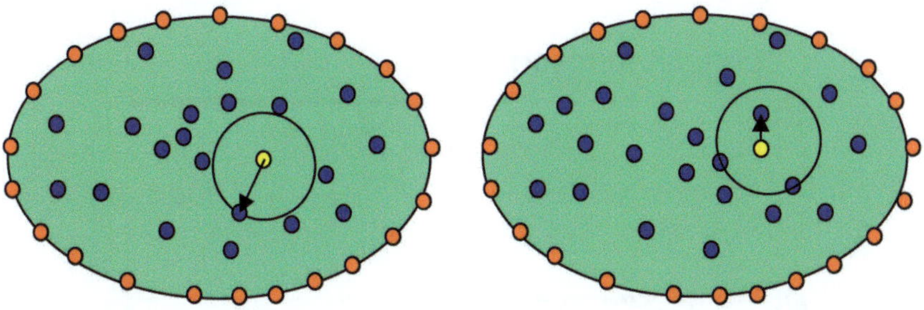

Fig. 6. Schematics of the definition of two quantities that measure the density of the data set. In the left case the fill distance is defined, which measures the maximum distance between two nodes in a sub-domain, and in the right case, the separation distance is defined, which measures the minimum distance between the two nodes in a sub-domain.

Each of the sub-domains ${}_l\omega$ includes ${}_lN$ grid-points of which ${}_lN_\Omega$ are in the domain and ${}_lN_\Gamma$ are on the boundary.

$$ {}_lN = {}_l N_\Omega + {}_l N_\Gamma. \tag{28} $$

For each of the sub-domains ${}_l\omega$, a separation distance is defined, which is the minimum distance between the two nodes, and the fill distance is defined, which is the maximum distance between the two nodes. In case the sub-domains are overlapping, the following is valid

$$ \sum_{l=1}^{L} {}_lN > N. \tag{29} $$

The coefficients can be calculated from the sub-domain nodes in two distinct ways. The first way is collocation (interpolation) and the second way is approximation by the least squares method (see Figures 7 and 8, respectively).

Approximation is needed where data smoothing rather than interpolation is needed. This is because data are often inaccurate, contain noise or – as happens sometimes in practical applications – are too plentiful and cannot and need not be reasonably all interpolated at once. Moreover, smoothing is almost always required as long as problems are ill-posed, which means that their solution for theoretical reasons, is extremely sensitive to even the smallest changes in input data.

4.2 Calculation of Coefficients by Collocation

Let us assume the known function values ${}_l\Phi_n$ in the nodes ${}_l\mathbf{p}_n$ of sub-domain ${}_l\omega$. The collocation implies

$$ \sum_{k=1}^{{}_lK} {}_l\psi_k({}_l\mathbf{p}_n)\, {}_l\alpha_k = \Phi({}_l\mathbf{p}_n);\; {}_lK = {}_lN. \tag{30} $$

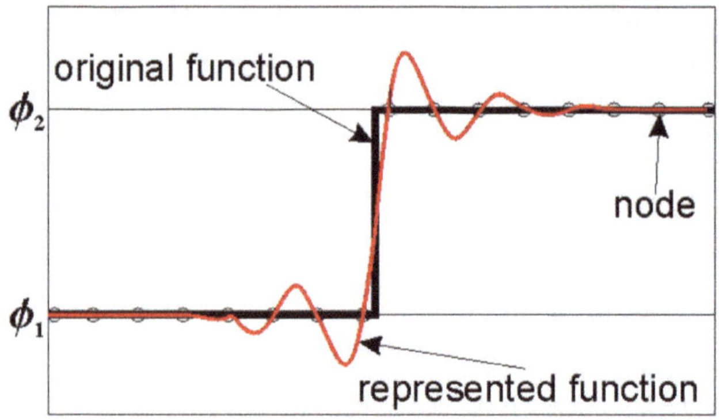

Fig. 7. A one-dimensional example of interpolation of the function.

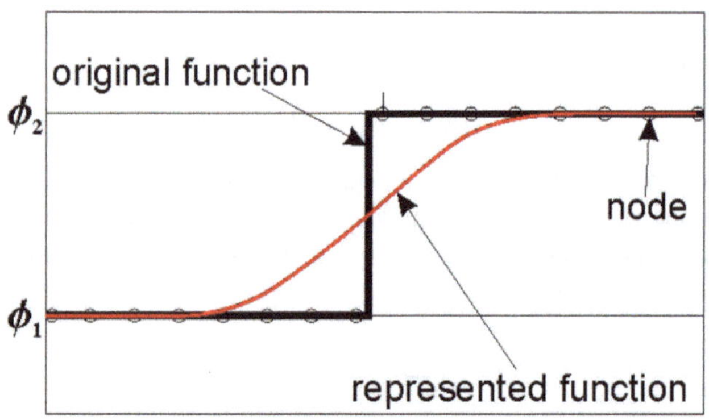

Fig. 8. A one-dimensional example of the least squares approximation of the function.

For the coefficients to be computable, the number of the shape functions has to match the number of the collocation points, and the collocation matrix has to be non-singular.

$$\sum_{k=1}^{{}_lK} {}_l\psi_k({}_l\mathbf{p}_n)\, {}_l\alpha_k = \Phi({}_l\mathbf{p}_n); \; {}_lK = {}_lN. \tag{31}$$

The system of Equations (31) can be written in matrix-vector notation

$${}_l\underline{\boldsymbol{\psi}}\, {}_l\boldsymbol{\alpha} = {}_l\boldsymbol{\Phi}, \tag{32}$$

with

$${}_l\underline{\psi}_{kn} = {}_l\psi_k({}_l\mathbf{p}_n), \tag{33}$$

$${}_l\Phi_n = \Phi({}_l\mathbf{p}_n). \tag{34}$$

The coefficients ${}_l\boldsymbol{\alpha}$ can be computed by inverting the system (32)

$$ {}_l\boldsymbol{\alpha} = {}_l\underline{\boldsymbol{\psi}}^{-1}\,{}_l\boldsymbol{\Phi}. \tag{35}$$

The matrices $\underline{\boldsymbol{\psi}}$ and $\underline{\boldsymbol{\psi}}^{-1}$ have dimension ${}_lN \times {}_lN$ and the vectors ${}_l\boldsymbol{\alpha}$ and ${}_l\boldsymbol{\Phi}$ have dimension ${}_lN$.

4.3 Calculation of Coefficients by Approximation

Let us assume the known function values ${}_l\Phi_n$ in the nodes ${}_l\mathbf{p}_n$ of sub-domain ω_l. The least squares approximation implies that the following functional should be minimized

$$\Im[{}_l\boldsymbol{\alpha}(\mathbf{p})] = \sum_{n=1}^{{}_lN} {}_l\upsilon_n(\mathbf{p}) \sum_{k=1}^{{}_lK} [\psi_k({}_l\mathbf{p}_n)\,{}_l\alpha_k(\mathbf{p}) - \Phi({}_l\mathbf{p}_n)]^2, \tag{36}$$

with respect to change of ${}_l\boldsymbol{\alpha}(\mathbf{p})$, i.e.

$$\frac{\partial}{\partial\,{}_l\boldsymbol{\alpha}(\mathbf{p})}\Im[{}_l\boldsymbol{\alpha}(\mathbf{p})] = 0. \tag{37}$$

The number of the nodes ${}_lN$ used in the approximation has to be greater or equal to the number of the shape functions ${}_lK$. Weight functions ${}_l\upsilon_n$ have been introduced in Equation (36) in addition to the shape functions. The weight functions measure the relative importance of the node. The coefficients ${}_l\boldsymbol{\alpha}(\mathbf{p})$ depend on the center of the weight function $\mathbf{p}$. The coefficients ${}_l\boldsymbol{\alpha}(\mathbf{p})$ are respectively calculated from the system of linear equations

$$ {}_l\,\underline{\widehat{\boldsymbol{\psi}}}\,(\mathbf{p})\,{}_l\boldsymbol{\alpha}(\mathbf{p}) = {}_l\,\underline{\breve{\boldsymbol{\psi}}}\,(\mathbf{p})\,{}_l\boldsymbol{\Phi}. \tag{38}$$

The matrices ${}_l\,\underline{\widehat{\boldsymbol{\psi}}}$ and ${}_l\,\underline{\breve{\boldsymbol{\psi}}}$ are of the dimensions ${}_lK \times {}_lK$ and ${}_lK \times {}_lN$ respectively. Their entries are

$$ {}_l\,\underline{\widehat{\psi}}_{ik}\,(\mathbf{p}) = \sum_{n=1}^{{}_lN} {}_l\upsilon_n(\mathbf{p})\,{}_l\psi_i({}_l\mathbf{p}_n)\,{}_l\psi_k({}_l\mathbf{p}_n), \tag{39}$$

$$ {}_l\,\underline{\breve{\psi}}_{in}\,(\mathbf{p}) = {}_l\upsilon_n(\mathbf{p})\,{}_l\psi_i({}_l\mathbf{p}_n). \tag{40}$$

The coefficients ${}_l\boldsymbol{\alpha}(\mathbf{p})$ in general depend on the weight function center $\mathbf{p}$. For practical reasons, their calculation is fixed to the central node of the sub-domain ${}_l\omega$, denoted by ${}_l\mathbf{p}$

$$ {}_l\,\underline{\widehat{\boldsymbol{\psi}}} \equiv {}_l\,\underline{\widehat{\boldsymbol{\psi}}}\,({}_l\mathbf{p}), \tag{41}$$

$$ {}_l\,\underline{\breve{\boldsymbol{\psi}}} \equiv {}_l\,\underline{\breve{\boldsymbol{\psi}}}\,({}_l\mathbf{p}), \tag{42}$$

$$ {}_{l}\boldsymbol{\alpha} \equiv {}_{l}\boldsymbol{\alpha}({}_{l}\mathbf{p}). \tag{43}$$

Similarly as in the case of collocation, an explicit expression for the calculation of the coefficients ${}_{l}\boldsymbol{\alpha}$ can be written

$$ {}_{l}\boldsymbol{\alpha} = {}_{l}\underline{\boldsymbol{\psi}}^{-1}\, {}_{l}\boldsymbol{\Phi}, \tag{44}$$

with

$$ {}_{l}\underline{\boldsymbol{\psi}}^{-1} = {}_{l}\underline{\widehat{\boldsymbol{\psi}}}^{-1}\, {}_{l}\underline{\breve{\boldsymbol{\psi}}}. \tag{45}$$

The main difference between collocation and approximation is as follows: In collocation, the representation of the function exactly satisfies the nodal values ${}_{l}\Phi_{n}$. In approximation, the nodal values are approximated in the least squares sense. The advantage of approximation compared to collocation is that it usually gives better estimation of the derivatives, particularly in cases with sharp gradients (see Figures 7 and 8). The drawback over collocation is obvious. The nodal values are not exactly satisfied in approximation.

4.4 From the Representation of the Function to the Representation of the Partial Derivatives

By taking into account the expressions for the calculation of the coefficients ${}_{l}\boldsymbol{\alpha}$ (by collocation or by approximation), the representation of function $\Phi(\mathbf{p})$ can be expressed as

$$ \Phi(\mathbf{p}) \approx \sum_{k=1}^{{}_{l}K} {}_{l}\psi_{k}(\mathbf{p}) \sum_{n=1}^{{}_{l}N} {}_{l}\underline{\psi}_{kn}^{-1}\, {}_{l}\Phi_{n} \tag{46}$$

The first partial derivatives of $\Phi(\mathbf{p})$ can be expressed as

$$ \frac{\partial}{\partial p_{\varsigma}}\Phi(\mathbf{p}) \approx \sum_{k=1}^{{}_{l}K} \frac{\partial}{\partial p_{\varsigma}}\psi_{k}(\mathbf{p}) \sum_{n=1}^{{}_{l}N} {}_{l}\underline{\psi}_{kn}^{-1}\, {}_{l}\phi_{n}; \quad \varsigma = x, y, z. \tag{47}$$

The second partial derivatives of $\Phi(\mathbf{p})$ can be expressed as

$$ \frac{\partial^{2}}{\partial p_{\varsigma} p_{\xi}}\Phi(\mathbf{p}) \approx \sum_{k=1}^{{}_{l}K} \frac{\partial^{2}}{\partial p_{\varsigma} p_{\xi}}\, {}_{l}\psi_{k}(\mathbf{p}) \sum_{n=1}^{{}_{l}N} {}_{l}\underline{\psi}_{kn}^{-1}\, {}_{l}\Phi_{n}; \quad \varsigma, \xi = x, y, z. \tag{48}$$

4.5 Selection of Global Representation Functions

Let us introduce the Cartesian coordinate system with base vectors $\mathbf{i}_{\varsigma}$; $\varsigma = x, y, z$ and coordinates p_{ς}; $\varsigma = x, y, z$, i.e.

$$ \mathbf{p} = \mathbf{i}_{\varsigma} p_{\varsigma}; \quad \varsigma = x, y, z. \tag{49}$$

The following six polynomials can be used in 2D to represent the quadratic basis:

$$\Psi_1 = 1,\ \psi_2 = p_x,\ \psi_3 = p_y,\ \psi_4 = p_x p_x,\ \psi_5 = p_x p_y,\ \psi_6 = p_y p_y. \tag{50}$$

The following additional four polynomials have to be used in 3D to represent the quadratic basis:

$$\psi_7 = p_z,\ \psi_8 = p_x p_z,\ \psi_9 = p_y p_z,\ \psi_{10} = p_z p_z. \tag{51}$$

The radial basis functions [4, 28, 33] such as multi-quadrics

$$\psi_k(\mathbf{p}) = [(\mathbf{p} - \mathbf{p}_k) \cdot (\mathbf{p} - \mathbf{p}_k) + c^2]^{1/2}, \tag{52}$$

or inverse multi-quadrics

$$\psi_k(\mathbf{p}) = [(\mathbf{p} - \mathbf{p}_k) \cdot (\mathbf{p} - \mathbf{p}_k) + c^2]^{1/2}, \tag{53}$$

can be used in 2D or 3D. The calculation of the coefficients ${}_l\boldsymbol{\alpha}$ can be made by collocation or by approximation.

4.6 Selection of Weight Functions

As a weight function typically a radial basis function is chosen. A typical example is the polynomial-like shape function

$$ {}_l\upsilon_n(\mathbf{p}) = \begin{cases} \left[\frac{(\mathbf{p} - {}_l\mathbf{p}_n)\cdot(\mathbf{p} - {}_l\mathbf{p}_n) - {}_l\sigma_n}{(\mathbf{p} - {}_l\mathbf{p}_n)\cdot(\mathbf{p} - {}_l\mathbf{p}_n)\ {}_l\sigma_n}\right]; & |\mathbf{p} - {}_l\mathbf{p}_n| < \sigma_n, \\ 0; & |\mathbf{p} - {}_l\mathbf{p}_n| \geq \sigma_n, \end{cases} \tag{54}$$

or the Gaussian function

$$ {}_l\upsilon_n(\mathbf{p}) = \begin{cases} \exp(-{}_lc(\mathbf{p} - {}_l\mathbf{p}_n) \cdot (\mathbf{p} - {}_l\mathbf{p}_n)/\sigma_n^2; & |\mathbf{p} - {}_l\mathbf{p}_n| < \sigma_n, \\ 0; & |\mathbf{p} - {}_l\mathbf{p}_n| \geq \sigma_n. \end{cases} \tag{55}$$

The size of the circular (2D) or spherical (3D) support ${}_l\sigma_n$ is chosen to contain an appropriate number of nodes. The weight function shape factor ${}_lc_n$ depends on the mesh (non-)uniformity. In case of a uniform mesh, the best results are obtained with a large shape factor. In case of the non-uniform node arrangements, a smaller shape factor has to be chosen. The size of the shape factor depends on the non-uniformity of the node arrangement (see [23]).

4.7 Diffuse Approximation

A limited number of applications using the polynomial basis functions in connection with the least squares approximation and strong formulation has been developed under the name Diffuse Approximation (DA) [21, 25] for heat transfer and fluid flow problems.

4.8 Kansa Method

A broad class of meshfree methods in development today are based on Radial Basis Functions (RBFs) [4]. The RBF collocation method or Kansa method [11] is the simplest of them. This method has been further upgraded to symmetric collocation [10, 24], to modified collocation [8] and to indirect collocation [20]. The method has been already used in a broad spectrum of computational fluid dynamics problems [29] such as the solution of Navier–Stokes equations [17, 18] or porous media flow [28, 31] and the solution of solid-liquid phase change problems [12]. In contrast to advantages over mesh generation, all the listed attempts unfortunately fail to perform for large problems, because they produce fully populated matrices, sensitive to the choice of the free parameters in RBFs. Sparse matrices can be generated by the introduction of the compactly supported RBFs and the accuracy of such approach can be improved by the multilevel technique [7]. One of the possibilities for mitigating the large fully populated matrix problem is to employ the domain decomposition [19]. However, the domain decomposition re-introduces some sort of meshing which is not attractive. The concept of local collocation in the context of RBF-based solution of Poisson equation has been introduced in [14, 34]. For interpolation of the function value in a certain node the authors use only data in the (neighbouring) nodes that fall into the domain of influence of this node. The procedure results in a matrix that is of the same size as the matrix in the original Kansa method, however it is sparse. Circular domains of influence have been used in [14] and stencil-shaped domains in [34]. In [14], the one-dimensional and two-dimensional Poisson equation has been solved by using multiquadrics and inverse multiquadrics RBFs with a detailed analysis of the influence of the free parameter on the results. In [34], a class of linear and non-linear elasticity problems have been solved with a fixed free parameter. The differential quadrature method, that calculates the derivatives of a function by a weighted linear sum of functional values at its neighbouring nodes has been structured with the RBFs in [26]. Despite the local properties, the matrix still has a similar form as in [14, 34]. This paper reviews a new, even more simple meshfree solution procedure for solving the transport phenomena, which overcomes even the solution of the large sparse matrices.

5 Semi-Explicit Solution of the General Transport Equation

5.1 Reformulation

This part elaborates the semi-explicit solution of the general transport equation (Equation 1), subject to the initial condition presented in Equation (3) and the boundary conditions presented in Equations (4–6). The general transport equation can be transformed into following expression by taking into account the explicit discretization

$$\frac{\rho \mathsf{A}(\Phi) - \rho_0 \mathsf{A}_0}{\Delta t} + \nabla \cdot [\rho_0 \mathbf{v}_0 \mathsf{A}_0] = \nabla \cdot (\underline{\mathbf{D}}_0 \nabla \Phi_0) + S. \tag{56}$$

At time $t = t_0 + \Delta t$, the functions $\mathsf{A}(\Phi)$ and $S(\Phi)$ can be expanded as

$$\mathsf{A}(\Phi) \approx \bar{\mathsf{A}} + \frac{\partial \bar{\mathsf{A}}}{\partial \Phi}(\Phi - \bar{\Phi}), \tag{57}$$

$$S(\Phi) \approx \bar{S} + \frac{\partial \bar{S}}{\partial \phi}(\Phi - \bar{\Phi}), \tag{58}$$

$$\frac{\rho \bar{\mathsf{A}} + \rho \bar{\mathsf{A}}_{l\Phi}(\Phi - \bar{\Phi}) - \rho_0 \, \mathsf{A}_0}{\Delta t} + \nabla \cdot [\rho_0 \mathbf{v}_0 \, \mathsf{A}_0]$$
$$= \nabla \cdot (\underline{\mathbf{D}}_0 \nabla \Phi_0) + \bar{S} + \bar{S}_{l\Phi}(\Phi - \bar{\Phi}). \tag{59}$$

The unknown function value can be extracted from the above equation

$$\Phi = \frac{\rho_0 \, \mathsf{A}_0 - \rho \bar{\mathsf{A}} + \rho \frac{\partial \bar{\mathsf{A}}}{\partial \Phi} \bar{\Phi}}{\rho \frac{\partial \bar{\mathsf{A}}}{\partial \Phi} - \Delta \frac{\partial \bar{S}}{\partial \Phi}} + \frac{\Delta t \left[\nabla \cdot (\underline{\mathbf{D}}_0 \nabla \Phi_0) - \nabla \cdot (\rho_0 \mathbf{v}_0 \, \mathsf{A}_0) + \bar{S} - \frac{\partial \bar{S}}{\partial \Phi} \bar{\Phi} \right]}{\rho \frac{\partial \bar{\mathsf{A}}}{\partial \Phi} - \Delta t \frac{\partial \bar{S}}{\partial \Phi}}. \tag{60}$$

The solution of the above equation in grid-point $\mathbf{p}_n$ can be calculated as

$$\Phi_n = \frac{\rho_{n0} \, \mathsf{A}_{n0} - \rho_n \bar{\mathsf{A}}_n + \rho_n \frac{\partial \bar{\mathsf{A}}}{\partial \Phi_n} \bar{\Phi}_n}{\rho \frac{\partial \bar{\mathsf{A}}}{\partial \Phi_n} - \Delta t \frac{\partial \bar{S}}{\partial \Phi_n}} + \frac{\Delta t \left[\nabla \cdot (\underline{\mathbf{D}}_{0n} \nabla \Phi_{0n}) - \nabla \cdot (\rho_{0n} \mathbf{v}_{0n} \, \mathsf{A}_{0n}) + \bar{S}_n - \frac{\partial \bar{S}}{\partial \Phi_n} \bar{\Phi}_n \right]}{\rho \frac{\partial \bar{\mathsf{A}}}{\partial \Phi_n} - \Delta t \frac{\partial \bar{S}}{\partial \Phi_n}}. \tag{61}$$

The calculation of the convective term $\nabla \cdot (\rho_{0n} \mathbf{v}_{0n} \mathsf{A}_{0n})$ includes derivatives of the form (see Appendix)

$$\frac{\partial}{\partial p_\varsigma} \rho_{0n} \, v_{\varsigma 0n} \, \mathsf{A}_{0n} \approx \varsigma = x, y, z. \tag{62}$$

These derivatives can be evaluated as

$$\frac{\partial}{\partial p_\varsigma} \rho_{0n} \, v_{\varsigma 0n} \, \mathsf{A}_{0n} \approx \sum_{k=1}^{K} \frac{\partial}{\partial p_\varsigma} \psi_k(\mathbf{p}_n) \sum_{n=1}^{{}_lN} {}_l\psi_{kn}^{-1} \, {}_l(\rho_0 \, v_{\varsigma 0} \, \mathsf{A}_0)_n; \quad \varsigma = x, y, z. \tag{63}$$

The calculation of the diffusive term $\nabla \cdot (\underline{\mathbf{D}}_{0n}\ \nabla\ \Phi_{0n})$ includes derivatives of the form (see Appendix)

$$\frac{\partial}{\partial p_\varsigma} D_{0n\zeta\xi},\ \frac{\partial^2}{\partial p_\varsigma \partial p_\tau} D_{0n\zeta\xi},\ \frac{\partial}{\partial p_\varsigma} \Phi_{0n\zeta},\ \frac{\partial^2}{\partial p_\varsigma \partial p_\tau} \Phi_{0n\zeta};\quad \varsigma, \tau, \zeta, \xi = x, y, z. \quad (64)$$

These derivatives can be evaluated as

$$\frac{\partial}{\partial p_\varsigma} D_{0n\xi\zeta} \approx \sum_{k=1}^{K} \frac{\partial}{\partial p_\varsigma} \psi_k(\mathbf{p}_n) \sum_{n=1}^{{}_lN} {}_l\psi_{kn}^{-1}\ {}_lD_{0n\xi\zeta};\quad \varsigma = x, y, z, \quad (65)$$

$$\frac{\partial^2}{\partial p_\varsigma \partial p_\tau} D_{0n\xi\zeta} \approx \sum_{k=1}^{K} \frac{\partial^2}{\partial p_\varsigma \partial p_\tau} \psi_k(\mathbf{p}_n) \sum_{n=1}^{{}_lN} {}_l\psi_{kn}^{-1}\ {}_lD_{0n\xi\zeta};\quad \varsigma = x, y, z. \quad (66)$$

The solution procedure for the governing Equation (1) and the boundary conditions presented in Equations (4–6) now follow the below defined steps 1–5:

Step 1:
First, the initial conditions are set in the domain and boundary nodes and the derivatives required in the convective and diffusive terms are calculated from the known nodal values.

Step 2:
Equation (61) is employed to calculate the new values of the variable Φ_n at time $t_0 + \Delta t$ in the domain nodes.

What follows defines variable Φ_n at time $t_0 + \Delta t$ in the Dirichlet, Neumann, and Robin boundary nodes. For this purpose, the coefficients ${}_l\boldsymbol{\alpha}$ have to be determined from the new values in the domain and from the information on the boundary conditions.

Let us introduce domain, Dirichlet, Neumann, and Robin boundary indicators. These indicators are defined as

$$\Upsilon_{\Omega n} = \begin{cases} 1; & \mathbf{p}_n \in \Omega, \\ 0; & \mathbf{p}_n \notin \Omega, \end{cases} \qquad \Upsilon_{\Gamma n}^{D} = \begin{cases} 1; & \mathbf{p}_n \in \Gamma^D, \\ 0; & \mathbf{p}_n \notin \Gamma^D, \end{cases}$$

$$\Upsilon_{\Gamma n}^{N} = \begin{cases} 1; & \mathbf{p}_n \in \Gamma^N, \\ 0; & \mathbf{p}_n \notin \Gamma^N, \end{cases} \qquad \Upsilon_{\Gamma n}^{R} = \begin{cases} 1; & \mathbf{p}_n \in \Gamma^R, \\ 0; & \mathbf{p}_n \notin \Gamma^R. \end{cases} \quad (67)$$

Step 3: The collocation version
Consider the collocation version for the calculation of the coefficients. This implies

$$\sum_{k=1}^{{}_lN} {}_l\Upsilon_{\Omega n}\ {}_l\psi_k({}_l\mathbf{p}_n)\ {}_l\alpha_k + \sum_{k=1}^{{}_lN} {}_l\Upsilon_{\Gamma n}^{D}\ {}_l\psi_k({}_l\mathbf{p}_n)\ {}_l\alpha_k$$

$$+\sum_{k=1}^{{}_lN} {}_l\Upsilon_{\Gamma n}^{N} \frac{\partial}{\partial n_\Gamma} {}_l\psi_k({}_l\mathbf{p}_n)\,{}_l\alpha_k + \sum_{k=1}^{{}_lN} {}_l\Upsilon_{\Gamma n}^{R} \frac{\partial}{\partial n_\Gamma} {}_l\psi_k({}_l\mathbf{p}_n)\,{}_l\alpha_k$$

$$= {}_l\Upsilon_{\Omega n}\,{}_l\Phi_n + {}_l\Upsilon_{\Gamma n}^{D}\,{}_l\Phi_n^{D} + {}_l\Upsilon_{\Gamma n}^{N}\,{}_l\Phi_n^{N}$$

$$+ {}_l\Upsilon_{\Gamma n}^{R}\,{}_l\Phi_{\Gamma n}^{R}\left(\sum_{k=1}^{{}_lN} {}_l\psi_k({}_l\mathbf{p}_n)\,{}_l\alpha_k - {}_l\Phi_{\Gamma\,\mathrm{ref}\,n}^{R}\right). \tag{68}$$

The Robin boundary conditions have been represented by

$${}_l\Phi_{\Gamma n}^{R}({}_l\Phi_n - {}_l\Phi_{\Gamma\,\mathrm{ref}\,n}^{R}) \approx {}_l\Phi_{\Gamma n}^{R}\left(\sum_{k=1}^{{}_lN} {}_l\psi_k({}_l\mathbf{p}_n)\,{}_l\alpha_k - {}_l\Phi_{\Gamma\,\mathrm{ref}\,n}^{R}\right). \tag{69}$$

The calculation of the coefficients ${}_l\boldsymbol{\alpha}$ follows from the following system of linear equations

$${}_l\underline{\boldsymbol{\Psi}}\,{}_l\boldsymbol{\alpha} = {}_l\mathbf{b}, \tag{70}$$

with the system matrix coefficients

$${}_l\underline{\Psi}_{nk} = {}_l\Upsilon_{\Omega n}\,{}_l\psi_k({}_l\mathbf{p}_n) + {}_l\Upsilon_{\Gamma n}^{D}\,{}_l\psi_k({}_l\mathbf{p}_n) + {}_l\Upsilon_{\Gamma n}^{N}\frac{\partial}{\partial n_\Gamma}\,{}_l\psi_k({}_l\mathbf{p}_n)$$

$$+ {}_l\Upsilon_{\Gamma n}^{R}\left[\frac{\partial}{\partial n_\Gamma}\,{}_l\psi_k({}_l\mathbf{p}_n) - {}_l\Phi_{\Gamma n}^{R}\sum_{k=1}^{{}_lN} {}_l\psi_k({}_l\mathbf{p}_n)\right], \tag{71}$$

and with the augmented right-hand side vector

$${}_l\mathbf{b}_n = {}_l\Upsilon_{\Omega n}\Phi_n + {}_l\Upsilon_{\Gamma n}^{D}\Phi_n^{D} + {}_l\Upsilon_{\Gamma n}^{N}\Phi_n^{N} - {}_l\Upsilon_{\Gamma n}^{R}\,{}_l\Phi_{\Gamma n}^{R}\,{}_l\Phi_{\Gamma\,\mathrm{ref}\,n}^{R}. \tag{72}$$

Step 4: The approximation version

Consider the least squares approximation version for the calculation of the coefficients. This implies

$$\Im[{}_l\boldsymbol{\alpha}] = \sum_{n=1}^{{}_lN} {}_l\Upsilon_{\Omega n}\,{}_l\upsilon_n({}_l\mathbf{p}) \sum_{k=1}^{{}_lN} [{}_l\psi_k({}_l\mathbf{p}_n){}_l\alpha_k - {}_l\Phi_n]^2$$

$$+\sum_{n=1}^{{}_lN} {}_l\Upsilon_{\Gamma n}^{D}\,{}_l\upsilon_n({}_l\mathbf{p}) \sum_{k=1}^{{}_lK} [{}_l\psi_k({}_l\mathbf{p}_n){}_l\alpha_k - {}_l\Phi_n]^2$$

$$+\sum_{n=1}^{{}_lN} {}_l\Upsilon_{\Gamma n}^{N}\,{}_l\upsilon_n({}_l\mathbf{p}) \sum_{k=1}^{{}_lK} \left[\frac{\partial}{\partial n_\Gamma}\,{}_l\psi_k({}_l\mathbf{p}_n){}_l\alpha_k - {}_l\Phi_{\Gamma n}^{N}\right]^2 \tag{73}$$

$$+\sum_{n=1}^{N_l} {}_l\Upsilon_{\Gamma n}^{R}\,{}_l\upsilon_n({}_l\mathbf{p}) \sum_{k=1}^{{}_lK} \left[\frac{\partial}{\partial n_\Gamma}\,{}_l\psi_k({}_l\mathbf{p}_n){}_l\alpha_k - {}_l\Phi_{\Gamma n}^{R}({}_l\psi_k({}_l\mathbf{p}_n){}_l\alpha_k - {}_l\Phi_{\Gamma\,\mathrm{ref}\,n}^{R})\right]^2.$$

The same representation of the Robin boundary conditions is made as in the collocation case.

The minimisation of the functional $\Im[{}_l\boldsymbol{\alpha}]$ with respect to the coefficients ${}_l\boldsymbol{\alpha}$ gives the following system of linear equations

$$ {}_l\underline{\boldsymbol{\Psi}}\,{}_l\boldsymbol{\alpha} = {}_l\mathbf{b}, \tag{74}$$

with the system matrix coefficients

$$\begin{aligned} {}_l\underline{\Psi}_{ik} &= \sum_{n=1}^{{}_lN} {}_l\Upsilon_{\Omega n} \upsilon_n({}_l\mathbf{p})\, {}_l\psi_k(\mathbf{p}_n)\, {}_l\psi_k(\mathbf{p}_n) \\ &+ \sum_{n=1}^{{}_lN} {}_l\Upsilon_{\Gamma n}^D \upsilon_n({}_l\mathbf{p})\, {}_l\psi_i(\mathbf{p}_n)\, {}_l\psi_k(\mathbf{p}_n) \\ &+ \sum_{n=1}^{{}_lN} {}_l\Upsilon_{\Gamma n}^N \upsilon_n({}_l\mathbf{p}) \frac{\partial}{\partial n_\Gamma}\, {}_l\psi_i({}_l\mathbf{p}_n) \frac{\partial}{\partial n_\Gamma}\, {}_l\psi_k({}_l\mathbf{p}_n) \\ &+ \sum_{n=1}^{{}_lN} {}_l\Upsilon_{\Gamma n}^R \upsilon_n({}_l\mathbf{p}) \left({}_l\Phi_{\Gamma\,\mathrm{ref}\,n}^R \psi_i({}_l\mathbf{p}_n) + \frac{\partial}{\partial n_\Gamma}\, {}_l\psi_i({}_l\mathbf{p}_n) \right) \\ &\times \left({}_l\Phi_{\Gamma\,\mathrm{ref}\,n}^R\, {}_l\psi_k({}_l\mathbf{p}_n) + \frac{\partial}{\partial n_\Gamma}\, {}_l\psi_k({}_l\mathbf{p}_n) \right) \end{aligned} \tag{75}$$

and with the augmented right-hand side vector

$$\begin{aligned} {}_l\mathbf{b}_i &= \sum_{n=1}^{{}_lN} {}_l\Upsilon_{\Omega n}\, {}_l\upsilon_n({}_l\mathbf{p})\, {}_l\psi_i({}_l\mathbf{p}_n)\, {}_l\Phi_n + \sum_{n=1}^{{}_lN} {}_l\Upsilon_{\Gamma n}^D\, {}_l\upsilon_n({}_l\mathbf{p})\, {}_l\psi_i({}_l\mathbf{p}_n)\, {}_l\Phi_{\Gamma n}^D \\ &+ \sum_{n=1}^{{}_lN} {}_l\Upsilon_{\Gamma n}^N\, {}_l\upsilon_n({}_l\mathbf{p}) \frac{\partial}{\partial n_\Gamma}\, {}_l\psi_i({}_l\mathbf{p}_n)\, {}_l\Phi_{\Gamma n}^N \\ &+ \sum_{n=1}^{{}_lN} {}_l\Upsilon_{\Gamma n}^R\, {}_l\upsilon_n({}_l\mathbf{p}) \left(({}_l\Phi_{\Gamma n}^R)^2\, {}_l\Phi_{\Gamma\,\mathrm{ref}\,n}^R\, {}_l\psi_i({}_l\mathbf{p}_n) + {}_l\Phi_{\Gamma n}^R\, {}_l\Phi_{\Gamma\,\mathrm{ref}\,n}^R \frac{\partial}{\partial n_\Gamma}\, {}_l\psi_i({}_l\mathbf{p}_n) \right). \end{aligned} \tag{76}$$

After the coefficients are calculated, the unknown values of the unknown ${}_l\boldsymbol{\Phi}$ in the Dirichlet, Neumann, and Robin nodes at time $t_0 + \Delta t$ are determined from the global representation Equation (27).

Step 5

The iteration and steady-state checks are performed. In case the iteration criterion is passed, calculation of the new time-step is performed. In case the steady-state criterion is passed or the time of calculation exceeds the predetermined time of interest, the calculation is stopped.

The performance of the presented method for diffusion problems can be found in [27], for convection-diffusion problems in [36], and for moving boundary problems with moving interfaces and moving domains in [38]. The microscopic species transfer problems have been dealt with by Kovačević and Šarler [13], based on the dynamic r-adaptivity of the nodal arrangement.

6 Explicit Solution of the Coupled Mass and Momentum Transport Equations

This part discusses the special issues encountered in mesh-free solution of the coupled mass

$$\frac{\partial}{\partial t}\rho + \nabla \cdot (\rho \mathbf{v}) = 0, \tag{77}$$

and momentum conservation equation

$$\frac{\partial}{\partial t}(\rho \mathbf{v}) + \nabla \cdot (\rho \mathbf{v}\mathbf{v}) = -\nabla P + \nabla \cdot \underline{\mathbf{T}} + \mathbf{f}, \tag{78}$$

where P, $\underline{\mathbf{T}}$, and $\mathbf{f}$ represent the pressure, the deviatoric part of the stress tensor, and the body force, respectively. For the sake of brevity, initial conditions and only the Dirichlet velocity boundary conditions are assumed

$$\mathbf{v}(\mathbf{p}, t) = \mathbf{v}_0(\mathbf{p}, t); \quad \mathbf{p} \in \Omega + \Gamma, \tag{79}$$

$$\mathbf{v} = \mathbf{v}_\Gamma^D; \quad \mathbf{p} \in \Gamma^D. \tag{80}$$

We seek the solution of the pressure field at time t_0 and the pressure and velocity fields at time $t_0 + \Delta t$. The main goal is to formulate the equations that can be solved by the "local" methodology, described in the previous part.

The pressure field is solved by taking the divergence of the momentum conservation equation

$$\nabla^2 P = -\frac{\partial}{\partial t}\nabla \cdot (\rho \mathbf{v}) - \nabla \cdot [\nabla \cdot (\rho \mathbf{v}\mathbf{v})] + \nabla \cdot (\nabla \cdot \underline{\mathbf{T}}) + \nabla \cdot \mathbf{f}. \tag{81}$$

The pressure is actually calculated from a false transient of the following equation towards the steady-state:

$$\frac{\partial}{\partial t}P = -\nabla^2 P - \frac{\partial}{\partial t}\nabla \cdot (\rho \mathbf{v}) - \nabla \cdot [\nabla \cdot (\rho \mathbf{v}\mathbf{v})] + \nabla \cdot (\nabla \cdot \underline{\mathbf{T}}) + \nabla \cdot \mathbf{f}. \tag{82}$$

Neumann boundary conditions for the Pressure Poisson equation are obtained by multiplication of the momentum equation with the normal derivative

$$\frac{\partial P}{\partial n_\Gamma} = \left[-\frac{\partial}{\partial t}(\rho \mathbf{v}_\Gamma^D) - \nabla \cdot (\rho \mathbf{v}_\Gamma^D \mathbf{v}_\Gamma^D) + \nabla \cdot (\nabla \cdot \underline{\mathbf{T}}) + \nabla \cdot \mathbf{f} \right] \cdot \mathbf{n}_\Gamma. \tag{83}$$

The initial pressure is calculated as

$$P_0 = \bar{P}_0 + \left[-\nabla^2 \bar{P}_0 - \nabla \cdot [\nabla \cdot (\rho_0 \mathbf{v}_0 \mathbf{v}_0] + \nabla \cdot [\nabla \cdot \underline{\mathbf{T}}_0) + \nabla \cdot \mathbf{f}_0 \right] \Delta t_P, \tag{84}$$

where $\bar{P}$ represents the value from previous iteration and Δt_P the artificial pressure transient time-step. The initial value of $\bar{P}_0$ can be set to 0. After the calculation of the pressure field at time t_0, the new velocity field at time $t_0 + \Delta t$ is calculated from

$$\hat{\mathbf{v}} = \left[\frac{\rho_0 \mathbf{v}_0}{\Delta t} - \nabla \cdot (\rho \mathbf{v}_0 \mathbf{v}_0) - \nabla P_0 + \nabla \cdot \underline{\mathbf{T}} + \mathbf{f} \right] \frac{\Delta t}{\rho}. \tag{85}$$

The calculated velocity is denoted by $\hat{\mathbf{v}}$, because it does not comply with the mass conservation in general. In order to assure the compliance of the new velocity field with the mass conservation, the following pressure and velocity corrections are made. The calculated velocity is corrected by $\tilde{\mathbf{v}}$, which ensures

$$\nabla \cdot (\rho \mathbf{v}) = \nabla \cdot [\rho(\hat{\mathbf{v}} + \tilde{\mathbf{v}})] = -\frac{\partial \rho}{\partial t}. \tag{86}$$

Consider that the velocity correction occurs exclusively due to action of the pressure correction

$$\frac{\rho}{\Delta t} \tilde{\mathbf{v}} = -\nabla \tilde{P}, \tag{87}$$

$$\nabla^2 \tilde{P} = \frac{1}{\Delta t} \nabla \cdot (\rho \hat{\mathbf{v}}) + \frac{1}{\Delta t} \frac{\partial \rho}{\partial t} \tag{88}$$

with the pressure correction Neumann boundary conditions

$$\frac{\partial \tilde{P}}{\partial n_\Gamma} = 0. \tag{89}$$

The pressure correction is actually calculated from a false transient of the following equation towards the steady-state:

$$\frac{\partial \tilde{P}}{\partial t} = -\nabla^2 \tilde{P} + \frac{1}{\Delta t} \nabla \cdot (\rho \hat{\mathbf{v}}) + \frac{1}{\Delta t} \frac{\partial \rho}{\partial t}. \tag{90}$$

The pressure correction is calculated as

$$\tilde{P} = \bar{\tilde{P}} + \left[-\nabla^2 \tilde{P} + \frac{1}{\Delta t} \nabla \cdot (\rho \hat{\mathbf{v}}) + \frac{\rho - \rho_0}{\Delta t^2} \right] \Delta t_P. \tag{91}$$

After calculation of the pressure correction, the pressure and velocity fields are updated

$$P = P + \tilde{P}, \tag{92}$$

$$\mathbf{v} = \hat{\mathbf{v}} - \nabla \tilde{P} \frac{\Delta t}{\rho}, \tag{93}$$

and the solution procedure is advanced to a new time-step. The formulation, described here, for solving fluid flow problems, has been numerically implemented by Perko [23] in the context of Diffuse Approximation and by Divo and Kassab [9] in the context of local Kansa method.

7 Conclusions

This article reviews a new (very) simple meshfree formulation for solving a wide range of transport phenomena. The numerical tests [6], included in the cited references of the new method, show much higher accuracy of the method as compared with the classical FDM. The only exception observed is the solution at short times immediately after Dirichlet jump where similar numerical observation properties are observed [27]. The time-marching is performed in a simple explicit way. The governing equation is solved in its strong form. No polygonisation and integrations are needed. The developed method is almost independent of the problem dimension. The complicated geometry can easily be coped with. The method appears efficient, because it does not require a solution of a large system of equations like the original Kansa method. Instead, small systems of linear equations have to be solved in each time-step for each node and associated domain of influence, probably representing the most natural and automatic domain decomposition. This feature of the developed method represents its principal difference from the other related local approaches, where the resultant matrix is large and sparse [7, 14, 19, 34]. The method is simple to learn and simple to code. The method can cope with very large problems since the computational effort grows approximately linear with the number of the nodes. The local approach, described in this article could be extended in a straightforward way to tackle other types of partial differential equations. Despite the fact that the represented method behaves excellent, the underlying basic mathematical theory is still lacking. First systematic numerical experiments with uniform and non-uniform local collocation for very simple cases are expected to appear in [5].

Appendix

Calculation of the convective term in three-dimensional Cartesian coordinates

$$\nabla \cdot (\rho \mathbf{v} \Phi) = \frac{\partial}{\partial p_x}(\rho v_x \Phi) + \frac{\partial}{\partial p_y}(\rho v_y \Phi) + \frac{\partial}{\partial p_z}(\rho v_z \Phi).$$

Calculation of the diffusive term in three-dimensional Cartesian coordinates

$$\nabla \cdot (\underline{\mathbf{D}} \cdot \nabla \Phi) =$$

$$+ \frac{\partial D_{xx}}{\partial p_x}\frac{\partial \Phi_x}{\partial p_x} + \frac{\partial D_{xy}}{\partial p_x}\frac{\partial \Phi_y}{\partial p_y} + \frac{\partial D_{xz}}{\partial p_x}\frac{\partial \Phi_z}{\partial p_z} + D_{xx}\frac{\partial^2 \Phi_x}{\partial p_x^2} + D_{xy}\frac{\partial^2 \Phi_y}{\partial p_x p_y} + D_{xz}\frac{\partial^2 \Phi_z}{\partial p_x p_z}$$

$$+ \frac{\partial D_{yx}}{\partial p_y}\frac{\partial \Phi_x}{\partial p_x} + \frac{\partial D_{yy}}{\partial p_y}\frac{\partial \Phi_y}{\partial p_y} + \frac{\partial D_{yz}}{\partial p_y}\frac{\partial \Phi_z}{\partial p_z} + D_{yx}\frac{\partial^2 \Phi_x}{\partial p_x \partial p_y} + D_{yy}\frac{\partial^2 \Phi_y}{\partial p_y^2} + D_{yz}\frac{\partial^2 \Phi_z}{\partial p_x \partial p_z}$$

$$+ \frac{\partial D_{zx}}{\partial p_z}\frac{\partial \Phi_x}{\partial p_x} + \frac{\partial D_{zy}}{\partial p_z}\frac{\partial \Phi_y}{\partial p_y} + \frac{\partial D_{zz}}{\partial p_z}\frac{\partial \Phi_z}{\partial p_z} + D_{zx}\frac{\partial^2 \Phi_x}{\partial p_x \partial p_z} + D_{zy}\frac{\partial^2 \Phi_y}{\partial p_y \partial p_z} + D_{zz}\frac{\partial^2 \Phi_z}{\partial p_z^2}.$$

Acknowledgements

The author wishes to thank his students Igor Kovačević, Dr. Janez Perko, Dr. Miha Založnik, Robert Vertnik and Miran Dragar for assistance in preparing this article.

References

1. Atluri S.N. *The Meshless Method (MLPG) for Domain and BIE Discretizations*. Tech Science Press, Forsyth, 2004.
2. Atluri S.N. and Shen S. *The Meshless Method*. Tech Science Press, Encino, 2002.
3. Boyd J.P. *Chebishev and Fourier Spectral Methods*. Dover, New York, 2001.
4. Buhmann M.D. *Radial Basis Function: Theory and Implementations*. Cambridge University Press, Cambridge, 2003.
5. Bulinsky Z., Nowak A. and Šarler B. (2007) Numerical experiments with the local radial basi function collocation method. *Computers, Materials, Continua*, in review, 2007.
6. Cameron A.D., Casey J.A. and Simpson G.B. *Benchmark Tests for Thermal Analysis*. NAFEMS National Agency for Finite Element Methods and Standards, Department of Trade and Industry, National Engineering Laboratory, Glasgow, 1986.
7. Chen C.S., Ganesh M., Golberg M.A. and Cheng A.H.-D. Multilevel compact radial basis functions based computational scheme for some elliptic problems. *Computers and Mathematics with Application*, 43:359–378, 2002.
8. Chen W. New RBF collocation schemes and kernel RBFs with applications. *Lecture Notes in Computational Science and Engineering*, 26:75–86, 2002.
9. Divo E. and Kassab A.J. An efficient localized RBF meshless method for fluid flow and conjugate heat transfer. *ASME Journal of Heat Transfer*, in print, 2006.
10. Fasshauer G.E. (1997) Solving partial differential equations by collocation with radial basis functions. In: Mehaute A.L., Rabut C. and Schumaker L.L. (Eds), *Surface Fitting and Multiresolution Methods*, 1997, pp. 131–138.

11. Kansa E.J. Multiquadrics – A scattered data approximation scheme with applications to computational fluid dynamics – II. Solutions to parabolic, hyperbolic and elliptic partial differential equations. *Computers and Mathematics with Application*, 19:147–161, 1990.
12. Kovačević I., Poredoš A. and Šarler B. Solving the Stefan problem by the RBFCM. *Numer. Heat Transfer, Part B: Fundamentals*, 44:575–599, 2003.
13. Kovačević I. and Šarler B. Solution of a phase-field model for dissolution of primary particles in binary aluminium alloys by an r-adaptive mesh-free method. *Materials Science and Engineering*, 413/414A:423–428, 2005.
14. Lee C.K., Liu X. and Fan S.C. Local muliquadric approximation for solving boundary value problems. *Computational Mechanics*, 30:395–409, 2003.
15. Liu G.R. *Mesh Free Methods*. CRC Press, Boca Raton, FL, 2003.
16. Liu G.R. and Gu Y.T. *An Introduction to Meshfree Methods and Their Programming*. Springer, Dordrecht, 2005.
17. Mai-Duy N. and Tran-Cong T. Numerical solution of differential equations using multiquadrics radial basis function networks. *International Journal for Numerical Methods in Engineering*, 23:1807–1829, 2001.
18. Mai-Duy N. and Tran-Cong T. Numerical solution of Navier–Stokes equations using multiquadric radial basis function networks. *Neural Networks*, 14:185–199, 2001.
19. Mai-Duy N. and Tran-Cong T. Mesh-free radial basis function network methods with domain decomposition for approximation of functions and numerical solution of Poisson's equations. *Engineering Analysis with Boundary Elements*, 26:133–156, 2002.
20. Mai-Duy N. and Tran-Cong T. Indirect RBFN method with thin plate splines for numerical solution of differential equations. *Computer Modeling in Engineering and Sciences*, 4:85–102, 2003.
21. Nayroles B., Touzot G. and Villon P. The diffuse approximation. *C.R. Acad. Sci. Paris* 313-II:293–296, 1991.
22. Özisik M.N. *Finite Difference Methods in Heat Transfer*. CRC Press, Boca Raton, FL, 1994.
23. Perko J. *Modelling of Transport Phenomena by the Diffuse Approximate Method*. Ph.D. Thesis, School of Applied Sciences, Nova Gorica Polytechnic, Nova Gorica, 2005.
24. Power H. and Barraco W.A. Comparison analysis between unsymmetric and symmetric RBFCMs for the numerical solution of PDE's. *Computers and Mathematics with Applications*, 43:551–583, 2002.
25. Sadat H. and Prax C. Application of the diffuse approximation for solving fluid flow and heat transfer problems. *International Journal of Heat and Mass Transer*, 39:214–218, 1996.
26. Shu C., Ding H. and Yeo K.S. Local radial basis function-based differential quadrature method and its application to solve two-dimensional incompressible Navier–Stokes equations. *Computer Methods in Applied Mechanics and Engineering*, 192:941–954, 2003.
27. Šarler B. and Vertnik R. Meshfree explicit local radial basis function collocation method for diffusion problems. *Computers and Mathematics with Applications*, 51:1269–1282, 2006.
28. Šarler B., Jelić N., Kovačević I., Lakner M. and Perko J. Axisymmetric multiquadrics. *Engineering Analysis with Boundary Elements*, 30:137–142, 2006.
29. Šarler B. A radial basis function collocation approach in computational fluid dynamics. *Computer Modeling in Engineering and Sciences*, 7:185–193, 2005.

30. Šarler B., Vertnik R. and Perko J. Application of diffuse approximate method in convective-diffusive solidification problems. *Computers, Materials, Continua*, 2(1):77–84, 2005.
31. Šarler B., Perko J. and Chen C.S. Radial basis function collocation method solution of natural convection in porous media. *International Journal of Numerical Methods for Heat and Fluid Flow*, 14:187–212, 2004.
32. Šarler B. Towards a mesh-free computation of transport phenomena. *Engineering Analysis with Boundary Elements*, 26:731–738, 2002.
33. Šarler B., Kovačević I. and Chen C.S. A mesh-free solution of temperature in direct-chill cast slabs and billets. In: Mammoli A.A. and Brebbia C.A. (Eds), *Moving Boundaries VI*. WIT Press, Southampton, 2004, pp. 271–280.
34. Tolstykh A.I. and Shirobokov D.A. On using radial basis functions in a "finite difference" mode with applications to elasticity problems. *Computational Mechanics*, 33:68–79, 2003.
35. Versteeg H.K. and Malalasekera W. *Computational Fluid Dynamics: The Finite Volume Method*. Prentice Hall, Harlow, 1995.
36. Vertnik R. and Šarler B. Meshless local radial basis function collocation method for convective-diffusive solid-liquid phase change problems. *International Journal of Numerical Methods for Heat and Fluid Flow* (Special Issue: European Congress on Computational Methods in Applied Sciences and Engineering, ECCOMAS 2004, Jyväskylä, 24–28 July 2004), 16:617–640, 2005.
37. Vertnik R., Perko J. and Šarler B. Solution of temperature field in DC cast aluminium alloy billet by the Diffuse Approximate Method. *Materials and Technology*, 38:257–261, 2004.
38. Vertnik R., Založnik M. and Šarler B. Solution of transient direct chill aluminium billet casting problem with simultaneous material and interphase moving boundaries. *Engineering Analysis with Boundary Elements*, 30: 847–855, 2006.
39. Wang J.G. and Liu G.R. On the optimal shape parameter of radial basis functions used for 2-D meshless method. *Computer Methods in Applied Mechanics and Engineering*, 26:2611–2630, 2002.
40. Wrobel L.C. *The Boundary Element Method – Volume 1: Applications in Thermofluids and Acoustics*. Wiley, New York, 2001.
41. Zienkiewicz O.C. and Taylor R.L. *The Finite Element Method*. Butterworth-Heinemann, Oxford, 2002.

Computation of Static Deformations and Natural Frequencies of Shear Deformable Plates by an RBF-Pseudospectral Method with an Optimal Shape Parameter

A.J.M. Ferreira and G.E. Fasshauer

[1] *Departamento de Engenharia Mecânica e Gestão Industrial, Faculdade de Engenharia da Universidade do Porto, Rua Dr. Roberto Frias, 4200-465 Porto, Portugal; E-mail: ferreira@fe.up.pt*
[2] *Department of Applied Mathematics, Illinois Institute of Technology, Chicago, IL 60616, U.S.A.; E-mail: fasshauer@iit.edu*

Abstract. A study of static deformations and free vibration of shear flexible isotropic and laminated composite plates is presented. The analysis is based on a new numerical scheme where collocation by radial basis functions is viewed as a pseudospectral method to produce highly accurate results. A cross-validation technique is used to optimize the shape parameter for the basis functions. Numerical results for isotropic and symmetric laminated composite plates are presented and discussed for various thickness-to-length ratios.

Key words: Radial basis functions, pseudospectral method, free vibrations, plates.

1 Introduction

The static and vibration analysis of plates is an important subject in the design of mechanical, civil and aerospace applications. While most classical theories usually neglect the transverse shear and the rotatory inertia, these effects are really non-negligible due to the thickness of most parts of beams and plates. Therefore the first-order theory, also known as Mindlin–Reissner theory for plates should be used in a general analysis.

The analysis of static deformations and free vibration of Mindlin plates is best performed by numerical techniques. The differential quadrature method [1–3], the boundary characteristic orthogonal polynomials [4] and the pseudospectral method [5] were used in recent years. A good survey on thick plate vibration analysis was performed by Liew and colleagues [36]. The finite element method also proved to be very adequate for this type of problems (see, e.g., [6–8]). More recently the analysis

V.M.A. Leitão et al. (eds), Advances in Meshfree Techniques, 283–310.

of Mindlin plates by Kansa's non-symmetric radial basis function (RBF) collocation method was performed by Ferreira [10–17].

In the present work we illustrate for the first time the application of radial basis functions in a pseudospectral framework and an optimization technique to the automatic selection of the shape parameter to the analysis of Mindlin plates. The method proves to be elegant and of very high quality when compared to analytical solutions and Kansa's collocation method [10]. Kansa's collocation method needs the computation of coefficients first while the present method does not. This is an advantage for many engineering problems, particularly for structural dynamics. Some early experiments with Kansa's method [10–14, 16] show its good ability for solving engineering problems, but there are other possibilities such as the method presented in this paper.

2 RBF-PS Methods

Pseudospectral (PS) methods (see [24] or [25] for an introduction to the subject) are known as highly accurate solvers for partial differential equations (PDEs). Generally speaking, one represents the spatial part of the approximate solution $\hat{u}$ of a given PDE by a linear combination of certain smooth basis functions $\phi_j,\ j = 1, \ldots, N$, i.e.,

$$\hat{u}(x) = \sum_{j=1}^{N} c_j \phi_j(x), \quad x \in \mathbb{R}. \tag{1}$$

Traditionally, polynomial basis functions are used, and therefore the formulation above is a univariate one. This leads to the well-known limitation for PS methods: for higher space dimensions their use is pretty much limited to tensor-product grids. In this paper, however, we will use radial basis functions (RBFs) instead of polynomials. This opens up the possibility to work with irregular grids, and on irregular geometries while maintaining a degree of accuracy similar to that obtained with PS methods.

The usual approach to solving PDEs with an RBF collocation method is frequently referred to as *Kansa's method*. For this method one also starts with an expansion of the form (1), now with $\boldsymbol{x} \in \mathbb{R}^s$. However, one then imposes the boundary conditions for the PDE, and forces the PDE and its boundary conditions to be satisfied at a set of collocation points. This leads to a system of linear equations which is solved for the expansion coefficients c_j in (1). Having these coefficients, one can then evaluate the approximate solution $\hat{u}$ at any point $\boldsymbol{x}$ via (1). Thus, with Kansa's collocation method we end up with an approximate solution that is given in terms of a (continuous) function. For more details see, e.g., [26].

Recently Fornberg et al. (see, e.g., [18, 19]) and Schaback [20] showed that certain limiting cases of radial basis functions correspond to polynomial interpolants.

This new insight has led to the idea of using pseudospectral methods combined with radial basis functions to solve PDEs (see, e.g., [21–23]). It is this numerical approach that we use for the eigenvalue analysis presented below. For the sake of completeness we summarize the main ideas of this approach.

Consider the linear elliptic PDE problem

$$\mathcal{L}u = f \quad \text{in } \Omega \tag{2}$$

with Dirichlet boundary condition

$$u = g \quad \text{on } \Gamma = \partial\Omega. \tag{3}$$

For the pseudospectral approach we start with an expansion of the form

$$\hat{u}(\boldsymbol{x}) = \sum_{j=1}^{N} c_j \varphi(\|\boldsymbol{x} - \boldsymbol{\xi}_j\|), \qquad \boldsymbol{x} \in \Omega \subseteq \mathbb{R}^s, \tag{4}$$

where the points $\boldsymbol{\xi}_j$, $j = 1, \ldots, N$, are the *centers* of the basis functions $\phi_j = \varphi(\| \cdot - \boldsymbol{\xi}_j\|)$ and φ is one of the usual radial basic functions such as the inverse multiquadric

$$\varphi(r) = \frac{1}{\sqrt{1 + (\varepsilon r)^2}}, \tag{5}$$

the multiquadric

$$\varphi(r) = \sqrt{1 + (\varepsilon r)^2}, \tag{6}$$

the Gaussian

$$\varphi(r) = e^{-(\varepsilon r)^2}, \tag{7}$$

or a Wendland compactly supported function such as

$$\varphi(r) = (1 - \varepsilon r)_+^8 \left(32(\varepsilon r)^3 + 25(\varepsilon r)^2 + 8\varepsilon r + 1\right). \tag{8}$$

The first three of these functions are infinitely differentiable, the Wendland function is six times continuously differentiable. The inverse multiquadric, Gaussian and Wendland function are positive definite, while the multiquadric is conditionally negative definite. Note that all of our examples contain a positive shape parameter ε. For (inverse) multiquadrics our notation differs from another popular one for which the shape parameter is denoted by c (not to be confused with the coefficients c_j in the expansion (4)), e.g., $\varphi(r) = 1/\sqrt{r^2 + c^2}$. However, the two formulations are equivalent if we set $\varepsilon = 1/c$. For the Wendland function the shape parameter determines the size of the support radius (since the $_+$ notation indicates that the function is identically equal to zero outside a sphere of radius r/ε). The advantage of our representation is that all RBFs behave similarly under changes of ε. In particular,

$\varepsilon \to 0$ always leads to "flat" basic functions, and it is exactly for this limiting case that the connection to polynomials mentioned at the beginning of this section arises. To be precise, since the compactly supported Wendland functions have possess only a limited amount of smoothness they will not be able to provide the full spectral accuracy that polynomials and the other infinitely smooth basic functions are able to. However, the experiments below show that they still provide very high accuracy, and moreover behave in a more stable way than the other basic functions which proved to be beneficial for our eigenvalue analysis.

If we evaluate (4) at a set of *collocation points* $\boldsymbol{x}_i$, $i = 1, \ldots, N$, then we get

$$\hat{u}(\boldsymbol{x}_i) = \sum_{j=1}^{N} c_j \varphi(\|\boldsymbol{x}_i - \boldsymbol{\xi}_j\|), \qquad i = 1, \ldots, N,$$

or in matrix-vector notation

$$\boldsymbol{u} = A\boldsymbol{c}, \tag{9}$$

where $\boldsymbol{c} = [c_1, \ldots, c_N]^T$ is the coefficient vector, the *evaluation matrix* A has entries $A_{ij} = \varphi(\|\boldsymbol{x}_i - \boldsymbol{\xi}_j\|)$, and $\boldsymbol{u} = [\hat{u}(\boldsymbol{x}_1), \ldots, \hat{u}(\boldsymbol{x}_M)]^T$ is a vector of values of the approximate solution at the collocation points.

An important feature of pseudospectral methods is the fact that one usually is content with obtaining an approximation to the solution on a discrete set of grid points $\boldsymbol{x}_i$, $i = 1, \ldots, N$ instead of at an arbitrary point $\boldsymbol{x}$, as in the popular non-symmetric RBF collocation approach (or Kansa's method). One of several ways to implement the pseudospectral method is via so-called *differentiation matrices*, i.e., one finds a matrix L such that at the grid points $\boldsymbol{x}_i$ we have

$$\mathbf{u}_{\mathcal{L}} = L\mathbf{u}. \tag{10}$$

Here $\mathbf{u} = [\hat{u}(\boldsymbol{x}_1), \ldots, \hat{u}(\boldsymbol{x}_N)]^T$ is the vector of values of $\hat{u}$ at the grid points mentioned above, and $\mathbf{u}_{\mathcal{L}}$ is the vector of values of the "derivatives" of u at the same points.

Therefore, instead of computing the coefficients $\boldsymbol{c}$ by solving a collocation system – as is done in the standard RBF collocation approach (Kansa's method) – we want to use the differentiation matrix L so that in the end we will have a discrete version of the PDE in the form

$$L\boldsymbol{u} = \boldsymbol{f}, \tag{11}$$

where $\boldsymbol{u}$ is as above, and $\boldsymbol{f}$ is the vector of values of the right-hand side f of (2) evaluated at the collocation points.

Since the differential equation for our problem is $\mathcal{L}u = f$, we will apply the differential operator $\mathcal{L}$ to the approximate solution $\hat{u}$ as given by (4). By linearity we get

$$\mathcal{L}\hat{u}(\boldsymbol{x}) = \sum_{j=1}^{N} c_j \mathcal{L}\varphi(\|\boldsymbol{x} - \boldsymbol{\xi}_j\|).$$

Evaluation of this formula at the collocation points $\boldsymbol{x}_i$ yields a system of linear algebraic equations which can be written in matrix-vector notation as

$$L\boldsymbol{u} = A_{\mathcal{L}}\boldsymbol{c}, \tag{12}$$

where $\boldsymbol{u}$ and $\boldsymbol{c}$ are as in (9) above, and the matrix $A_{\mathcal{L}}$ has entries $\mathcal{L}\varphi(\|\boldsymbol{x} - \boldsymbol{\xi}_j\|)\big|_{\boldsymbol{x}=\boldsymbol{x}_i}$. The coefficients – which we do not explicitly compute – are given by (9), i.e., $\boldsymbol{c} = A^{-1}\boldsymbol{u}$, so that we formally get

$$L\boldsymbol{u} = A_{\mathcal{L}}A^{-1}\boldsymbol{u}, \tag{13}$$

and we see that the differentiation matrix L is given by $L = A_{\mathcal{L}}A^{-1}$. The name differentiation matrix is due to the fact that L takes the vector $\boldsymbol{u}$ of function values to the vector $L\boldsymbol{u} = \boldsymbol{u}_{\mathcal{L}}$ of "derivative" values (cf. (10)).

Note that this matrix involves inverting the *standard RBF interpolation matrix A* which is known to be non-singular for all distributions of centers $\boldsymbol{\xi}_j$ and (coinciding) collocation points $\boldsymbol{x}_i$. This property of A ensures that (at this point in our discussion) we will not run into the problems of possible non-invertibility of the collocation matrix encountered in the popular Kansa method.

Also note that we have not yet enforced the boundary conditions. This, however, is – for the Dirichlet case we are considering here – an absolutely trivial matter. We simply replace those rows of L corresponding to the boundary collocation points (at which we want to enforce the boundary conditions) by standard unit vectors with a one in the diagonal position and zeros elsewhere, and replace the corresponding $f(\boldsymbol{x}_k)$ on the right-hand side by $g(\boldsymbol{x}_k)$ (cf. (3)). It is obvious that this works since the resulting product of (boundary) row k of L with the vector $\boldsymbol{u}$ now corresponds to $\hat{u}(\boldsymbol{x}_k) = g(\boldsymbol{x}_k)$ (see, e.g., [25]). One can show that the resulting matrix L_{BC} which also enforces the boundary condition is very closely related to the Kansa matrix, i.e., after a possible permutation of rows we obtain

$$L_{BC}\boldsymbol{u} = \begin{bmatrix} \tilde{A}_{\mathcal{L}} \\ \tilde{A} \end{bmatrix} A^{-1}\boldsymbol{u},$$

where the block matrix on the right-hand side is exactly Kansa's matrix.

To obtain a numerical approximation to the solution of the elliptic problem (2)–(3) we actually need to compute

$$\boldsymbol{u} = L_{BC}^{-1}\begin{bmatrix} \boldsymbol{f} \\ \boldsymbol{g} \end{bmatrix} = \left[\begin{bmatrix} \tilde{A}_{\mathcal{L}} \\ \tilde{A} \end{bmatrix} A^{-1}\right]^{-1}\begin{bmatrix} \boldsymbol{f} \\ \boldsymbol{g} \end{bmatrix}, \tag{14}$$

which is the solution of the fully discretized problem (including both the differential operator and the boundary conditions). Assuming invertibility of the two matrix factors this gives

$$\boldsymbol{u} = A\begin{bmatrix} \tilde{A}_{\mathcal{L}} \\ \tilde{A} \end{bmatrix}^{-1}\begin{bmatrix} \boldsymbol{f} \\ \boldsymbol{g} \end{bmatrix}.$$

This is the same end result as one obtains if the approximate solution for the non-symmetric collocation method (Kansa's method) is evaluated at the collocation points. Note, that in this formulation we now do require invertibility of the Kansa matrix (just as in Kansa's method).

However, as noted above, we do not work with the individual matrices A, $\tilde{A}_{\mathcal{L}}$, and $\tilde{A}$, but instead use only the differentiation matrix L_{BC}, so that

$$\boldsymbol{u} = L_{BC}^{-1} \begin{bmatrix} \boldsymbol{f} \\ \boldsymbol{g} \end{bmatrix}$$

as stated in (14). Moreover, the coefficient vector $\boldsymbol{c}$ is never computed. This can be especially beneficial in time-dependent problems.

3 Finding an "Optimal" Shape Parameter

As mentioned above, a small shape parameter $\varepsilon \to 0$ will always lead to "flat" basic functions. In fact, the shape parameter ε can be used to influence the accuracy of our numerical method: smaller values of ε generally lead to higher accuracy. However, it is known that there exists a so-called trade-off principle (for infinitely smooth RBFs), i.e., high accuracy can only be achieved at the cost of low numerical stability or vice versa (see, e.g., [27]). This means that it is very difficult to get near the polynomial limit in practice. On the other hand, the optimal value of ε, i.e., the value that produces the smallest error, is usually a positive value [19]. For the Wendland functions we use below the trade-off linked to the variation of ε balances higher accuracy (for small ε) against numerical efficiency (since higher values or ε lead to increasingly sparse matrices). Numerical stability is not so much of an issue with these functions, and that is why we use them here.

A popular strategy for estimating the parameter of a model based on the given data is known as *cross validation*. In [28] Rippa describes an algorithm that corresponds to a variant of cross validation known as *"leave-one-out" cross validation* (LOOCV). This method is rather popular in the statistics literature where it is also known as PRESS (Predictive REsidual Sum of Squares). In this algorithm an "optimal" value of ε for the RBF interpolation problem is selected by minimizing the error for a fit to the data based on an interpolant for which one of the centers was "left out". This method takes into account the dependence of the error on the data function. Therefore, the predicted "optimal" shape parameter is usually close to the actual optimum value (which we can only find if we know the exact solution of the interpolation problem). We will adapt Rippa's strategy to find the "optimal" shape parameter ε of the basic function used in the RBF-PS method.

First, we explain how the LOOCV method was used in [28] for the interpolation problem. Specifically, if $\mathcal{P}_f^{[k]}$ is the radial basis function interpolant to the data $\{f_1, \ldots, f_{k-1}, f_{k+1}, \ldots, f_N\}$, i.e.,

$$\mathcal{P}_f^{[k]}(\boldsymbol{x}) = \sum_{\substack{j=1 \\ j \neq k}}^{N} c_j^{[k]} \varphi(\|\boldsymbol{x} - \boldsymbol{x}_j\|),$$

and if E_k is the error

$$E_k = f_k - \mathcal{P}_f^{[k]}(\boldsymbol{x}_k),$$

then the quality of the fit is determined by the norm of the vector of errors $E = [E_1, \ldots, E_N]^T$ obtained by removing in turn one of the data points and comparing the resulting fit with the (known) value at the removed point. The norm of E as a function of ε will serve as a *cost function* for the shape parameter.

While a naive implementation of the leave-one-out algorithm is rather expensive (on the order of N^4) Rippa shows that the algorithm can be simplified to a single formula

$$E_k = \frac{c_k}{A_{kk}^{-1}}, \tag{15}$$

where c_k is the kth coefficient in the interpolant $\mathcal{P}_f$ based on the full data set, and A_{kk}^{-1} is the kth diagonal element of the inverse of the corresponding interpolation matrix. This results in $\mathcal{O}(N^3)$ computational complexity. Note that all entries in the error vector E can be computed in a single statement in Matlab if we vectorize the component formula (15) (see line 4 in Program 3.1). In order to find a good value of the shape parameter as quickly as possible we can use the Matlab function `fminbnd` to find the minimum of the cost function for ε.

Thus, we can implement the cost function in the subroutine `CostEpsilon.m` displayed in Program 3.1. Here the pseudoinverse of A was used to ensure maximum stability in the solution of the linear system. The cost is computed via the 2-norm.

Program 3.1 `CostEpsilon.m`

```
1  function ceps = CostEpsilon(ep,rbf,r,rhs)
2  A = rbf(ep,r);
3  invA = pinv(A);
4  EF = (invA*rhs)./diag(invA);
5  ceps = norm(EF(:));
```

The calling sequence for `CostEpsilon` will look something like

```
[ep,fval] = fminbnd(@(ep) CostEpsilon(ep,rbf,DM,rhs),mine,maxe);
```

where `mine` and `maxe` define the interval to search in for the optimal ε value, and `DM` is a *distance matrix* with entries $\|\boldsymbol{x}_i - \boldsymbol{\xi}_j\|$ used to evaluate the RBF (in the interpolation setting).

The original algorithm in Rippa's paper [28] was intended for the interpolation problem. Therefore, in the context of RBF-PS methods we use a modification of the basic routine `CostEpsilon` which we call `CostEpsilonDRBF` (see Program 3.2

below). Instead of finding an optimal ε for the interpolation problem $Ac = f$ we now need to optimize the choice of ε for the matrix problem (*c.f.* (13))

$$L = A_{\mathcal{L}} A^{-1} \iff LA = A_{\mathcal{L}} \iff A^T L^T = (A_{\mathcal{L}})^T.$$

For simplicity we illustrate the procedure with a first-order derivative $\frac{\partial}{\partial x}$. In this case we will write A_x instead of the generic $A_{\mathcal{L}}$. Any other differential operator can be implemented analogously. As long as the differential operator is of odd order we will have to provide both a distance matrix and a *difference matrix*. For differential operators of even order such as the Laplacian a distance matrix will suffice. For more details see [26].

Program 3.2 `CostEpsilonDRBF.m`

```
% ceps = CostEpsilonDRBF(ep,r,dx,rbf,dxrbf)
% Provides the "cost of epsilon" function for LOOCV optimization
% of shape parameter
% Input: ep, values of shape parameter
%        r, dx, distance and difference matrices
%        rbf, dxrbf, definition of rbf and its derivative
1  function ceps = CostEpsilonDRBF(ep,r,dx,rbf,dxrbf)
2  [m,n] = size(r);
3  A = rbf(ep,r);    % = A^T since A is symmetric
4  rhs = dxrbf(ep,r,dx)';    % A_x^T
5  invA = pinv(A);
6  EF = (invA*rhs)./repmat(diag(invA),1,m);
7  ceps = norm(EF(:));
```

Note that `CostEpsilonDRBF.m` is very similar to `CostEpsilon.m` (*c.f.* Program 3.1). Now, however, we compute a right-hand side *matrix* corresponding to the transpose of A_x. Therefore, the denominator – which remains the same for all right-hand sides (see formula (15)) – needs to be cloned on line 6 via the `repmat` command. The cost of ε is now again the (Frobenius) norm of the matrix `EF`. Other measures for the error may also be appropriate. For the standard interpolation setting Rippa compared the use of the ℓ_1 and ℓ_2 norms (see [28]) and concluded that the ℓ_1 norm yields more accurate "optima". For the RBF-PS problems to be presented here we have observed very good results with the ℓ_2 (or Frobenius) norm, and therefore that is what is used on line 7 of Program 3.2.

A program that calls `CostEpsilonDRBF` and computes the differentiation matrix (with optimal ε) is given by

Program 3.3 `DRBF.m`

```
% [D,x] = DRBF(N,rbf,dxrbf)
% Computes the differentiation matrix D for 1-D derivative
```

```
% using Chebyshev points and LOOCV for optimal shape parameter
% Input: N, number of points -1
%        rbf, dxrbf function handles for rbf and its derivative
% Calls on: DistanceMatrix, DifferenceMatrix
% Requires: CostEpsilonDRBF
function [D,x] = DRBF(N,rbf,dxrbf)
if N==0, D=0; x=1; return, end
x = cos(pi*(0:N)/N)';     % Chebyshev points
mine = .1; maxe = 10;     % Shape parameter interval
r = DistanceMatrix(x,x);
dx = DifferenceMatrix(x,x);
ep = fminbnd(@(ep) CostEpsilonDRBF(ep,r,dx,rbf,dxrbf),mine,maxe);
fprintf('Using epsilon = %f\n', ep)
A = rbf(ep,r);
DA = dxrbf(ep,r,dx);  D = DA/A;
```

4 Free Vibration Analysis of Symmetric Laminated Plates

4.1 Governing Equations

Based on the FSDT (first-order shear deformation theory), the transverse displacement $w(x, y)$ and the rotations $\theta_x(x, y)$ and $\theta_y(x, y)$ about the $y-$ and $x-$axes are independently interpolated due to uncoupling between inplane displacements and bending displacements for symmetrically laminated plates. The equations for isotropic plates can be readily obtained from the following. The equations of motion for the free vibration of laminated plates [31, 32] are:

$$D_{11}\frac{\partial^2\theta_x}{\partial x^2}+D_{16}\frac{\partial^2\theta_y}{\partial x^2}+(D_{12}+D_{66})\frac{\partial^2\theta_y}{\partial x\partial y}+2D_{16}\frac{\partial^2\theta_x}{\partial x\partial y}+D_{66}\frac{\partial^2\theta_x}{\partial y^2}+D_{26}\frac{\partial^2\theta_y}{\partial y^2}+ \\ -kA_{45}\left(\theta_y+\frac{\partial w}{\partial y}\right)-kA_{55}\left(\theta_x+\frac{\partial w}{\partial x}\right)=I_2\frac{\partial^2\theta_x}{\partial t^2} \quad (16)$$

$$D_{16}\frac{\partial^2\theta_x}{\partial x^2}+D_{66}\frac{\partial^2\theta_y}{\partial x^2}+(D_{12}+D_{66})\frac{\partial^2\theta_x}{\partial x\partial y}+2D_{26}\frac{\partial^2\theta_y}{\partial x\partial y}+D_{26}\frac{\partial^2\theta_x}{\partial y^2}+D_{22}\frac{\partial^2\theta_y}{\partial y^2}+ \\ -kA_{44}\left(\theta_y+\frac{\partial w}{\partial y}\right)-kA_{45}\left(\theta_x+\frac{\partial w}{\partial x}\right)=I_2\frac{\partial^2\theta_y}{\partial t^2} \quad (17)$$

$$\frac{\partial}{\partial x}\left[kA_{45}\left(\theta_y+\frac{\partial w}{\partial y}\right)+kA_{55}\left(\theta_x+\frac{\partial w}{\partial x}\right)\right]+$$
$$\frac{\partial}{\partial y}\left[kA_{44}\left(\theta_y+\frac{\partial w}{\partial y}\right)+kA_{45}\left(\theta_x+\frac{\partial w}{\partial x}\right)\right]+P=I_0\frac{\partial^2 w}{\partial t^2}, \quad (18)$$

where D_{ij} and A_{ij} are the bending and shear stiffness components, k is the shear correction factor, and I_i are the mass inertias defined as [31], and P is the external load applied to the plate, as illustrated in Figure 1.

$$I_0=\int_{-\frac{h}{2}}^{\frac{h}{2}}\rho dz,\ I_2=\int_{-\frac{h}{2}}^{\frac{h}{2}}\rho z^2 dz. \quad (19)$$

Here ρ and h denote the density and the total thickness of the composite plate, respectively. The bending moments M_x, M_y and M_{xy} and the shear forces Q_x and Q_y are expressed as functions of the displacement gradients and the material constitutive equations by

$$M_x=D_{11}\frac{\partial\theta_x}{\partial x}+D_{12}\frac{\partial\theta_y}{\partial y}+D_{16}\left(\frac{\partial\theta_x}{\partial y}+\frac{\partial\theta_y}{\partial x}\right) \quad (20)$$

$$M_y=D_{12}\frac{\partial\theta_x}{\partial x}+D_{22}\frac{\partial\theta_y}{\partial y}+D_{26}\left(\frac{\partial\theta_x}{\partial y}+\frac{\partial\theta_y}{\partial x}\right) \quad (21)$$

$$M_{xy}=D_{16}\frac{\partial\theta_x}{\partial x}+D_{26}\frac{\partial\theta_y}{\partial y}+D_{66}\left(\frac{\partial\theta_x}{\partial y}+\frac{\partial\theta_y}{\partial x}\right) \quad (22)$$

$$Q_x=kA_{55}\left(\theta_x+\frac{\partial w}{\partial x}\right)+kA_{45}\left(\theta_y+\frac{\partial w}{\partial y}\right) \quad (23)$$

$$Q_y=kA_{45}\left(\theta_x+\frac{\partial w}{\partial x}\right)+kA_{55}\left(\theta_y+\frac{\partial w}{\partial y}\right). \quad (24)$$

For free vibration problems we assume a harmonic solution in terms of the displacements w, θ_x, θ_y in the form

$$w(x,y,t)=W(x,y)e^{i\lambda t} \quad (25)$$

$$\theta_x(x,y,t)=\Psi_x(x,y)e^{i\lambda t} \quad (26)$$

$$\theta_y(x,y,t)=\Psi_y(x,y)e^{i\lambda t}, \quad (27)$$

where λ is the frequency of natural vibration. Substituting the harmonic expansion into the equations of motion (16-18) we obtain the following equations in terms of the amplitudes W, Ψ_x, Ψ_y

$$D_{11}\frac{\partial^2\Psi_x}{\partial x^2}+D_{16}\frac{\partial^2\Psi_y}{\partial x^2}+(D_{12}+D_{66})\frac{\partial^2\Psi_y}{\partial x\partial y}+2D_{16}\frac{\partial^2\Psi_x}{\partial x\partial y}+D_{66}\frac{\partial^2\Psi_x}{\partial y^2}+D_{26}\frac{\partial^2\Psi_y}{\partial y^2}-$$
$$-kA_{45}\left(\Psi_y+\frac{\partial W}{\partial y}\right)-kA_{55}\left(\Psi_x+\frac{\partial W}{\partial x}\right)=-I_2\lambda^2\Psi_x \quad (28)$$

$$D_{16}\frac{\partial^2\Psi_x}{\partial x^2}+D_{66}\frac{\partial^2\Psi_y}{\partial x^2}+(D_{12}+D_{66})\frac{\partial^2\Psi_x}{\partial x\partial y}+2D_{26}\frac{\partial^2\Psi_y}{\partial x\partial y}+D_{26}\frac{\partial^2\Psi_x}{\partial y^2}+D_{22}\frac{\partial^2\Psi_y}{\partial y^2}-$$
$$-kA_{44}\left(\Psi_y+\frac{\partial W}{\partial y}\right)-kA_{45}\left(\Psi_x+\frac{\partial W}{\partial x}\right)=-I_2\lambda^2\Psi_y \quad (29)$$

$$\frac{\partial}{\partial x}\left[kA_{45}\left(\Psi_y+\frac{\partial W}{\partial y}\right)+kA_{55}\left(\Psi_x+\frac{\partial W}{\partial x}\right)\right]+$$
$$+\frac{\partial}{\partial y}\left[kA_{44}\left(\Psi_y+\frac{\partial W}{\partial y}\right)+kA_{45}\left(\Psi_x+\frac{\partial W}{\partial x}\right)\right]=-I_0\lambda^2 W \quad (30)$$

4.2 Boundary Conditions

The boundary conditions for an arbitrary edge with simply supported, clamped or free edge conditions are as follows:

(a) Simply supported
- SS1, $w = 0$; $M_n = 0$; $M_{ns} = 0$.
- SS2, $w = 0$; $M_n = 0$; $\theta_s = 0$.

(b) Clamped, $w = 0$; $\theta_n = 0$; $\theta_s = 0$.
(c) Free, $Q_n = 0$; $M_n = 0$; $M_{ns} = 0$.

In previous equations, the subscripts n and s refer to the normal and tangential directions of the edge, respectively; M_n, M_{ns} and Q_n represent the normal bending moment, twisting moment and shear force on the plate edge; θ_n and θ_s represent the rotations about the tangential and normal coordinates at the plate edge. The stress resultants on an edge whose normal is represented by $\mathbf{n} = (n_x, n_y)$ can be expressed as

$$M_n = n_x^2 M_x + 2n_x n_y M_{xy} + n_y^2 M_y, \quad (31)$$

$$M_{ns} = (n_x^2 - n_y^2)M_{xy} - n_x n_y(My - Mx), \quad (32)$$

$$Q_n = n_x Q_x + n_y Q_y, \quad (33)$$

$$\theta_n = n_x\theta_x + n_y\theta_y, \quad (34)$$

$$\theta_s = n_x\theta_y - n_y\theta_x, \quad (35)$$

where n_x and n_y are the direction cosines of a unit normal vector at a point at the laminated plate boundary [31, 32].

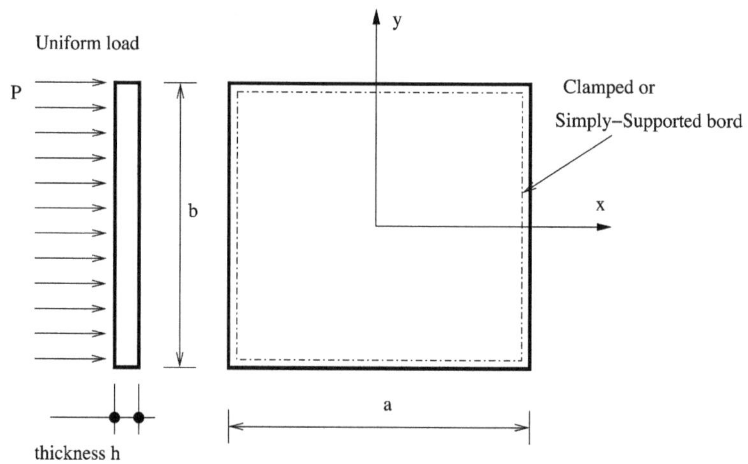

Fig. 1. Plate geometry.

Note that we can easily analyze Mindlin isotropic plates by considering

$$D_{11} = D_{22} = D = \frac{Eh^3}{12(1-\nu^2)}, \quad D_{12} = \nu D_{11}, \quad D_{66} = \frac{Gh^3}{12},$$

$$A_{55} = A_{44} = kGh \quad \text{and} \quad D_{16} = D_{26} = A_{45} = 0,$$

where E is the modulus of elasticity and ν Poisson's ratio of an isotropic material.

4.3 Shear Correction Factors

At layer interfaces, continuity of transverse shear stresses is required, for laminates with distinct materials across the thickness direction. According to the first-order shear deformation assumptions, the transverse shear deformation is constant through the thickness, which is a coarse approximation to the actual variation even for a homogeneous cross-section. For homogeneous cross-sections, the shear deformation is commonly accepted to be a parabolic function of z. Therefore a shear correction factor k must be introduced to approximate on an average basis the transverse deformation energy. Assuming a heterogeneous plate free of tangential tractions, the equilibrium equation in the x direction can be expressed as

$$\frac{\partial \sigma_x}{\partial x} + \frac{\partial \tau_{xy}}{\partial y} + \frac{\partial \tau_{xz}}{\partial z} = 0 \tag{36}$$

Assuming, for simplicity, cylindrical bending, then

$$
\begin{aligned}
\tau_{xz} &= -\int_{-h/2}^{z} \frac{\partial \sigma_x}{\partial x} dz = -\int_{-h/2}^{z} \frac{\partial M_x}{\partial x} \frac{D_1(z)}{R_1} z dz \\
&= -\frac{Q_x}{R_1} \int_{-h/2}^{z} D_1(z) z dz = \frac{Q_x}{R_1} g(z)
\end{aligned}
\tag{37}
$$

where Q_x is the shear force on the xz plane; $R_1 = \int_{-h/2}^{h/2} D_1(z) z^2 dz$ is the flexural plate stiffness in the x direction; z is the coordinate through the thickness; $g(z) = -\int_{-h/2}^{z} D_1(z) z dz$ is the shear shape function.

The function $g(z)$ that shapes the shear stress diagram is independent of loadings, becoming the well known parabolic function $g(z) = [D_1 h^2/8][1 - 4(z/h)^2]$ for the case of a homogeneous cross-section. The strain energy component is given as

$$
w_s = \int_{-h/2}^{h/2} \frac{\tau_{xz}^2}{G_{13}(z)} dz = \frac{Q_x^2}{R_1^2} \int_{-h/2}^{h/2} \frac{g^2(z)}{G_{13}(z)} dz \tag{38}
$$

where $G_{13}(z)$ is the shear modulus, variable through the thickness, in the xz plane. The strain energy component, under the assumption of constant shear strain, is given as

$$
\overline{w}_s = \int_{-h/2}^{h/2} \overline{\gamma}_{xz} G_{13}(z) \overline{\gamma}_{xz} dz = \frac{Q_x^2}{h^2 \overline{G}_1^2} h\overline{G}_1 = \frac{Q_x^2}{h\overline{G}_1} \tag{39}
$$

where

$$
h\overline{G}_1 = \int_{-h/2}^{h/2} G_{13}(z) dz \tag{40}
$$

and $\overline{\gamma}_{xz}$ is the mean value of the shear strains. Therefore it is now possible to evaluate the correction factor k_1 in the xz plane to be

$$
k_1 = \frac{\overline{w}_s}{w_s} = \frac{R_1^2}{h\overline{G}_1 \int_{-h/2}^{h/2} g^2(z)/G_{13}(z) dz} \tag{41}
$$

For k_2 we proceed the same way. This can be applied to symmetric or non-symmetric cross-sections [44]. Here we use the same correction factor ($k = k_1 = k_2$).

For numerical implementation, all integrals are replaced by summation over the layer thicknesses in the case of composite laminated structures with different material layers [44].

5 Discretization of the Equations of Motion

The equations of motion can now be discretized according to the radial basis function method as

$$
\begin{aligned}
&D_{11}\sum_{j=1}^{N}\alpha_j^{\Psi_x}\frac{\partial^2\phi_j}{\partial x^2}+D_{16}\sum_{j=1}^{N}\alpha_j^{\Psi_y}\frac{\partial^2\phi_j}{\partial x^2}+(D_{12}+D_{66})\sum_{j=1}^{N}\alpha_j^{\Psi_y}\frac{\partial^2\phi_j}{\partial x\partial y}+\\
&\quad+2D_{16}\sum_{j=1}^{N}\alpha_j^{\Psi_x}\frac{\partial^2\phi_j}{\partial x\partial y}+D_{66}\sum_{j=1}^{N}\alpha_j^{\Psi_x}\frac{\partial^2\phi_j}{\partial y^2}+D_{26}\sum_{j=1}^{N}\alpha_j^{\Psi_y}\frac{\partial^2\phi_j}{\partial y^2}-\\
&\qquad-kA_{45}\left(\sum_{j=1}^{N}\alpha_j^{\Psi_y}\phi_j+\sum_{j=1}^{N}\alpha_j^{W}\frac{\partial\phi_j}{\partial y}\right)-\\
&\qquad-kA_{55}\left(\sum_{j=1}^{N}\alpha_j^{\Psi_x}\phi_j+\sum_{j=1}^{N}\alpha_j^{W}\frac{\partial\phi_j}{\partial x}\right)=-I_2\omega^2\sum_{j=1}^{N}\alpha_j^{\Psi_x}\phi_j
\end{aligned}
\tag{42}
$$

$$
\begin{aligned}
&D_{16}\sum_{j=1}^{N}\alpha_j^{\Psi_x}\frac{\partial^2\phi_j}{\partial x^2}+D_{66}\sum_{j=1}^{N}\alpha_j^{\Psi_y}\frac{\partial^2\phi_j}{\partial x^2}+(D_{12}+D_{66})\sum_{j=1}^{N}\alpha_j^{\Psi_x}\frac{\partial^2\phi_j}{\partial x\partial y}+\\
&\quad+2D_{26}\sum_{j=1}^{N}\alpha_j^{\Psi_y}\frac{\partial^2\phi_j}{\partial x\partial y}+D_{26}\sum_{j=1}^{N}\alpha_j^{\Psi_x}\frac{\partial^2\phi_j}{\partial y^2}+D_{22}\sum_{j=1}^{N}\alpha_j^{\Psi_y}\frac{\partial^2\phi_j}{\partial y^2}-\\
&\qquad-kA_{44}\left(\sum_{j=1}^{N}\alpha_j^{\Psi_y}\phi_j+\sum_{j=1}^{N}\alpha_j^{W}\frac{\partial\phi_j}{\partial y}\right)-\\
&\qquad-kA_{45}\left(\sum_{j=1}^{N}\alpha_j^{\Psi_x}\phi_j+\sum_{j=1}^{N}\alpha_j^{W}\frac{\partial\phi_j}{\partial x}\right)=-I_2\omega^2\sum_{j=1}^{N}\alpha_j^{\Psi_y}\phi_j
\end{aligned}
\tag{43}
$$

$$
\begin{aligned}
&\frac{\partial}{\partial x}\left[kA_{45}\left(\sum_{j=1}^{N}\alpha_j^{\Psi_y}\phi_j+\sum_{j=1}^{N}\alpha_j^{W}\frac{\partial\phi_j}{\partial y}\right)+kA_{55}\left(\sum_{j=1}^{N}\alpha_j^{\Psi_x}\phi_j+\sum_{j=1}^{N}\alpha_j^{W}\frac{\partial\phi_j}{\partial x}\right)\right]+\\
&+\frac{\partial}{\partial y}\left[kA_{44}\left(\sum_{j=1}^{N}\alpha_j^{\Psi_y}\phi_j+\sum_{j=1}^{N}\alpha_j^{W}\frac{\partial\phi_j}{\partial y}\right)+kA_{45}\left(\sum_{j=1}^{N}\alpha_j^{\Psi_x}\phi_j+\sum_{j=1}^{N}\alpha_j^{W}\frac{\partial\phi_j}{\partial x}\right)\right]+P\\
&\qquad=-I_0\omega^2\sum_{j=1}^{N}\alpha_j^{W}\phi_j
\end{aligned}
\tag{44}
$$

where N represents the total number of points of the structure. The vector α^{Ψ_x} corresponds to the vector of unknowns related to rotation Ψ_x. For free vibration problems, P is set to zero.

6 Calculation of Transverse Shear Stresses

In the following examples, transverse shear stresses are calculated through the equilibrium equations. For the calculation of τ_{xz} we consider first the equilibrium equation

$$\frac{\partial \sigma_x}{\partial x} + \frac{\partial \tau_{xy}}{\partial y} + \frac{\partial \tau_{xz}}{\partial z} = 0 \tag{45}$$

or

$$\tau_{xz} = -\int_{-h/2}^{z} \left(\frac{\partial \sigma_x}{\partial x} + \frac{\partial \tau_{xy}}{\partial y} \right) dz \tag{46}$$

By integrating through the k laminate layers we can then obtain the transverse shear stresses at any point in z direction, by using the derivatives of radial basis functions. It will be clearer in the following examples.

7 Numerical Examples

7.1 Square Isotropic Thick Plate under Uniform Load

In this first example we consider a simply supported square isotropic plate (side length a and thickness h) under uniform pressure. The modulus of elasticity (E) and the Poisson's ratio (ν) are taken as 10920 and 0.25, respectively. The normalized transverse displacement, $\overline{w}$, and the normalized normal x-stress, $\overline{\sigma}_{xx}$ are obtained as

$$\overline{w} = \frac{Eh^3 10^2 w_{\max}}{qa^4} \qquad \overline{\sigma}_{xx} = \frac{\sigma_{xx} h^2}{qa^2}$$

In Table 1 the normalized results are compared with finite element results of Reddy [8] and exact (analytical) results [9]. Results are also compared with a previous higher-order formulation with RBFs presented by Ferreira et al. [13]. We use 9×9, 13×13, 17×17 and 21×21 Chebyshev grids for both the internal and boundary nodes, with an optimal shape parameter together with the Wendland compactly supported function

$$\varphi_{3,3}(r) = (1 - \varepsilon r)_+^8 (32(\varepsilon r)^3 + 25(\varepsilon r)^2 + 8\varepsilon r + 1). \tag{47}$$

The boundary conditions are of SS2 type and the collocation points we consider lie on Chebyshev grids. For all a/h ratios the present formulation is better than [13] and Reddy [9] when compared with exact solutions, for transverse displacements and normal xx stress. The results converge rapidly to exact values.

Table 1. Square isotropic plate under uniform load.

$\frac{a}{h}$	Method	$\overline{w}$	$\overline{\sigma}_{xx}$
10	Reddy [8]	4.770	0.2899
	exact [9]	4.791	0.2762
	Ferreira et al. [13] (N=21)	4.7866	0.2777
	present (9 × 9)	4.7349	0.2681
	present (13 × 13)	4.7881	0.2754
	present (17 × 17)	4.7908	0.2762
	present (21 × 21)	4.7911	0.2762
20	Reddy [8]	4.570	0.2683
	exact [9]	4.625	0.2762
	Ferreira et al. [13] (N=21)	4.6132	0.2761
	present (9 × 9)	4.570	0.2596
	present (13 × 13)	4.6206	0.2749
	present (17 × 17)	4.6246	0.2761
	present (21 × 21)	4.6253	0.2762
50	Reddy [8]	4.496	0.2667
	exact [9]	4.579	0.2762
	Ferreira et al. [13] (N=21)	4.5753	0.2762
	present (9 × 9)	4.9485	0.3078
	present (13 × 13)	4.5615	0.2806
	present (17 × 17)	4.5758	0.2766
	present (21 × 21)	4.5784	0.2762
100	Reddy [8]	4.482	0.2664
	exact [9]	4.572	0.2762
	Ferreira et al. [13] (N=21)	4.5737	0.2764
	present (9 × 9)	5.3071	0.3322
	present (13 × 13)	4.4498	0.2665
	present (17 × 17)	4.5597	0.2745
	present (21 × 21)	4.5755	0.2752

7.2 Four Layer [0°/90°/90°/0°] Square Cross-Ply Laminated Plate under Sinusoidal Load

A simply supported square laminated plate of side a and thickness h is composed of four equal layers oriented at [0°/90°/90°/0°]. The plate is subjected to a sinusoidal vertical pressure of the form:

$$p_z = P \sin\left(\frac{\pi x}{a}\right) \sin\left(\frac{\pi y}{a}\right)$$

with the origin of the coordinate system located at the lower left corner on the mid-plane.

The orthotropic material properties are given by

Table 2. [0°/90°/90°/0°] square laminated plate under sinusoidal load.

$\frac{a}{h}$	Method	$\overline{w}$	$\overline{\sigma}_{xx}$	$\overline{\sigma}_{yy}$	$\overline{\tau}_{zx}$	$\overline{\tau}_{xy}$
4	3 strip [39]	1.8939	0.6806	0.6463	0.2109	0.0450
	HSDT [41]	1.8937	0.6651	0.6322	0.2064	0.0440
	FSDT [40]	1.7100	0.4059	0.5765	0.1398	0.0308
	elasticity [42]	1.954	0.720	0.666	0.270	0.0467
	Ferreira et al. [13] (N=21)	1.8864	0.6659	0.6313	0.1352	0.0433
	present (9 × 9)	1.7103	0.4339	0.6150	0.2542	0.0308
	present (13 × 13)	1.7109	0.4336	0.6145	0.2598	0.0309
	present (17 × 17)	1.7110	0.4335	0.6144	0.2607	0.0309
	present (21 × 21)	1.7110	0.4335	0.6144	0.2610	0.0309
10	3 strip [39]	0.7149	0.5589	0.3974	0.2697	0.0273
	HSDT [41]	0.7147	0.5456	0.3888	0.2640	0.0268
	FSDT [40]	0.6628	0.4989	0.3615	0.1667	0.0241
	elasticity [42]	0.743	0.559	0.403	0.301	0.0276
	Ferreira et al. [13] (N=21)	0.7153	0.5466	0.4383	0.3347	0.0267
	present (9 × 9)	0.6633	0.5324	0.3853	0.3104	0.0241
	present (13 × 13)	0.6638	0.5321	0.3854	0.3174	0.0240
	present (17 × 17)	0.6638	0.5321	0.3854	0.3186	0.0242
	present (21 × 21)	0.6638	0.5321	0.3854	0.3189	0.0242
100	3 strip [39]	0.4343	0.5507	0.2769	0.2948	0.0217
	HSDT [41]	0.4343	0.5387	0.2708	0.2897	0.0213
	FSDT [40]	0.4337	0.5382	0.2705	0.1780	0.0213
	elasticity [42]	0.4347	0.539	0.271	0.339	0.0214
	Ferreira et al. [13] (N=21)	0.4365	0.5413	0.3359	0.4106	0.0215
	present (9 × 9)	0.4343	0.5663	0.2852	0.3319	0.0203
	present (13 × 13)	0.4344	0.5734	0.2889	0.3399	0.0213
	present (17 × 17)	0.4346	0.5738	0.2884	0.3425	0.0214
	present (21 × 21)	0.4347	0.5739	0.2884	0.3433	0.0214

$$E_1 = 25.0E_2 \qquad G_{12} = G_{13} = 0.5E_2 \qquad G_{23} = 0.2E_2 \qquad \nu_{12} = 0.25$$

In Table 2 the present method is compared with a finite strip formulation by Akhras [39, 40] who used three strips, an analytical solution by Reddy [41] using a higher-order formulation and an exact three dimensional solution by Pagano [42]. The present solution is also compared with another higher-order solution by the authors [13]. The in-plane displacements, the transverse displacements, the normal stresses and the in-plane and transverse shear stresses are presented in normalized form as

$$\overline{w} = \frac{10^2 w_{\max} h^3 E_2}{Pa^4} \qquad \overline{\sigma}_{xx} = \frac{\sigma_{xx} h^2}{Pa^2} \qquad \overline{\sigma}_{yy} = \frac{\sigma_{yy} h^2}{Pa^2}$$

$$\overline{\tau}_{zx} = \frac{\tau_{zx} h}{Pa} \qquad \overline{\tau}_{xy} = \frac{\tau_{xy} h^2}{Pa^2}$$

The membrane stresses were evaluated at the following locations: $\overline{\sigma}_{xx}(a/2, b/2, \frac{h}{2})$, $\overline{\sigma}_{yy}(a/2, b/2, \frac{h}{4})$, and $\overline{\sigma}_{xy}(a, b, -\frac{h}{2})$. The transverse shear stresses are calculated using the equilibrium equations at locations $\overline{\tau}_{zx}(0, b/2, 0)$.

We used four nodal grids, with 9×9, 13×13, 17×17 and 21×21 points. The present shear deformation theory discretized with Wendland compact support functions present good results in line of the FSDT used by other authors. For higher thicknesses the need for higher order theories becomes quite clear.

As mentioned before we can evaluate the transverse shear stresses at any point in the z direction.

In this example, we can calculate explicitly τ_{xz} as

$$\tau_{xz} = -\int_{-h/2}^{z} \left(\frac{\partial \sigma_x}{\partial x} + \frac{\partial \tau_{xy}}{\partial y} \right) dz = -\int_{-h/2}^{z} \left(z e_{11}^{k} \frac{\partial^2 \theta_x}{\partial x^2} + z e_{12}^{k} \frac{\partial^2 \theta_y}{\partial x \partial y} \right) dz \quad (48)$$

with e_{11}^{k} being one of the components of the elasticity matrix for the k layer, given by

$$e_{11}^{k} = \frac{E_1^k}{1 - \nu_{12}^2} \quad (49)$$

where E_1^k is the modulus of elasticity in the composite fiber direction for the material in the k-th layer, and ν_{12} the major Poisson's ratio. Other components are obtained in a similar way. For this laminate the $z = 0$ stress is given by

$$\tau_{xz}(z=0) = \frac{\partial^2 \theta_x}{\partial x^2} \left[\frac{3h^2}{32} e_{11}^{(1)} + \frac{h^2}{32} e_{11}^{(2)} \right] + \\ + \frac{\partial^2 \theta_x}{\partial x \partial y} \left[\frac{3h^2}{32} e_{12}^{(1)} + \frac{h^2}{32} e_{12}^{(2)} \right] + \frac{\partial^2 \theta_y}{\partial y^2} \left[\frac{3h^2}{32} e_{22}^{(1)} + \frac{h^2}{32} e_{22}^{(2)} \right] \quad (50)$$

7.3 Three Layer Square Sandwich Plate under Uniform Load

The well-known sandwich example of Srinivas [37] is presented and discussed. It considers a simply supported square sandwich plate under uniform pressure. The ratio of side to thickness, a/h is taken as 10. The sandwich laminate considers two outside layers (skins) of thickness $h_1 = h_3 = 0.1h$ and one inner layer (core) of thickness $h_2 = 0.8h$. The skin orthotropic properties are obtained by multiplying an integer, R, by the core orthotropic properties, given by

$$\overline{Q}_{core} = \begin{bmatrix} 0.999781 & 0.231192 & 0 & 0 & 0 \\ 0.231192 & 0.524886 & 0 & 0 & 0 \\ 0 & 0 & 0.262931 & 0 & 0 \\ 0 & 0 & 0 & 0.266810 & 0 \\ 0 & 0 & 0 & 0 & 0.159914 \end{bmatrix}$$

Table 3. Square laminated plate under uniform load, $R = 5$.

Method	$\overline{w}$	$\overline{\sigma}_{xx}^1$	$\overline{\sigma}_{xx}^2$	$\overline{\sigma}_{xx}^3$	$\overline{\tau}_{xz}^1$	$\overline{\tau}_{xz}^2$
	(a/2,a/2,0)	(a/2,a/2,h/2)	(a/2,a/2,2h/5)	(a/2,a/2,2h/5)	(0,a/2,0)	(0,a/2,-2h/5)
HSDT [38]	256.13	62.38	46.91	9.382	3.089	2.566
FSDT [38]	236.10	61.87	49.50	9.899	3.313	2.444
CLT	216.94	61.141	48.623	9.783	4.5899	3.386
Ferreira & Barbosa [43]	258.74	59.21	45.61	9.122	3.593	3.593
Ferreira (N=15) [12]	257.38	58.725	46.980	9.396	3.848	2.839
exact [37]	258.97	60.353	46.623	9.340	4.3641	3.2675
Ferreira et al. [13] (HSDT)(N=21)	257.1100	60.3660	47.0028	9.4006	4.5481	2.3910
Ferreira (layerwise) [15] (N=21)	257.5231	59.9675	46.2906	9.2581	4.0463	2.3901
present (9 × 9)	257.65	58.9449	47.1559	9.4312	3.8626	2.4523
present (13 × 13)	259.17	59.0723	47.2578	9.4516	4.0282	2.6002
present (17 × 17)	259.26	59.1129	47.2904	9.4581	4.0508	2.6232
present (21 × 21)	259.27	59.1189	47.2951	9.4590	4.0554	2.6280

Table 4. Square laminated plate under uniform load, $R = 10$.

Method	$\overline{w}$	$\overline{\sigma}_{xx}^1$	$\overline{\sigma}_{xx}^2$	$\overline{\sigma}_{xx}^3$	$\overline{\tau}_{xz}^1$	$\overline{\tau}_{xz}^2$
	(a/2,a/2,0)	(a/2,a/2,h/2)	(a/2,a/2,2h/5)	(a/2,a/2,2h/5)	(0,a/2,0)	(0,a/2,-2h/5)
HSDT [38]	152.33	64.65	51.31	5.131	3.147	2.587
FSDT [38]	131.095	67.80	54.24	4.424	3.152	2.676
CLT	118.87	65.332	48.857	5.356	4.3666	3.7075
Ferreira & Barbosa [43]	159.402	64.16	47.72	4.772	3.518	3.518
Ferreira (N=15) [12]	158.55	62.723	50.16	5.01	3.596	3.053
exact [37]	159.38	65.332	48.857	4.903	4.0959	3.5154
Ferreira et al. [13] (HSDT)(N=21)	154.6581	65.3809	49.9729	4.9973	3.5280	2.3984
Ferreira (layerwise) [15] (N=21)	158.3799	64.8462	48.4434	4.8443	3.9237	2.8809
present (9 × 9)	158.92	63.3129	50.6503	5.0650	3.8274	2.6466
present (13 × 13)	159.56	63.1135	50.4908	5.0491	3.9542	2.7979
present (17 × 17)	159.60	63.1318	50.5055	5.0505	3.9706	2.8216
present (21 × 21)	159.61	63.1358	50.5087	5.0509	3.9739	2.8267

The skin properties are obtained by

$$\overline{Q}_{skin} = R\overline{Q}_{core}$$

In this example we considered four nodal grids, as in previous example. Results are compared with exact results of Srinivas [37] and finite element results of Pandya and Kant [38]. We also compare results with a shell finite formulation by Ferreira and Barbosa [43] and first order shear formulation with multiquadrics by Ferreira [12]. The present solution is also compared with another higher-order solution with

Table 5. Square laminated plate under uniform load, $R = 15$.

Method	$\overline{w}$	$\overline{\sigma}^1_{xx}$	$\overline{\sigma}^2_{xx}$	$\overline{\sigma}^3_{xx}$	$\overline{\tau}^1_{xz}$	$\overline{\tau}^2_{xz}$
	(a/2,a/2,0)	(a/2,a/2,h/2)	(a/2,a/2,2h/5)	(a/2,a/2,2h/5)	(0,a/2,0)	(0,a/2,-2h/5)
HSDT [38]	110.43	66.62	51.97	3.465	3.035	2.691
FSDT [38]	90.85	70.04	56.03	3.753	3.091	2.764
CLT	81.768	69.135	55.308	3.687	4.2825	3.8287
Ferreira & Barbosa [43]	121.821	65.650	47.09	3.140	3.466	3.466
Ferreira (N=15) [12]	121.184	63.214	50.571	3.371	3.466	3.099
exact [37]	121.72	66.787	48.299	3.238	3.9638	3.5768
Ferreira et al. [13] (HSDT)(N=21)	114.6442	66.9196	50.3230	3.3549	3.0213	2.2750
Ferreira (layerwise) [15] (N=21)	120.9883	66.2911	47.8992	3.1933	3.8311	3.2562
present (9 × 9)	121.51	64.0033	51.2026	3.4135	3.8104	2.6892
present (13 × 13)	121.89	63.6445	50.9156	3.3944	3.9214	2.8412
present (17 × 17)	121.93	63.6517	50.9213	3.3948	3.9353	2.8651
present (21 × 21)	121.93	63.6547	50.9238	3.3949	3.9382	2.8702

Table 6. Natural frequencies of a CCCC square Mindlin/Reissner plate with $h/a = 0.1, k = 0.8601, \nu = 0.3$.

Mode no.	m	n	9 × 9	13 × 13	17 × 17	Rayleygh-Ritz [33]	Liew et al. [34]
1	1	1	1.5967	1.5911	1.5913	1.5940	1.5582
2	2	1	3.0157	3.0393	3.0388	3.0390	3.0182
3	1	2	3.0157	3.0393	3.0390	3.0390	3.0182
4	2	2	4.1762	4.2641	4.2620	4.2650	4.1711
5	3	1	4.9579	5.0290	5.0244	5.0350	5.1218
6	1	3	5.0333	5.0756	5.0645	5.0350	5.1594
7	3	2	5.9067	6.0890	6.0769		6.0178
8	2	3	5.9067	6.0890	6.0771		6.0178
9	4	1	7.4848	7.4323	7.4113		7.5169
10	1	4	7.9839	7.4323	7.4129		7.5169

multiquadrics by Ferreira et al. [13] and a layerwise formulation by Ferreira [15]. In-plane displacements, transverse displacement and stresses are normalized through factors

$$\overline{w} = w\frac{0.999781}{hq} \qquad \overline{\sigma}_{\alpha\alpha} = \frac{\sigma_{\alpha\alpha}}{q} \qquad \overline{\tau}_{\alpha\beta} = \frac{\tau_{\alpha\beta}}{q}$$

The locations are given in Tables 3 to 5, where various values of R are considered (5, 10 and 15). The present formulation produces results that are in excellent agreement with all higher order formulations and with exact results.

Table 7. Natural frequencies of a CCCC square Mindlin/Reissner plate with $h/a = 0.01$, $k = 0.8601$, $\nu = 0.3$.

Mode no.	m	n	9×9	13×13	17×17	Rayleygh-Ritz [33]	Liew et al. [34]
1	1	1	0.1011	0.1846	0.1753	0.1754	0.1743
2	2	1	0.2764	0.3787	0.3575	0.3576	0.3576
3	1	2	0.2764	0.3787	0.3575	0.3576	0.3576
4	2	2	0.4817	0.5615	0.5280	0.5274	0.5240
5	3	1	1.1218	0.6525	0.6433	0.6402	0.6465
6	1	3	1.1307	0.6596	0.6462	0.6432	0.6505
7	3	2	1.5977	0.7722	0.8136		0.8015
8	2	3	1.5977	0.7722	0.8137		0.8015
9	4	1	2.9106	0.8870	1.0440		1.0426
10	1	4	3.9583	1.0703	1.0440		1.0426

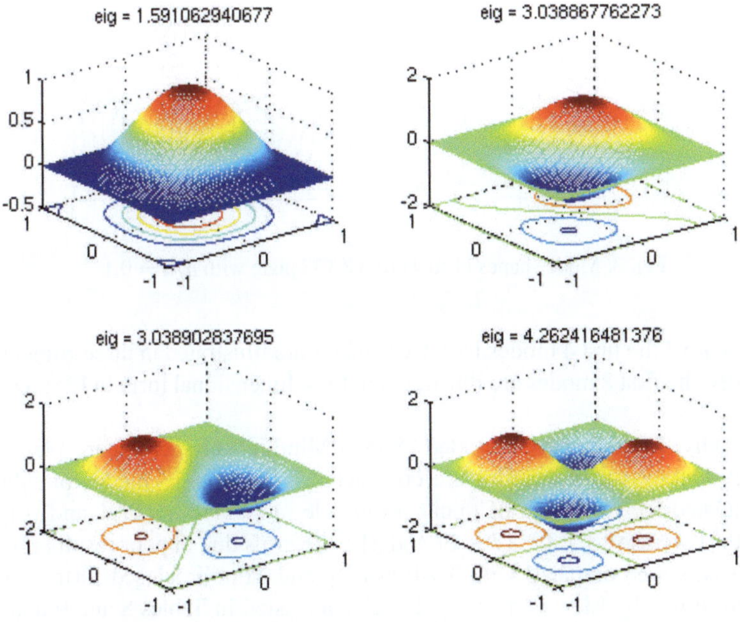

Fig. 2. Mode shapes (1 to 4) for CCCC plate with $h/a = 0.1$.

7.4 Natural Frequencies of Mindlin Plates

In Tables 6 and 7 we compare natural frequencies for fully clamped plates with Ritz solution [33] and a competitive meshless solution by Liew et al. [34]. Results show excellent agreement with analytical solution for both h/a cases.

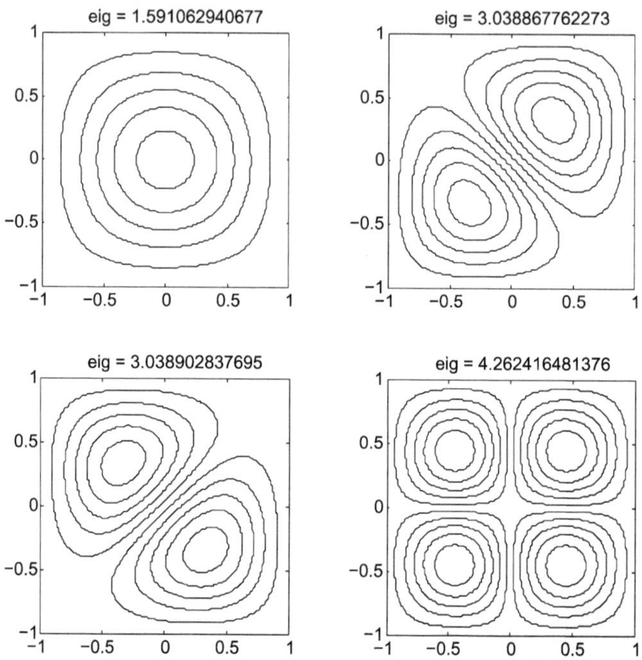

Fig. 3. Mode shapes (1 to 4) for CCCC plate with $h/a = 0.1$.

In Figure 2 the first 4 modes for CCCC plates are illustrated in three-dimensional form. Also the first 8 modes are illustrated in two-dimensional form in Figures 3 and 4.

Secondly, fully simply supported (SSSS) Mindlin/Reissner square plates with different thickness-to-side ratios are considered. The first 13 modes of vibration are calculated for two cases of thickness-to-side ratios $h/a = 0.01$ and 0.1. For $h/a = 0.01$ we use 13×13, 17×17 and 21×21 grids due to poor results for 9×9 grid. Results are compared with 3D-Elasticity and Mindlin closed form solutions [35], and results by Liew et al. [34]. Results are listed in Tables 8 and 9 and show excellent agreement with closed form solutions.

In Figure 5 the first 4 modes for SSSS plates are illustrated in three-dimensional form.

7.5 Natural Frequencies of Composite Plates

We now consider square laminated plates, where all layers of the laminate are assumed to be of the same thickness, density and made of the same linearly elastic composite material. The following material parameters of a layer are used:

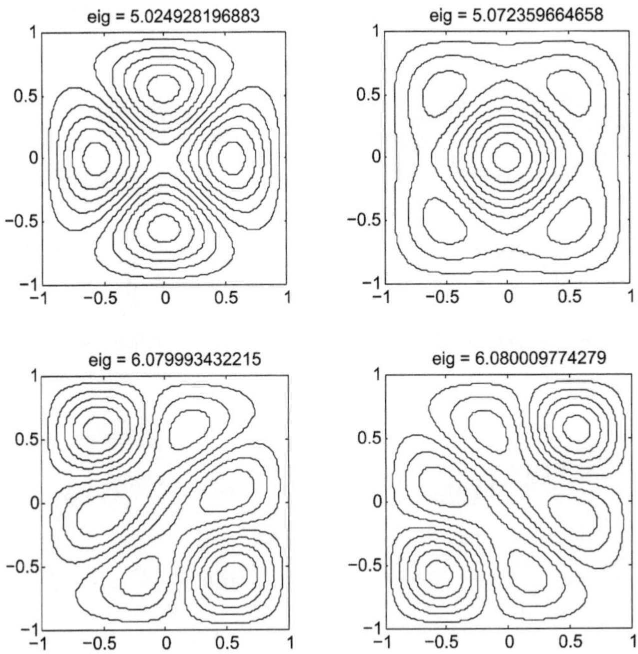

Fig. 4. Mode shapes (5 to 8) for CCCC plate with $h/a = 0.1$.

Table 8. Natural frequencies of a SSSS square Mindlin/Reissner plate with $h/a = 0.1$, $k = 0.833$, $\nu = 0.3$ (* – closed form).

Mode no.	m	n	9 × 9	13 × 13	17 × 17	3D *	Mindlin *	Liew et al. [34]
1	1	1	0.9309	0.9303	0.9303	0.9320	0.9300	0.9220
2	2	1	2.2140	2.2195	2.2193	2.2260	2.2190	2.2050
3	1	2	2.2140	2.2195	2.2193	2.2260	2.2190	2.2050
4	2	2	3.4157	3.4062	3.4057	3.4210	3.4060	3.3770
5	3	1	4.1208	4.1509	4.1495	4.1710	4.1490	4.1390
6	1	3	4.1254	4.1510	4.1495	4.1710	4.1490	4.1390
7	3	2	5.2197	5.2087	5.2059	5.2390	5.2060	5.1700
8	2	3	5.2197	5.2087	5.2059	5.2390	5.2060	5.1700
9	4	1	6.8810	6.5309	6.5207	-	6.5200	6.5240
10	1	4	6.8810	6.5309	6.5207	-	6.5200	6.5240

$$\frac{E_1}{E_2} = 10, 20, 30 \text{ or } 40;\ G_{12} = G_{13} = 0.6E_2;\ G_3 = 0.5E_2;\ \nu_{12} = 0.25$$

The subscripts 1 and 2 denote the directions normal and transverse to the fiber direction in a lamina, which may be oriented at an angle to the plate axes. The ply

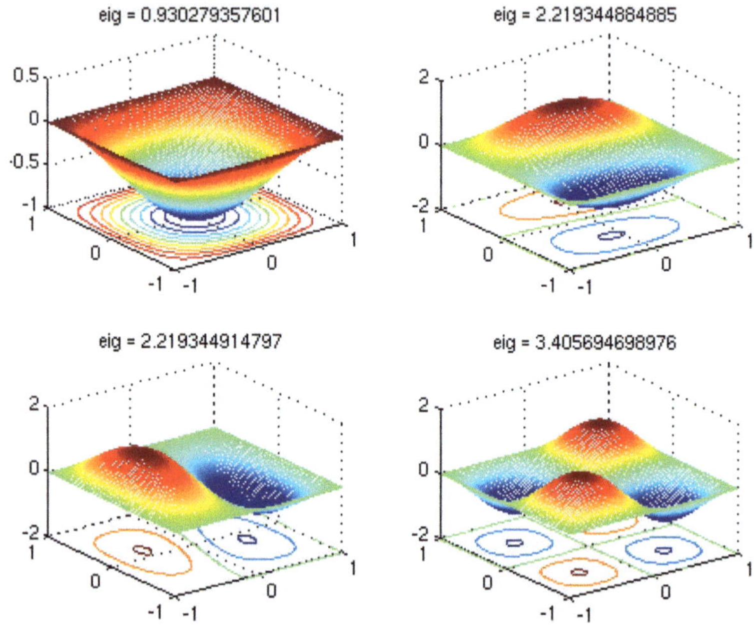

Fig. 5. Mode shapes (1 to 4) for SSSS plate with $h/a = 0.1$.

Table 9. Natural frequencies of a SSSS square Mindlin/Reissner plate with $h/a = 0.01$, $k = 0.833$, $\nu = 0.3$ (* – closed form).

Mode no.	m	n	13 × 13	17 × 17	21 × 21	Mindlin *	Liew et al. [34]
1	1	1	0.0965	0.0963	0.0963	0.0963	0.0961
2	2	1	0.2417	0.2407	0.2363	0.2406	0.2419
3	1	2	0.2417	0.2407	0.2393	0.2406	0.2419
4	2	2	0.3884	0.3851	0.3847	0.3848	0.3860
5	3	1	0.4775	0.4818	0.4809	0.4809	0.4898
6	1	3	0.4788	0.4819	0.4864	0.4809	0.4898
7	3	2	0.6290	0.6267	0.6243	0.6249	0.6315
8	2	3	0.6290	0.6267	0.6243	0.6249	0.6315
9	4	1	0.8145	0.8238	0.8290	0.8167	0.8447
10	1	4	0.8145	0.8238	0.8390	0.8167	0.8447

angle of each layer is measured from the global x-axis to the fiber direction. Unless otherwise stated, the simply supported boundary condition is taken to be the hard type SS2 condition. In all examples we use a shear correction factor $k = \pi^2/12$, as proposed in [32].

Table 10. The normalized fundamental frequency of the simply-supported cross-ply laminated square plate [0°/90°/90°/0°] ($\bar{w} = (wa^2/h)\sqrt{\rho/E_2}, h/a = 0.2$).

Method	Grid	E_1/E_2			
		10	20	30	40
Liew [32]		8.2924	9.5613	10.320	10.849
Exact (Reddy)[31]		8.2982	9.5671	10.326	10.854
Present	9×9	8.2694	9.5320	10.2856	10.8097
	13×13	8.2670	9.5297	10.2835	10.8077
	17×17	8.2669	9.5296	10.2833	10.8076

The example considered is a simply supported square plate of the cross-ply lamination [0°/90°/90°/0°]. The thickness and length of the plate are denoted by h and a, respectively. The thickness-to-span ratio $h/a = 0.2$ is employed in the computation. Table 10 lists the fundamental frequency of the simply supported laminate made of various modulus ratios of E_1/E_2. It is found that the results are in very close agreement with the values of [31] and the meshfree results of Liew [32] based on the FSDT. The relative errors between the analytical and present solutions are below 1.0%.

The influence of mixed boundary conditions and span-to-thickness ratio is now considered. The plate is simply supported along the edges parallel to the x-axis while the other two edges are subjected to simply supported (S) and clamped (C) boundary conditions. Notations SS, SC, CC refer to the boundary conditions of the two edges parallel to the y-axis only. The three layer cross-ply [0°/90°/0°] square laminate is here considered with $E_1/E_2 = 40$. The plate is discretized with 9×9, 13×13 and 17×17 grids. In Table 11, results are compared with Liew [32] and Reddy [31]. It is found that the present results are in excellent agreement with exact results and very close to Liew's results. The boundary conditions play no essential role in the quality of solution. For SS and SC conditions the difference of present results to exact are below 1%. For CC conditions difference to exact results are below 0.6% except for $a/h = 100$ where error is 1.7%.

8 Conclusions

A study of static deformations and free vibration of Mindlin plates was presented. Equations of motion and boundary conditions for Mindlin plates were presented using a first-order shear deformation theory for plates with a shear correction algorithm. Numerical results were presented and discussed for various thickness-to-length ratios. The analysis is based on a new numerical scheme, where collocation by radial basis functions are viewed in the sense of pseudospectral methods in order to produce highly accurate results. This method allows the use of non-rectangular geometries and promotes a more flexible framework for pseudospectral methods. Furthermore a

Table 11. The normalized fundamental frequency of the 3-layer [0°/90°/0°] laminated square plate with various boundary conditions and span to thickness ratios ($\bar{w} = (wa^2/h)\sqrt{\rho/E_2}$, $E_1/E_2 = 40$).

a/h	Method	Grid	SS	SC	CC
2	Liew [32]		5.205	5.210	5.257
	Exact (Reddy) [31]		5.205	5.211	5.257
	Present	9 × 9	5.2052	5.1846	5.2301
		13 × 13	5.2047	5.1842	5.2297
		17 × 17	5.2046	5.1841	5.2296
5	Liew [32]		10.290	10.647	11.266
	Exact (Reddy) [31]		10.290	10.646	11.266
	Present	9 × 9	10.2900	10.5964	11.2068
		13 × 13	10.2882	10.5949	11.2057
		17 × 17	10.2881	10.5948	11.2056
10	Liew [32]		14.767	17.176	19.669
	Exact (Reddy) [31]		14.767	17.175	19.669
	Present	9 × 9	14.7688	17.1106	19.5777
		13 × 13	14.7638	17.1061	19.5733
		17 × 17	14.7635	17.1059	19.5733
100	Liew [32]		18.769	28.164	40.004
	Exact (Reddy) [31]		18.891	28.501	40.743
	Present	9 × 9	18.8224	28.1989	39.6371
		13 × 13	18.8311	28.2375	40.0557
		17 × 17	18.8262	28.2360	40.0610

cross-validation technique was used to optimize the shape parameter in radial basis functions. Compact support functions were used in this paper, showing high stability and convergence of results, when compared with existing finite element or meshless schemes.

The present method is a simple yet powerful alternative to other finite element or meshless methods in the static deformation and free vibration analysis of isotropic and laminated composite plates.

References

1. P. A. A. Laura and R. H. Gutierrez, Analysis of vibrating Timoshenko beams using the method of differential quadrature. *Shock and Vibration*, 1(1):89–93, 1993.
2. C. W. Bert and M. Malik, Differential quadrature method in computational mechanics: A review. *Applied Mechanics Review*, 49(1):1–28, 1996.

3. K. M. Liew, J. B. Han and Z. M. Xiao, Vibration analysis of circular Mindlin plates using differential quadrature method. *Journal of Sound and Vibration*, 205(5):617–630.
4. K. M. Liew, K. C. Hung and M. K. Lim, Vibration of Mindlin plates using boundary characteristic orthogonal polynomials. *Journal of Sound and Vibration*, 182(1):77–90, 1995.
5. J. Lee and W. W. Schultz, Eigenvalue analysis of Timoshenko beams and axisymmetric Mindlin plates by the pseudospectral method. *Journal of Sound and Vibration*, 269:609–621.
6. K. Bathe, *Finite Element Procedures in Engineering Analysis*. Prentice-Hall, Englewood Cliffs, NJ, 1982.
7. W. Weaver and P. Johnson, *Structural Dynamics by Finite Elements*. Prentice-Hall, Englewood Cliffs, NJ, 1987.
8. J. N. Reddy, *Introduction to the Finite Element Method*. McGraw-Hill, New York, 1993.
9. J. N. Reddy, *Energy Principles and Variational Methods in Applied Mechanics*, 2nd Edition. John Wiley, New York, 2002.
10. A. J. M. Ferreira, Free vibration analysis of Timoshenko beams and Mindlin plates by radial basis functions. *International Journal of Computational Methods*, 2(1):15–31, 2005.
11. A. J. M. Ferreira, C. M. C. Roque and P. A. L. S. Martins, Radial basis functions and higher-order theories in the analysis of laminated composite beams and plates. *Composite Structures*, 66:287–293.
12. A. J. M. Ferreira, A formulation of the multiquadric radial basis function method for the analysis of laminated composite plates. *Composite Structures*, 59:385–392, 2003.
13. A. J. M. Ferreira, C. M. C. Roque and P. A. L. S. Martins, Analysis of composite plates using higher-order shear deformation theory and a finite point formulation based on the multiquadric radial basis function method. *Composites: Part B*, 34:627–636, 2003.
14. A. J. M. Ferreira, Thick composite beam analysis using a global meshless approximation based on radial basis functions. *Mechanics of Advanced Materials and Structures*, 10:271–284, 2003.
15. A. J. M. Ferreira, Analysis of composite plates using a layerwise deformation theory and multiquadrics discretization. *Mechanics of Advanced Materials and Structures*, in press, 2005.
16. A. J. M. Ferreira, Polyharmonic (thin-plate) splines in the analysis of composite plates. *International Journal of Mechanical Sciences*, 46:1549–1569.
17. A. J. M. Ferreira, R. C. Batra, C. M. C. Roque, L. F. Qian and P. A. L. S. Martins, Static analysis of functionally graded plates using third-order shear deformation theory and a meshless method, *Composite Structures*, 69:449–457, 2005.
18. T. A. Driscoll and B. Fornberg, Interpolation in the limit of increasingly flat radial basis functions. *Comput. Math. Appl.*, 43:413–422, 2002.
19. E. Larsson and B. Fornberg, Theoretical and computational aspects of multivariate interpolation with increasingly flat radial basis functions. *Comput. Math. Appl.*, 49:103–130, 2005.
20. R. Schaback, Multivariate interpolation by polynomials and radial basis functions. *Constr. Approx.*, 21:293–317, 2005.
21. G. E. Fasshauer, RBF collocation methods as pseudospectral methods. In: *Boundary Elements XXVII*, A. Kassab, C. A. Brebbia, E. Divo, and D. Poljak (Eds.). WIT Press, Southampton, 2005, pp. 47–56.
22. G. E. Fasshauer, RBF collocation methods and pseudospectral methods, preprint, 2004.

23. R. B. Platte and T. A. Driscoll, Eigenvalue stability of radial basis function discretizations for time-dependent problems. *Comput. Math. Appl.*, 51(8): 1251–1268, 2006.
24. B. Fornberg, *A Practical Guide to Pseudospectral Methods*. Cambridge University Press, 1998.
25. L. N. Trefethen, *Spectral Methods in Matlab*. SIAM, Philadelphia, PA, 2000.
26. G. E. Fasshauer, *Meshfree Approximation Methods with Matlab*. World Scientific Publishers, Singapore, to appear.
27. R. Schaback, On the efficiency of interpolation by radial basis functions. In: *Proceedings of Surface fitting and multiresolution methods*, A. Le Méhauté, C. Rabut, and L. L. Schumaker (Eds), Vanderbilt University Press, 1997, pp. 309–318.
28. S. Rippa, An algorithm for selecting a good value for the parameter c in radial basis function interpolation. *Adv. Comput. Math.*, 11:193–210, 1999.
29. A. H.-D. Cheng, and J. J. S. P. Cabral, Direct solution of certain ill-posed boundary value problems by collocation method. In: *Boundary Elements XXVII*, A. Kassab, C. A. Brebbia, E. Divo, and D. Poljak (Eds). WIT Press, Southampton, 2005, pp. 35–44.
30. C. Canuto, M. Y. Hussaini, A. Quateroni and T. A. Zang, *Spectral Methods in Fluid Dynamics*. Springer Verlag, Berlin, 1998.
31. J. N. Reddy, *Mechanics of Laminated Composite Plates*. CRC Press, New York, 1997.
32. K. M. Liew, Y. Q. Huang and J. N. Reddy, Vibration analysis of symmetrically laminated plates based on FSDT using the moving least squares differential quadrature method. *Comp. Meth. Appl. Mech. Eng.*, 192:2203–2222, 2003.
33. D. J. Dawe and O. L. Roufaeil, Rayleigh–Ritz vibration analysis of Mindlin plates, *Journal of Sound and Vibration*, 69(3):345–359, 1980.
34. K. M. Liew, J. Wang, T. Y. Ng and M. J. Tan, Free vibration and buckling analyses of shear-deformable plates based on FSDT meshfree method. *Journal of Sound and Vibration*, 276:997–1017, 2004.
35. E. Hinton, *Numerical Methods and Software for Dynamic Analysis of Plates and Shells*. Pineridge Press, Swansea, 1988.
36. K. M. Liew, Y. Xiang, S. Kitipornchai, Research on thick plate vibration: A literature survey. *Journal of Sound and Vibration*, 180(1):163–176, 1995.
37. S. Srinivas, A refined analysis of composite laminates. *Journal of Sound and Vibration*, 30:495–507, 1973.
38. B. N. Pandya, T. Kant, Higher-order shear deformable theories for flexure of sandwich plates-finite element evaluations. *Int. J. Solids and Structures*, 24:419–451, 1988.
39. G. Akhras, M. S. Cheung, W. Li, Finite strip analysis for anisotropic laminated composite plates using higher-order deformation theory. *Computers & Structures*, 52(3):471–477, 1994.
40. G. Akhras, M. S. Cheung, W. Li, Static and vibrations analysis of anisotropic laminated plates by finite strip method. *Int. J. Solids Struct.*, 30(22):3129–3137, 1993.
41. J. N. Reddy, A simple higher-order theory for laminated composite plates. *J. Appl. Mech.*, 51:745–752, 1984.
42. N. J. Pagano, Exact solutions for rectangular bidirectional composites and sandwich plates. *J. Compos. Mater.*, 4:20–34, 1970.
43. A. J. M. Ferreira, J. T. Barbosa, Buckling behaviour of composite shells. *Composite Structures*, 50:93–98, 2000.
44. A. J. M. Ferreira, Analysis of laminated composite and sandwich shells. Ph.D. Thesis, FEUP, Porto, Portugal, 1997 [in Portuguese].

Author Index

Subject Index

Computational Methods in Applied Sciences

1. J. Holnicki-Szulc and C.A.M. Soares (eds.): *Advances in Smart Technologies in Structural Engineering*. 2004 ISBN 3-540-22331-2
2. J.A.C. Ambrósio (ed.): *Advances in Computational Multibody Systems*. 2005 ISBN 1-4020-3392-3
3. E. Oñate and B. Kröplin (eds.): *Textile Composites and Inflatable Structures*. 2005 ISBN 1-4020-3316-8
4. J.C. García Orden, J.M. Goicolea and J. Cuadrado (eds.): *Multibody Dynamics*. Computational Methods and Applications. 2007 ISBN 1-4020-5683-4
5. V.M.A. Leitão, C.J.S. Alves and C.A. Duarte (eds.): *Advances in Meshfree Techniques*. 2007 ISBN 1-4020-6094-6

www.springer.com

MIX
Papier aus verantwortungsvollen Quellen
Paper from responsible sources
FSC® C105338

If you have any concerns about our products,
you can contact us on
ProductSafety@springernature.com

In case Publisher is established outside the EU,
the EU authorized representative is:
Springer Nature Customer Service Center GmbH
Europaplatz 3, 69115 Heidelberg, Germany

Printed by Libri Plureos GmbH
in Hamburg, Germany